THE WORLD PETROLEUM INDUSTRY

THE WORLD PETROLEUM INDUSTRY

The Market for Petroleum and Petroleum Products in the 1980's

Stuart Sinclair

EUROMONITOR PUBLICATIONS LIMITED

Euromonitor Publications Ltd.
87-88 Turnmill Street, London EC1M 5QU

THE WORLD PETROLEUM INDUSTRY
The Market for Petroleum and
Petroleum Products in the 1980's

British Library Cataloguing in Publication Data
Sinclair, Stuart W.
   The world petroleum industry
   1. Petroleum industry and trade
   1. Title
   338.2'7282   HD9560.5

   ISBN 0-86338-027-1

CONTENTS

LIST OF DIAGRAMS

LIST OF TABLES IN TEXT

LIST OF TABLES IN TEXT: continued

LIST OF TABLES IN TEXT: continued

LIST OF TABLES IN TEXT: continued

LIST OF TABLES IN TEXT: continued

LIST OF TABLES IN TEXT: continued

FACT FILE

FACT FILE   List of Tables

FACT FILE List of Tables

Chapter One

INTRODUCTION

THIS study of the petroleum and petroleum products markets appears at a time of considerable interest both in the future of oil as an energy source and in the future of the products derived from it. After declining, in inflation-adjusted terms, for the best part of two decades, oil prices then soared, fell away again, soared again and are, at the time of this writing, tending to drift down once more. Where they will go during the rest of the 1980s, and how this will affect the markets for the products drawn from oil, are obviously subjects of major concern to producers and consumers alike.

This introductory chapter presents a brief overview of the issues which have been, and will continue to be, central to any discussion of oil, oil trade, oil refining and the production, consumption and trade of products derived from crude oil. It begins with a short recapitulation of the way the international oil market has evolved during the past decade. It looks at how demand for oil has changed in developed countries since the early 1970s, then looks at patterns of demand in LDCs (less developed countries). It then assesses the ways in which the major institutions involved in the business—OPEC as an agency, and the oil companies—have changed, and the political climate within which they have changed, and which in turn they have had a hand in shaping.

Oil in the Last Decade

In recent years the oil industry has been subject to an enormous amount of attention. After nearly three decades of virtually uninterrupted recovery from the ravages of the Second World War, the developed countries were made aware in 1973 of the extent to which their dependence on crude oil as an energy source had grown. The relationship between the affluence attained throughout many rich countries, and the ready availability of fuels, was then made only too clear as oil prices quadrupled and physical shortages—in the US—appeared. After a brief though relatively severe recession, economic growth picked up again in the mid-1970s and prosperity, as well as confidence in continued energy availability, returned. The fall of the Shah of Iran in early 1979, and the series of oil supply interruptions and price increases precipitated by that event, then shattered the complacency into which many had fallen since 1973. As the posted or official price of a barrel of crude oil leapt upwards, chasing the 'spot' or free market price on the Rotterdam market towards $42 in 1980, analysts, government ministers, and industrialists began to prepare themselves for a period in which patterns of energy use would begin to look very different. The warnings of bodies such as the Club of Rome, which in the early 1970s had forecast the inevitable exhaustion of some of the world's natural resources, began to be taken seriously, notwithstanding the fact that nothing material had changed regarding estimates of the world's oil reserves during the intervening period. A new era of violent political upheaval in the Middle East began to take shape in political scientists' minds, with the collapse of the Shah's régime in Iran seen as merely the precursor to a series of revolutions, of which the House of Saud in Saudi Arabia was likely to be the most obvious future victim. Indeed, the profession of political risk analyst suddenly came to the forefront of social science, with companies anxious to recruit such people in the hope of thus being better prepared for future catastrophes. The recession, which began in most Western European countries in mid-1980 and in the US slightly earlier, was, like its predecessor in 1974-75, blamed by most politicians on the second oil crisis. A number of economists were not quite so sure, and tended to take the view that while a massive transfer of purchasing power away from developed countries' economies and towards a clutch of oil-exporting nations did not by any means help sustain the level of aggregate demand, certain deficiencies, especially in the framing and conduct of monetary policy in developed countries, had also contributed to the recession. In the private sector, those firms most immediately affected were the heavy energy users, such as pulp and paper firms and the steel industry; there, substantial unit cost increases had either to be absorbed or passed on as far as the recession would allow. For firms making goods whose use depended on energy prices—and automobiles using gasoline are the obvious instance—the second oil shock appeared to be little short of a calamity. Some firms, such as those in Japan, were already well-versed in the production of small, inexpensive and fuel-efficient cars. In Europe, for similar reasons (high excise taxes on gasoline being chief among them) small cars were also a long-established line. But in the US only General Motors had much to offer for discerning purchasers of fuel-efficient cars: luckily, it had the X-car line ready to be introduced slightly before the oil price rises. Ford and Chrysler were less fortunate.

Consistent with conventional supply and demand economics, a quick response to the higher oil prices came from oil firms. With oil now fetching so much more, their natural reaction was to look for more of it. Between 1979 and 1980 the number of wells being drilled soared. Early evidence of this strong response came from the soaring spot rates for hiring drilling rigs. Indeed, by 1980 there was scarcely a country in the world in which oil exploration was not going ahead. The rig count fell again thereafter, however, with activity in the US particularly hard hit. Oil exploration and

2

drilling did pick up a little in the later part of 1983, however, although it still stood well below its levels of a year earlier, which were in turn depressed from earlier highs. The Hughes Tool count of rig activity, taken in the industry as the definitive account of exploration, showed 1,199 rotary drilling rigs in operation outside the USA at the end of 1983. This compares to 1,476 in use a year earlier.

1982 and 1983 saw the pendulum swing the other way again. Far from proceeding onwards towards $100/barrel, as some analysts had been forecasting in 1980, the oil price began to slip—not only in a real terms, as before, but in nominal terms too—with the OPEC market price settling at $28. What few forecasters appeared to have anticipated was the severity of the 1980-83 recession, and the extent of the natural demand reaction to nearly a decade of rising energy prices. Here at last, as long predicted by such commentators as Milton Friedman, was the market at work, with a higher price calling forth both an enhanced volume of supply and a sharply lower volume of demand. Moreover, what began to be realized during 1983, as demand for OPEC oil slipped down to between 17 and 18 million barrels per day from its peak of 33 million barrels per day in 1979, was that there was probably a good deal of energy conservation effort still to show up in the statistics. Even when aggregate demand recovered, according to this view, total energy needs would be materially less than in the past due to the energy-saving technologies and practices put into place during harder times.

If much of this thinking pointed to a low-energy future for the industrialized world, there was less hope for other countries, specifically the oil-importing developing countries.

Most models of world energy demand to the end of the century agree that it is the Third World which will see the lion's share of future growth. This is particularly true of the semi-industrialized middle-income countries, such as Brazil, which alone is already importing more oil than the whole of developing Africa. South Korea, Argentina, Hong Kong and Singapore are also big net oil importers.

Within the Third World, however, there is another group of countries that tends to have been overlooked as far as research into oil demands is concerned—the 13 members of OPEC itself. Within the Organisation's Secretariat, it is only recently that there has been much interest in forecasting the likely evolution of OPEC's oil needs.

An early pointer was given by Dr Al-Janabi, of the Secretariat, at a conference in London in mid-1980. At that time he argued that 'what is missed by practically all analysts is the speed with which domestic consumption of petroleum products in OPEC countries is growing.'' From consumption of some 2.3 million barrels per day of products in 1980, OPEC now believes that 6.3 million bpd may be required by 1990 and possibly 17m bpd by the end of the century. This means that as much as half of total Third World oil needs by 1990 could be attributable to demand within OPEC itself. Is this realistic?

First, it is clear that even in the 1960s, demand for refined oil products within the countries which now constitute OPEC was already rapidly increasing. Their 1967-79 annual average growth rate of 12.5 per cent was considerably higher than that seen elsewhere in the Third World. The highest rate of increase elsewhere, that of Africa, was considerably lower, at 6.3 per cent annually; for the world the growth rate was only 5.5 per cent.

Other figures confirm that many OPEC members are emerging as substantial energy users. UN data for the period 1974-79 indicates that commercial energy use in four Middle Eastern OPEC states (Iraq, Libya, Kuwait and Saudi Arabia) grew by 10.4 per cent annually—against an average growth rate of 6.4 per cent for the most industralised Third World countries and a mere 2.5 per cent for the advanced economies.

For the future, the dominant influence on OPEC energy requirements generally and on refined products demand will be the pace, extent and nature of the countries' industrialisation programmes.

To a greater or lesser extent all the members of OPEC are engaged upon the expansion of their industrial capacity. In this they share an objective with virtually all developing countries, which together accounted for an estimated 10.9 per cent share of world industrial output in 1980.

The UN target of a 25 per cent share by the year 2000 (the so-called ''Lima'' target of UNIDO, the United Nations Industrial Development Organisation) may be overly ambitious, and in any case will tend not to have much relevance for the least industrialised within OPEC (such as Gabon or Nigeria). However, given that much of the industrialisation foreseen within OPEC is likely to be highly energy-intensive (such as aluminium smelting and oil-based downstream activity) there is every reason to expect a continued strong growth of oil needs.

In the past, the tendency has been for the share of industry within GDP to begin to accelerate at per capita income levels between $265-520 and $521-1,075, before slowing down to peak and then even decline. While some OPEC countries, especially those with small populations, are arguably already beyond this stage, the larger members states still have this phase to come. In principle, therefore, the OPEC Secretariat's forecast

4

that its demand for refined products will grow at 10 per cent a year between 1980 and 1990 seems reasonable. Indeed, in view of the 12 per cent annual growth recorded between 1975 and 1980—despite the collapse of demand in Iran after 1978—that estimate may even be modest.

The implications of this forecast depend critically on the way in which each OPEC member views the relationship between crude oil output, refining, home consumption and then exports of refined and unrefined volume. During the 1970s, years of below-average oil output (such as 1975) saw the ratio of home consumption to output rise considerably. In 1980, the year of the lowest crude output by the Organisation as a whole since 1971, domestic usage reached a record 8.4 per cent of output. This compares with only 3 per cent of output used domestically in the late 1960s.

Elsewhere in the Third World, oil products' use scarcely faltered after the big price increases of the 1970s. Policy set by central and local government certainly had a lot to do with this: not only did many governments in LDCs not tax gasoline and other petroleum products particularly severely—so as to induce careful use of it—some actually subsidized, directly or indirectly, the use of oil-derived products. This paradox will be examined at some length later in the book; for the time being what needs to be noted is that oil use is often not constrained at all by public policy. This in turn has implications for the likely future volume of demand.

The Oil Companies: Which Way Forward?

A further issue which arose during the 1970s concerned the future pattern of involvement of the oil companies in the industry. Given that most of the main oil-exporting countries' governments had, during the 1970s, arrogated to themselves far more authority in dealing with their oil reserves and with oil production policy generally than before, the involvement of the traditional suppliers in many countries was tending to be pre-empted. At the same time, economic forces in the developed countries were tending to make the conventional pattern of vertical integration—from well-head, through refining and distribution to retail sale—less attractive for the oil majors. Other pieces of the familiar picture—for instance, the oil companies' erstwhile practice of owning or chartering their own tanker fleets to carry crude oil and refined products between locations—were also being questioned as volumes shrank, interest-rates rose and a rather slimmer business structure, with assets concentrated on the highest-return segments of the industry, began to emerge.

There is, of course, no question that the oil companies possess unparalleled

5

expertise in their areas. In exploration, drilling, production management, marketing, financing and other functions they usually define the state of the art. But how OPEC member governments will wish to work with them, and in what regions and functions, is not yet entirely clear. There is little doubt that in both style and substance the oil companies may undergo considerable change before the century is out.

Not the least of the forces tending to undermine the oil majors' hegemony in the industry is the growing desire of many oil agencies in the OPEC members to involve themselves more closely in downstream activities. Indeed, intrinsic to the growth in this confidence has been a desire to add more value to crude oil before it is exported. Oil agencies in OPEC members were, by the mid-1970s, acutely aware of the fact that they acounted for trivial proportions of the world's oil refining capacity and, further downstream, the world's petrochemicals industry.

Other elements of the traditional vertically integrated pattern of the business, notably bulk oil transport in ships, were also observed to lie wholly or chiefly in developed country firms' hands. Indeed, in 1977, the 13 OPEC members accounted for a mere 6% of world oil refining capacity, as against 60% located in developed countries, 18% in the centrally planned economies, and 16% elsewhere. In the same year OPEC member agencies accounted for only 3% of the world's oil tanker fleet, measured in tonnes of capacity and judged by flag of origin. This contrasted with 54% for the developed countries and—due to the familiar anomaly of flags of convenience—38% in Liberia, Panama, and Singapore.

Increasingly frequently, OPEC spokesmen announced dissatisfaction with this state of affairs. Just as developing country governments and negotiators had for long argued the case for greater processing of raw materials prior to export in order to reap the benefits accruing from having more processing and manufacturing carried out within one's borders, so too did OPEC member spokesmen take up the theme that it was logical, rational and moral that the countries in which oil was found should also refine it and process it. Some speeches on this subject left little doubt as to the moral element in this option; "industrialization is as ferocious a battle as war, but we will finally triumph," was the observation of Saudi industry minister Ghazi al-Ghosaibi in mid-1978.

More recently, however, with OPEC members accounting (in 1982) for 8% of world refining capacity and a growing—albeit still small—proportion of ethylene capacity, doubts have come to be expressed over the extent to which the economic case for locating large refineries and petrochemical plants in OPEC countries is truly appealing. The huge glut

6

of refining and simpler bulk petrochemical capacity which overtook the European and North American markets in 1981 to 1983 raised questions about the desirability of having new facilities erected elsewhere. In some countries—Indonesia, for instance—the ambitious plans once laid for the downstream processing of oil have been substantially curtailed.

But this was largely on an emergency budget cutback basis; there did not seem to have been much reconsideration of the fundamental case for having petrochemical plants in the country. In those countries apparently still determined to press ahead—this largely means Saudi Arabia, Iran and Iraq—the case is rather more complicated, and depends on the availability of associated natural gas reserves to be used as the feedstock for the petrochemical plants. The extent to which OPEC capacity will integrate with existing refinery output is discussed later in the book. Central to this development is the economic structure of the OPEC countries.

OPEC countries were, particularly in the immediate aftermath of the 1973 and 1979 oil price rises, cast as countries of enormous power and capricious action. Talk of petrodollar surpluses spilling around the world helped create the impression that a new generation of wealthy states, much like Switzerland as Kuwait, had emerged. Nothing could, of course, be further from the truth. Most of the people in the OPEC member countries are desperately poor. Income distribution tends to be highly unequal, so that a small élite of wealthy aristocrats or functionaries gains disproportionate attention. But the governments of these countries—even the richest, such as Kuwait or Saudi Arabia—face enormous tasks in building up their states' infrastructure, public services, communications and—not least—expertise in international affairs.

Figures showing growth-rates for the countries' exports over long time periods indicate that petroleum-exporting countries' exports grew at an annual average rate of 14.0% over the 1950-80 period, and at rates of 7.3% in the 1950-60 decade, 7.9% in the next decade and 31.7% in the last, decisive, 1970-80 decade. By contrast, the corresponding figures for the other developing countries were 8.8% for the 30-year period, and 1.3%, 6.8% and 20.8% in the three decades. Imports by value grew at roughly corresponding rates, at 12.1% over the 30-year period for oil exporting countries and 9.4% for the other developing countries.

Oil and Politics Always Mix

As the price of oil stabilized and began to drop, so too did the degree of interest in the political dimension of the oil market. During the 1973-81 period analysts had come to see the supply of crude oil as being as much determined by political factors as anything else—a view which probably contributed to their underestimation of the purely market reaction which would later overwhelm the business. Many western analysts were convinced that the Middle East was so unstable a region that oil supplies would almost certainly be subject to frequent disruption. Not only would there be domestic upheavals—with Saudi Arabia the most likely candidate for a revolution, followed by Iraq, Iran, Egypt and others—but the extent of the Soviet Union's interest in the region would soon propel it to invading, in one way or another. This vision of the oil market was supplemented by one in which sales of oil from nationalized oil agencies would be—at least partially—dependent upon 'acceptable' policies being pursued by the client. As the most obvious example, it would be necessary to adopt an appropriate posture with respect to the subject of Israel.

To a large extent, of course, this has not happened. In the first place, exporting countries' oil agencies were often happy to supply to whomever wished to buy. South Africa, for instance, had little difficulty in securing deliveries from the National Iranian Oil Company. Second, even if a country should fall from favour for a while, complicated third-party shipping arrangements could be devised so as to disguise the true identity of the ultimate recipient of the oil. In addition, by the time the oil majors had relinquished a degree of control to national oil agencies in some countries, conditions of supply and demand had changed so as to make forgoing sales on the grounds of politics very unlikely.

This is not to say, however, that fears of political upheavals jeopardizing oil supplies have waned completely. There remain very real grounds for concern. Not only can it be argued that a number of the main oil-exporting states are intrinsically unstable as regards their internal political equilibrium, but it has also recently been argued that, from 1987 in particular, as certain governments' revenue begins to contract in line with the diminished real value of their oil royalties, strains will arise as public expenditure commitments are cut back. How might one begin to analyse the politics of OPEC, then?

Despite having existed for twenty years, OPEC members as political entities are still only hazily understood. Studies of many kinds have been made in the last decade, from multi-equation econometric models to political risk analyses, yet still the search for evidence of clear political objectives or coherent priorities between the thirteen member states' often conflicting interests has gone more or less unrewarded. Among the many economists' views that have held their adherents are those interpreting OPEC behaviour as a cartel; those stressing the price-leadership of a small sub-group of members with particularly large proven reserves; and the 'swing producer' view, whereby so-called 'moderate' members underwrite the price level that best suits them by adjusting the degree of their excess capacity. Close examination shows each of these views to have its merits and its frailties.

Yet a further view is one which focusses particularly upon OPEC's potential as a catalyst for bringing about political change for the benefit of the non-oil exporting developing countries. Two strands of this view exist. One is the argument that OPEC must for its own sake act increasingly in the interests of the non-oil LDCs because it depends for diplomatic support on critical foreign policy issues upon the latter. Thus the American theorist's opinion that "Saudi Arabia and Kuwait, whose regimes live in fear of overthrow, will be eager to seek political allies among the poor." The second strain of this theory asserts that OPEC members will give increasing weight to non-oil LDCs' interests because they want to—rather than have to—due to genuine interests in parts of the new international economic order.

Which, then, of these theories is most likely to be correct? And what does this imply for OPEC oil pricing policy? Will oil pricing be shaped with non-oil LDCs' interests in mind, and what priority would that criterion assume?

OPEC agencies have tended to reiterate some of the arguments—by now familiar—about the essential congruence of interests between OPEC and the rest of the Third World. The early theme, popular in OPEC press statements immediately after 1973, and amplified by those members—like Algeria—with more ambitious aims for a new international economic order was that OPEC was the first of a series of international commodity schemes which would redress the balance between primary product exporting countries and the others.

Given the lack of substance to this early promise, however, the ground of the OPEC diplomatic offensive has shifted somewhat in recent years. The more recent favoured position is that OPEC's own success, while not directly replicable in other raw materials, can nonetheless still be used to assist the non-oil LDCs via various aid schemes, oil price concessions, and by more imaginative and judicious use of oil supply threats to create new negotiating levers for the benefit of the unindustrialised world generally.

This recent change of position is illustrated in a speech of René Ortiz, OPEC Secretary General, in which he observed that "oil power... has not, thus far, been brought to bear... in international relations." This parallels the view of Fadhill Al-Chalabi, ex-OPEC Assistant Secretary General, whose recent book reworks the earlier theme that OPEC pricing actions not only serve as "a model for correcting the prices of other materials," but also "effectively strengthened inter-Third World solidarity."

The view that the Middle-Eastern OPEC states require the continuing diplomatic support of those countries they wooed away from Israeli aid and alignment in the early 1970s now appears to be rather too simple. Just as many of these states did not begin voting with the Arabs because they were convinced of the intrinsic merits of the Arabs' case, so too for the Arabs the active support of the European countries—now within sight—has become a far more valuable objective. While anything is worth acquiring at a price, the diplomatic support has in this case carried a price in terms of a substantial aid programme. That price may have become excessive. In real terms OPEC multilateral aid peaked in 1977; the objections from several black African leaders that their states were receiving insufficient aid to compensate them for oil price rises may have prompted a re-evaluation of the value to OPEC of these allies. More recently, the protestations of the hurriedly formed OPIC, the Organisation of Petroleum Importing Countries, at the fifth conference of UNCTAD in 1979, prompted chiefly a regional response from Venezuela (in concert with Mexico, a non-member of OPEC). As is clear from the diplomatic activity seen in the second half of 1982 and the early part of 1983, if the Israelis were ever truly 'out' of Africa, they are now fully back in. This hesitant start with which the proclaimed unity between oil client and exporter has opened is itself instructive, for it confirms that OPEC has at no stage accepted that higher oil prices have damaged the non-oil states' economies. As Ibrahim Shihata, Director General of the OPEC Fund, has argued, "the principle of compensation has not been invoked in respect of any other commodity." In contrast, OPEC's statements argue that by its oil price policy, and its preceding takeover of much of the multinationals' oil operations, the number of structural problems or injustices facing the Third World has been reduced. Thus to the extent that OPEC members collectively do not identify their actions as having harmed the non-oil LDCs' economies, price rises are unlikely in the future to be constrained substantially by the possibility that further rises will erode diplomatic support from these non-oil states.

Just as significant, perhaps, is the fact that any effort to give substance to the proposed elements of a new economic order—insofar as this entailed OPEC members contributing funds to LDCs for energy acquisition, or preferential oil pricing, for instance—would probably founder on the heterogeneity of OPEC itself. For the commonality of purpose needed for a significant effort along lines such as those listed is unlikely to be mustered. This is not to deny that, individually, certain OPEC states might be prepared to make substantial gestures; they conceivably

10

would. But by acting as individuals over the only important variable they can as yet periodically manipulate—the oil price—each would threaten the basis of their very success.

There may be some shared interests between OPEC and the non-oil LDCs, but they are likely to continue to fail to be coherently expressed while the world is interpreted through such different eyes.

Questions for the 1980s

Coming back to the question of how all these forces might affect the oil producers, governments and the consumers of oil products, one needs to bear in mind the factors already outlined which bear on supply and demand in crude oil. Related to these are a host of essentially political questions about the likelihood of interruptions to oil supply. Next is the question of oil refining, the process whereby products are derived, in the required volumes, from crude. Refiners have faced times just as uncertain as oil explorers and traders in the last decade. For just as demand for oil products has oscillated in sympathy with the price of crude oil, demand for oil refining capacity has also moved erratically. No longer can new capacity be brought onstream with an assured market for its products; demand has been too erratic for that. In fact, demand has simply been too low; in the first quarter of 1983, for instance, US refiners averaged only 66% capacity utilization. Figures for refinery closures confirm that recent years have been awkward. In 1982, nearly 1.2 million bpd of oil refining capacity was scrapped or mothballed in the US alone. No fewer than 57 refineries were closed. In 1981, by contrast, 23 refineries with aggregate capacity of 451,000 bpd closed. Although the closures have a certain logic to them—most operators are taking this opportunity of getting rid of their oldest and least economic facilities, as any firm tries to in a recession—they reflect serious financial problems for many firms.

By the middle of 1983, however, the picture was beginning to clear up a little. Unexpectedly rapid growth in the American economy, despite still very high interest rates, was encouraging greater personal car and air travel, while the recovery in the volume of industrial production was pushing up industrial fuel and business transport use. The refinery utilization rate in the US rose by 4.9% over its mid-1982 level to reach 78.3% in August 1983, and, by reopening a refinery previously closed and mothballed by Texaco at Tulsa, Sinclair Oil signalled that, for the time being at least the worst was over.

Not that gasoline demand had grown spectacularly—in July 1983 demand was still only averaging 6.858 million bpd, up 0.9% on the same period in 1982. Sales in the first five months of the year, at 6.412 million barrels/day were still 10.3% below the level reached in the same period of 1978. Moreover, for 1983 as a whole the Independent Petroleum Association of America forecast average sales only fractionally above their 1982 level, at 6.552 million barrels/day. Taking all refined products together, 1983 demand was expected by most analysts to run at approximately 15.3 million barrels/day, or about 0.3% above the 1982 level.

Meanwhile at the retail end of the oil industry—selling gasoline to consumers at service stations—some rather different questions have come to preoccupy participants. There is as yet no end in sight to the era of tight margins for gasoline retailers, and years such as 1982, which saw the biggest ever fall in retail gasoline prices in America (10 cents/gallon), do not help matters. Since 1972, the year when the biggest number of service stations existed in the US, some 80,000 units have disappeared. This is equivalent to nearly one-third of the sites. Some of these have been relinquished voluntarily, and have been absorbed into larger-volume sites; others have simply gone bankrupt, victim to high interest rates, decelerating and sometimes falling gasoline demand, and shifting patterns of travel.

Given this era of closure, what many operators have done to stave off competition is look outside the conventional confines of the industry to bolster sales and attract customers from other locations. They have co-operated with small retail stores (convenience stores, in trade jargon) and are increasingly diversifying into such enterprises as instant photographic print shops, as well as the more accustomed facilities such as car washes and repairs.

A period for consolidation now seems to be ahead, with most of the majors investing in slimming down the number of outlets they supply, concentrating them geographically, boosting the throughput of those which remain and diversifying their revenue sources away from reliance upon gasoline as far as possible.

It can be seen, then, that for oil-exporting countries' oil agencies, oil explorers, oil-importing countries' governments, multi-national oil firms, independent refiners and gasoline retailers, the world looks like an uncertain place. Each of these individuals or organisations must plan in expectation of a host of factors—few of them in his or their own control—coming about. Some, like a fall in the spot price of crude oil, may be welcome news to some (say, refiners) but bad news for others (finance ministers in oil-exporting countries, for instance). Some events, say a coup d'état in an oil-exporting country, may, in the short term at least, benefit some of those in the business—active spot-market speculators,

12

say, or those holding substantial oil inventories which can be upvalued on paper. In the longer term, however, events of that sort are likely to be unwelcome to virtually all of those in the business. The only thing which binds them all together is the unknown—the fact that in their industry, change has become the norm and equilibrium the unexpected.

★   ★   ★

The rest of the book is organized as follows. Chapter Two provides an overview of the oil products business in the major areas of the world, looking at patterns of consumption, output and trade. The main products are identified, along with their uses and the competition they face from competing products. This leads to a longer consideration, in Chapter Three, of the process whereby crude oil is transformed into products—refining. The refining business has faced many problems in the last decade after a period of rapid growth. Not only has the slump in demand for oil products seriously affected capacity utilization and thus profitability in the business, but refiners have also been faced with the need to ready themselves for new patterns of crude oil availability and an ever-changing pattern of product demand. If this were not enough, certain refiners have also faced problems arising from the intended entry of some OPEC-based oil firms into the refining business.

Chapter Four examines the marketing of refined oil products in the major markets—the US, Europe and elsewhere. It examines the forces acting on the retail end of the business as a counterpart to the earlier investigation, in Chapters Two and Three, of the upstream businesses. Chapter Five looks at the question of policy as it affects petroleum products. No business operates in a vacuum as far as political intervention is concerned, least of all the oil business, and many public policy decisions impinge on the gasoline and other products businesses. Chapter Six begins a detailed analysis of the future of the oil products business by looking at the likely future availability of the refiner's most crucial input—crude oil. Factors likely to play a part in determining oil demand are assessed in Chapter Seven, along with the resurgence of interest in the idea of a long-term stategy, which might guide OPEC countries' thinking on the oil industry and their place in it. Chapter Eight draws together some of the themes which have run through the book and looks forward to the 1990s.

Chapter Two

OIL AND OIL PRODUCTS: AN OVERVIEW

THIS chapter offers an overview of the world oil, oil refining and oil products business. It acts as an introduction to the scope of the book, and touches on some of the issues which will be explored at greater length later on.

World Oil Output

World oil production peaked in 1979. Since then it has been declining gently, to reach 53.19 million barrels/day in 1982. During the last decade oil output from communist areas has grown appreciably, from under 10 million barrels/day in 1970 to 14.71 million barrels/day in 1982 (see diagram 2.1). Other countries' oil output has also grown, while the 13 members of OPEC have suffered a very serious drop in output. To some extent this fall has reflected physical constraints—both Iran and Iraq have suffered from war-damage—but the bulk of the fall reflects output cuts made to forestall greater price falls than occurred. As Table 2.1 shows, OPEC members' output fell by no less than 17.2% between 1981 and 1982, with Saudi Arabia suffering a massive 34% fall to 6.364 million barrels/day. Diagram 2.2 shows oil output by region in 1982, and indicates that the area whose output fell most was the Middle East (down 21.1%) while European output grew by 10.7%. Among the countries whose output rose significantly were Iran, whose 50% output growth (to just under 2 million bpd) prevented the Middle East total falling even further; Mexico and the USA. Cognizant of the growing need to service her debts, Mexico's nationalized oil agency PEMEX boosted output by 18.8%, while British oil output rose 15.7% to average 2.13 million barrels per day. American production grew by 0.9% to 8.456 million barrels per day.

Substitution of other fuel sources for oil has been proceeding, often unnoticed, since the major oil price rises of 1973. As Table 2.2 shows, coal output has grown considerably over the last decade, rising from 2,141 million metric tons in 1970 to 2,861 million metric tons in 1982—an increase of 34%. Lignite output has also grown by 34% over the same period. National gas output was up from 38,440 petajoules in 1970 to 56,465 petajoules in 1982, while electricity output has increased very substantially. Electricity, other than that produced by hydro-power, is a secondary product, of course, and has in turn to be produced by burning oil, coal or some other materials. Over the same time-period the table shows crude oil output rising from 2,275 million metric tons in 1970 to

14

reach 3,127 million metric tons in 1979 before tailing off, by 1982 falling approximately to its 1975 level.

Table 2.1

OPEC OIL PRODUCTION IN 1982

Country	1982	Percent change from 1981
Abu Dhabi......	866	− 24.3
Algeria	775	− 4.6
Dubai	356	− 0.8
Ecuador	209	− 0.9
Gabon..........	141	− 6.6
Indonesia	1,344	− 16.2
Iran	1,998	+ 50.0
Iraq	914	− 0.3
Kuwait	657	− 29.0
Libya...........	1,203	+ 9.0
Neutral Zone ★ ..	305	− 17.8
Nigeria	1,289	− 10.0
Qatar...........	331	− 18.3
Saudi Arabia....	6,364	− 34.0
Sharjah.........	7	− 30.0
Venezuela.......	1,891	− 10.3
Total.........	18,650	− 17.2

★ Shared by Kuwait and Saudi Arabia
Source: Oil and Gas Journal

Diagram 2.1

HOW WORLD OIL PRODUCTION IS DECLINING

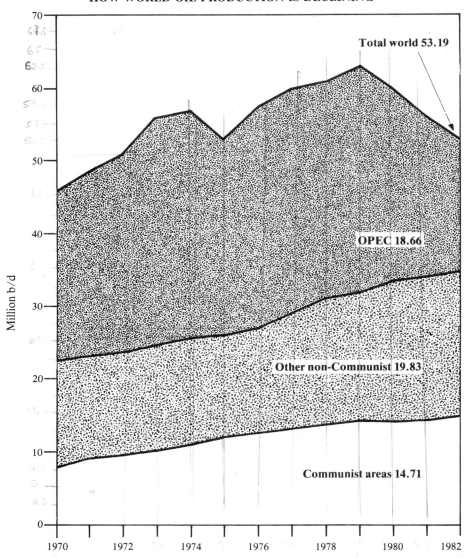

Source: Oil and Gas Journal

16

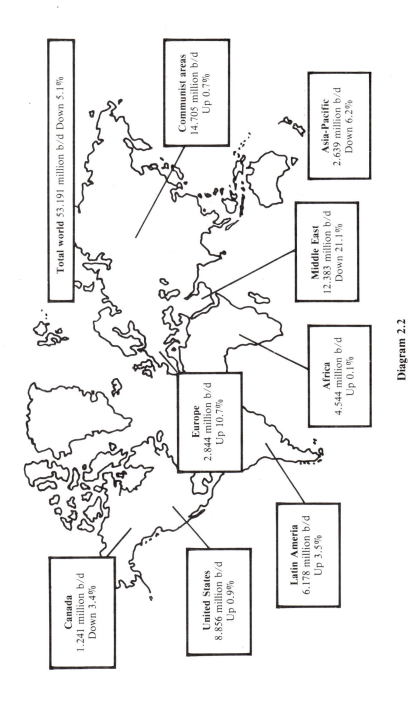

Total world 53.191 million b/d Down 5.1%

Communist areas
14.705 million b/d
Up 0.7%

Asia-Pacific
2.639 million b/d
Down 6.2%

Middle East
12.383 million b/d
Down 21.1%

Africa
4.544 million b/d
Up 0.1%

Europe
2.844 million b/d
Up 10.7%

Canada
1.241 million b/d
Down 3.4%

United States
8.856 million b/d
Up 0.9%

Latin Ameria
6.178 million b/d
Up 3.5%

Diagram 2.2

WORLD OIL PRODUCTION BY REGION, 1982

Source: Oil and Gas Journal

Table 2.2

WORLD OUTPUT OF ENERGY MATERIALS 1970-82

Year	coal	lignite	crude oil	natural gas	electricity
	million	metric tons		petajoules	10^9 kWh
1970	2,141	794	2,275	38,440	4,953
1972	2,164	813	2,548	43,391	5,697
1973	2,209	828	2,780	45,284	6,126
1974	2,245	848	2,789	45,383	6,313
1975	2,361	869	2,644	46,232	6,519
1976	2,420	896	2,875	48,444	6,985
1977	2,500	935	2,873	49,726	7,300
1978	2,540	951	3,011	51,742	7,688
1979	2,686	985	3,127	54,376	7,999
1980	2,732	1,005	2,975	54,921	8,224
1981	2,742	1,038	2,789	55,759	8,357
1982	2,861	1,062	2,625	56,465	8,188

Source: United Nations, *Monthly Bulletin of Statistics, 1983*.

Refining: An Introduction

After being brought to refineries in tankers or pipelines, crude oil is stored in groups (or 'farms') of tanks, which are made of steel. Up to 50,000 tons of oil may be stored in each tank. The primary process with which the crude is treated in the refinery is distillation. This entails having the various hydro-carbon compounds (oil being primarily composed of hydrogen and carbon) separated from one another. Heated crude oil enters a fractionating tower (as in diagram 2.3) where it is separated into streams of gasoline, kerosene, and so on, with the lighter products appearing at the top of the fractionating tower and the thicker products at the bottom. The fractions thereby obtained are called 'straight runs'. The oil is then treated in secondary stages, where specific products can be obtained from the straight runs. Among the more complicated of these secondary processes are those used to purify gasoline. Reforming is one such process. It entails taking naphtha, for instance, and changing its chemical structure, by heat and pressure with a catalyst present. Cracking is a process whereby heavy cuts from distillation are converted into lighter products.

Diagram 2.3

FRACTIONATING TOWER

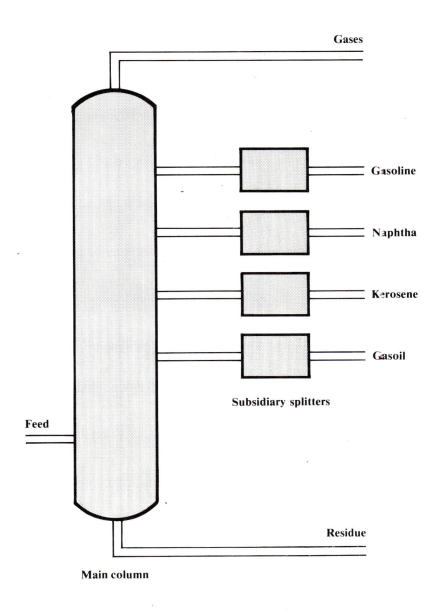

Source: OPEC

Diagram 2.4

A HYPOTHETICAL AVERAGE CRUDE

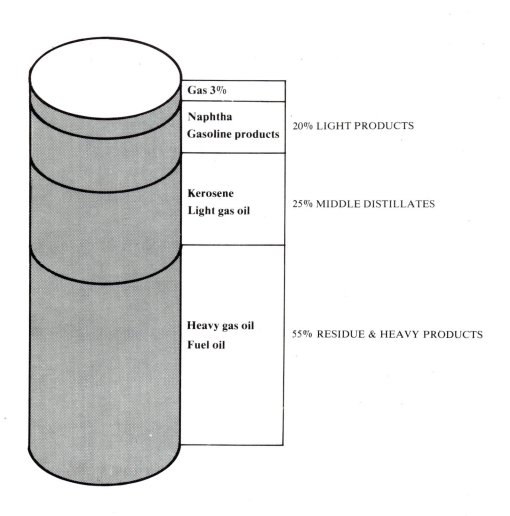

Gas 3%

Naphtha
Gasoline products

20% LIGHT PRODUCTS

Kerosene
Light gas oil

25% MIDDLE DISTILLATES

Heavy gas oil
Fuel oil

55% RESIDUE & HEAVY PRODUCTS

Source: OPEC

During refining, a number of impurities are removed from crudes. Sulphur is one which it is particularly important to remove, since too much sulphur in fuel oils can render them unacceptable for certain pollution standards. Among the other chemicals which may be present in crudes are nitrogen, salt and oxygen, nickel, vanadium, arsenic and chlorine. 'Sweetening' the crude entails adding chemical compounds to mitigate the unattractive aspects (such as smell or corrosiveness) of certain crudes. A crude which intrinsically has a low (say under 1%) sulphur content is called 'sweet'.

When crude oil is being treated in a refinery it yields products which can be considered under three headings: gas and gasoline, middle distillates, and fuel oil and residuals. Gas and gasoline (''white'' products) yield gas for domestic use, aviation fuel, gasoline for cars, and feedstocks for petrochemicals. These items constitute the lighter end of the barrel. Diagram 2.4 shows these items as part of a notional 'typical' barrel. The middle distillates include kerosene and light gasoil, heating oil, diesel oils and waxes. Naphtha, which is the essential input to petrochemicals plants, is obtained both from middle and from light distillate cuts. The bottom end of the barrel contains heavy fuel oils (used in such places as power stations and ships' furnaces) and asphalt and bitumen. Diagram 2.5 shows a selection of products ranged from light at the top to heavy at the bottom.

The products yielded by a notional 'typical' refinery are as follows:

Light products

● *Fuel gas* or *refinery gas*. These are produced in considerable quantities during the refining processes and are used as fuel for the refinery. They are an important feedstock for petrochemicals.

● *Liquid petroleum gas* (LPG). This is used as domestic bottled gas for cooking and heating, and in industry for such tasks as cutting metal.

● *Motor spirit* and *gasoline,* for cars and trucks. Known also as petrol, this forms one of the most important refinery products. The final fuel used in transport is a carefully blended mixture.

Refineries have to produce motor fuels with high octane rates which apply to what is called the anti-knock quality of gasoline. The anti-knock property of a gasoline prevents it from igniting too early under compression and ensures that it burns smoothly and easily in car engines. High octane rates, necessary for modern engines, are produced during the secondary refining process. The gasoline group

Diagram 2.5

OIL-PRODUCTS AND USES

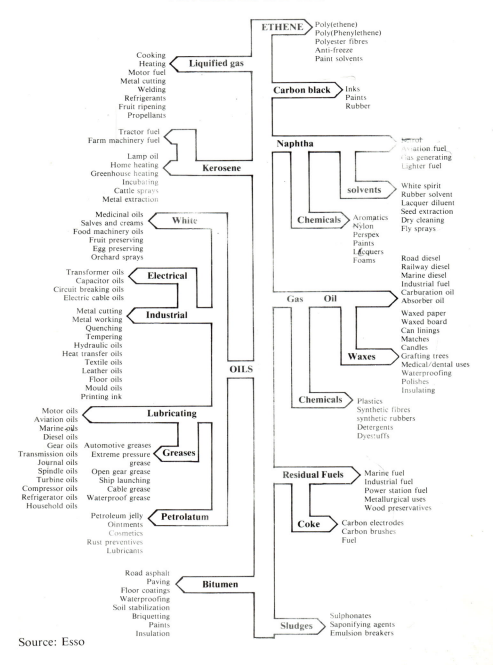

Source: Esso

of products also includes special solvents and industrial spirits used in paints, dry-cleaning and rubber solvents.

Middle distillates

- *Aviation* and *jet fuels* for piston and turbine-engined aircraft.

- *Gas oil* or *kerosene*. Sometimes known as paraffin, this product is used for central heating, domestic heating stoves, space heaters and in farm machinery.

- *Diesel fuels* also form part of the middle distillate gasoils group. Diesel fuels come in two broad groups, for high speed engines in cars and trucks requiring a high quality product, and lower quality heavier diesel fuel for slower engines, such as in marine engines or for stationary power plants.

Heavy products

- *Fuel oils*. These black oils come from the bottom end of the barrel and are the residue left over after cracking and distillation. They are largely used in power stations, ships and industrial furnaces. Fuel oil is often a blend of the lighter unused surplus products and heavy viscous residue oils.

Most large ships, such as tankers, are powered by bunker oils—usually a mixture of inexpensive fuel oils from the residue, suited to the large low speed marine engine. Bunker fuels also include some heavy diesel fuels such as that classified as Marine Diesel Oil.

The fuel oils cover a wide range of uses, from domestic heating and cooking oils to small scale commercial heating and large scale industrial heating. Market requirements vary enormously from region to region and according to quality. Depending on the market, standards are set according to viscosity and sulphur content. Kerosene and gasoil are often selectively blended to dilute the residues to meet the required market gradings, which in some parts of the world are strictly set according to environmental considerations.

In the US, one of the largest markets, fuel oils used for heating are graded from No. 1 Fuel Oil to No. 6 Fuel Oil and cover light distillate oils to medium distillate, heavy distillate, a blend of distillate and residue, and residue oil. For instance, No. 2 and No. 3 Fuel Oils refer to medium-to-light distillate grades used in domestic central heating, while No. 5 Fuel Oil refers to a medium heavy residual oil used for heating large commercial premises.

- *Bitumen*, sometimes referred to as asphalt, is used in roads and for roofing and waterproofing. It is produced to certain standards of hardness or softness in controlled vacuum distillation processes.

- *Lubricating oils* and *waxes*. Lubrication oils come in a huge variety of products, ranging from very fine fluid lubricants to heavy viscous oils and greases. Waxes are used for candles, paper, waterproofing of cartons, etc.

CH.1

IV *Consumption of Refined Products*

Figures from the United Nations are available which show demand by year for different types of refined oil product. In aggregate, world demand for refined products in the UN definition approached 2,000 million metric tonnes annually in the late 1970s, but fell away somewhat after peaking at 1,988 million metric tonnes in 1979.

Table 2.3

THE WORLD MARKET FOR REFINED PRODUCTS, 1976-80, ALL PRODUCTS
(thousand metric tons)

Year	Production	Imports	Consumption per capita (kg)
1976	1,859,578	235,940	422
1977	1,919,087	245,078	428
1978	1,947,793	234,399	426
1979	1,987,746	243,586	424
1980	1,913,948	236,838	400

Source: UN *Yearbook of Energy Statistics, 1982*

Breaking down the total into its constituent products, Table 2.4 shows production, imports and per capita consumption over a five-year period for six products. It is apparent that residual fuel oil is the single most

24

important product when measured by volume of production. Motor gasoline comes second. It is apparent from the table that only in the cases of aviation gasoline and residual fuel oil did output fall over the period shown; output of the other products rose.

Table 2.4
THE WORLD MARKET FOR REFINED PRODUCTS, 1976-80
(thousand metric tons)

Year	Production	Imports	Consumption per capita (kg)
(a) liquified petroleum gas			
1976	100,421	15,839	25
1977	105,568	18,028	25
1978	106,891	17,839	25
1979	115,603	23,289	27
1980	114,645	24,137	26
(b) aviation gasoline			
1976	3,325	1,132	1
1977	3,287	1,059	1
1978	3,396	1,010	1
1979	3,287	1,001	1
1980	3,153	931	1
(c) motor gasoline			
1976	612,994	30,351	148
1977	635,270	34,749	150
1978	657,838	33,919	153
1979	655,273	35,202	149
1980	642,469	34,789	143
(d) kerosene			
1976	114,897	6,296	27
1977	120,672	7,864	28
1978	123,277	7,746	28
1979	129,608	7,830	29
1980	123,559	7,344	27
(e) jet fuel			
1976	93,165	13,035	16
1977	96,651	14,962	17
1978	99,570	15,316	17
1979	105,866	16,180	17
1980	106,256	16,169	17
(f) residual fuel oil			
1976	934,776	169,287	205
1977	957,639	168,416	207
1978	956,821	158,569	202
1979	978,109	160,084	201
1980	923,866	153,468	186

Source: UN

Diagram 2.6 shows aggregate production of all six refined products, and indicates the preponderance of residual fuel oil and motor gasoline in the mix of output. It is clear from the diagram that there has been comparatively little change in the relative importance of the various products. Looked at in per capita terms, as diagram 2.7 shows, there has been a slight shift away from residual fuel oil (from 48.6% of total personal consumption in 1976 to 46.5% in 1980) towards liquid petroleum gas, kerosene and gasoline. These movements have, however, been slight.

Table 2.5 shows the importance of imports relative to total production for each product, averaged over the 1976-80 period. This can be taken as a measure of the import reliance of countries in each fuel's case. The table shows that international trade is far more intense in the aviation gasoline business than in kerosene or motor gasoline. Motor gasoline turns out in fact to be the least intensively-traded of the main petroleum products.

Table 2.5

IMPORTS AS % OF PRODUCTION OF REFINED PRODUCTS, WORLD AVERAGE, 1976-80

	%
All products	12.4
LPG	18.1
Aviation gasoline	31.1
Motor gasoline	5.3
Kerosene	6.0
Jet fuel	15.1
Residual fuel oil	17.1

Source: UN

Consumption of motor gasoline, the petroleum product given the most attention in this book, ranged in 1980 (the last year for which internationally comparable figures from the United Nations were available) from 1,244 kg per capita in the USA to 3 in Burundi. Table 2.6 shows per capita use for 1980. A major fall in US consumption was recorded after 1978, when consumption reached 1,458 kg/capita. Between 1978 and 1980 per capita usage in the US fell by 14.7%. Indeed, over the world as a whole, per capita use of gasoline fell from a peak of 153 kg/capita in 1978 to 143 in 1980. It tended to be the case that the richer countries were the ones in which consumption fell most dramatically. The case of the US has already been cited. Western Europe provides a

26

Diagram 2.6

WORLD PRODUCTION OF REFINED PRODUCTS, 1976-80
(% breakdown)

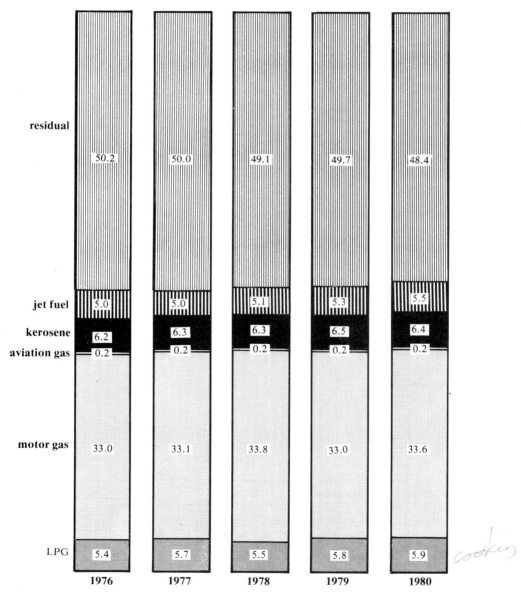

Source: Calculated from UN figures

27

Diagram 2.7

**CONSUMPTION OF REFINED
PETROLEUM PRODUCTS, 1976-80**

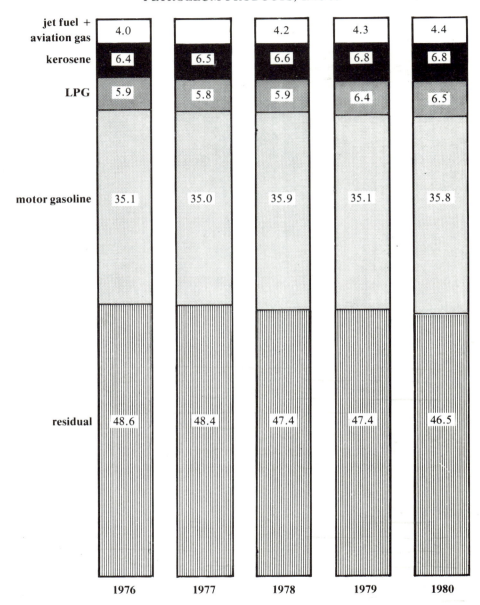

Source: Calculated from UN figures

28

Table 2.6

PER CAPITA CONSUMPTION OF MOTOR GASOLINE, 1980: SELECTED REGIONS (kg/year)

The world	143
Developed market economies	572
Developing market economies	36
Centrally planned economies	65
OPEC members	64
Africa	30
Africa excluding South Africa	23
South & Central America	127
Middle East	83
Far East, developing	9
Centrally planned Asia	11
West Europe	273
Other Europe	133
Centrally planned Europe	210
Oceania, developed	705
Oceania, developing	88
memo: USA	1,244

Source: UN

similar, though less marked, case of contracting demand (a 2.8% fall between 1979 and 1980). But in centrally planned Europe (consumption up by 4.5% between 1976 and 1980), developing Africa (consumption up by 21% over the same period) and the 13 OPEC members (a rise of 36%), the reverse phenomenon was observed. Chapter Five deals at length with aspects of gasoline (and petroleum) policy in different countries, but here it is worth noting a few points since public policy, at the stages of exploration, production, exporting, transporting, stockpiling and consuming, is so intimately involved in the industry.

First, an important distinction between different groups of countries is that after 1973 most governments in developed countries enacted tax increases to decelerate the rate of growth of gasoline demand in their economies. On the other hand many developing country governments failed to make this move, or made it but blunted its effect through inadequate execution. Some developing country governments continued to subsidize the consumption of gasoline. In centrally planned economies, subsidies coupled with very low prices at retail level, and few serious incentives to diminish oil use, all contributed to growing demand. Only well into the 1980s was the profligate use of oil in the USSR, for instance, even begun to be checked.

Second, as the volume of consumption used rises, a margin of arbitrary or discretionary fuel use inevitably arises. Driving slightly less briskly, heating a house to a slightly lower temperature and being more careful about sheer wastefulness are relatively easy options open to consumers when encouraged, through price and non-price means, to conserve. On the other hand when fuel use is extremely low in absolute terms, that margin for easy savings ('costless' savings, one might say) is necessarily reduced.

In industrial uses, other energy sources have been substituted for oil, often to an impressive extent. In the US utility sector, for instance, facilities burned 50% less oil in 1982 than in 1972 while generating 440% more power from nuclear sources. Hydroelectric power from Canada is being used far more intensively than before 1973: American imports doubled over the 1978-82 period and may double again by 1985.

Just how profound the reversal of decades of upward oil consumption has been is shown in tables 2.7 to 2.12. These show data on oil intensity, which is to say the volume of oil used per $1,000 of GDP generated in various countries and regions of the world. In the case of all the OECD countries taken together (the major developed economies), oil intensity peaked at 3.35 in 1973, exactly as one would expect, in retrospect, and since then has fallen to reach 2.31 in 1982. Looking at total energy intensity, which embraces not only oil, but also coal, nuclear fuels, etc., one sees that the OECD intensity index peaked at a level of 6.11 in 1970, before falling to 4.73 by 1982. The next column in the same table shows indices with different base years, ranging from 1970 to 1976. Taking 1973 as a base, it appears that the oil index had fallen by 1982 to 69.09 and the energy index overall had fallen to 78.95.

The table on the centrally planned economies shows, for oil intensity, a much later peak, in 1978, and only a gradual falling off, to the level of 3.19 by 1982. Compared to the earlier table, the record of the CPEs is of longer growth in oil intensity, from a slightly lower base in 1950, a less abrupt and more diffuse reaction to the price rises of 1973, and an arrival at a significantly higher level of oil intensity than in the OECD economies by 1982. In terms of energy intensity, the story is similar, with a peak coming rather earlier than with oil alone (in this case, in 1960) but with a prolonged and gradual decline in energy intensity. By 1982 it still stood at 10.30, however, a level more than twice as high as in the OECD countries. Looking at the base 1973 series, it appears that oil use per $1,000 of GDP in these countries has fallen only to 95.57 in the last decade—around one-seventh the size of the reaction in the OECD countries.

30

Table 2.7

ENERGY & OIL INTENSITY IN OECD COUNTRIES 1950-1982

	Energy demand	Oil demand	GDP	Energy intensity	Oil intensity
	1,000 b/doe[1]		$ million	boe/$1,000 GDP[2]	
1950	25,176	8,058	1,390,932	6.61	2.11
1960	34,827	14,754	2,146,710	5.92	2.51
1970	58,602	31,172	3,499,180	6.11	3.25
1971	60,305	32,317	3,622,540	6.08	3.26
1972	63,078	35,004	3,821,420	6.02	3.34
1973	66,676	37,229	4,059,930	5.99	3.35
1974	64,768	35,578	4,085,580	5.79	3.18
1975	62,600	34,139	4,077,570	5.60	3.06
1976	66,615	36,566	4,284,280	5.68	3.12
1977	67,578	37,570·	4,436,500	5.56	3.09
1978	69,047	38,273	4,610,550	5.47	3.03
1979	71,454	38,893	4,756,220	5.48	2.98
1980	68,851	35,704	4,811,620	5.22	2.71
1981	65,949	52,731	4,882,780	4.93	2.45
1982	63,050	30,810	4,862,900	4.73	2.31

Source: OPEC

Notes: [1]1,000 barrels per day oil equivalent

[2]Barrels oil equivalent per $1,000 GDP

Table 2.8

ENERGY & OIL INTENSITY IN CENTRALLY PLANNED ECONOMIES 1950—1982

	Energy demand	Oil demand	GDP	Energy intensity	Oil intensity
	1,000 b/doe		$ million	boe/$1,000 GDP	
1950	6,271	881	182,990	12.51	1.76
1960	16,804	2,892	437,150	14.03	2.41
1970	25,098	6,891	817,830	11.20	3.08
1971	26,956	7,607	866,550	11.35	3.20
1972	28,143	8,347	911,100	11.27	3.34
1973	29,655	9,088	993,234	10.90	3.34
1974	31,046	9,819	1,051,200	10.78	3.41
1975	32,785	10,465	1,107,380	10.81	3.45
1976	34,182	11,024	1,166,420	10.70	3.45
1977	36.073	11,585	1,230,450	10.70	3.44
1978	38,021	12,335	1,291,900	10.74	3.48
1979	38,825	12,514	1,330,440	10.65	3.43
1980	39,373	12,831	1,376,790	10.44	3.40
1981	39,811	12,703	1,408,812	10.31	3.29
1982	41,045	12,717	1,454,290	10.30	3.19

Source: OPEC

Table 2.9

ENERGY & OIL INTENSITY IN LESS DEVELOPED COUNTRIES 1950-82

	Energy demand	Oil demand	GDP	Energy intensity	Oil intensity
	1,000 b/doe		$ million	boe/$1,000 GDP	
1950	1,659	964	202,915	2.98	1.73
1960	3,362	2,072	324,664	3.78	2.33
1970	6,803	4,636	549,304	4.52	3.08
1971	7,318	5,082	581,718	4.59	3.19
1972	7,766	5,394	623,055	4.55	3.16
1973	8,577	6,034	665,288	4.71	3.31
1974	8,833	6,105	703,708	4.58	3.17
1975	9,046	6,100	730,133	4.52	3.05
1976	9,786	6,706	769,250	4.64	3.18
1977	10,326	7,053	813,387	4.63	3.17
1978	10,925	7,428	859,339	4.64	3.16
1979	11,767	7,931	906,889	4.74	3.19
1980	12,291	8,132	950,744	4.72	3.12
1981	12,667	8,246	972,918	4.75	3.09
1982	13,072	8,425	997,904	4.78	3.08

Source: OPEC

Turning now to the table on the less developed countries (LDCs), it is clear that the general upward trend was only briefly interrupted by the oil price rises of 1973. Although oil intensity has fallen to only 93% of its 1973 level, overall energy use per $1,000 of GDP has risen since that year, by around 1.6%. The comments made in the chapter on policy have a bearing on this. OPEC is shown as a seperate group of countries in the next table. The figures contained there make it clear that from a much lower starting point in 1950, the OPEC countries' oil intensity had grown by 1982 to more than two and a half times its starting point. Oil use, relative to the base of 1973, has grown by no less than 36% over the decade to 1982—by far the biggest rise seen in any group of countries. Overall energy use is also clearly growing very quickly too, although by no means as fast as oil use itself.

32

Table 2.10

ENERGY & OIL INTENSITY IN OPEC COUNTRIES 1950-1982

	Energy demand	Oil demand	GDP	Energy intensity	Oil intensity
	1,000 b/doe		$ million	boe/$1,000 GDP	
1950	228	162	54,559	1.53	1.09
1960	806	605	87,294	3,37	2.53
1970	1,511	1,028	172,878	3.19	2.17
1971	1,631	1,138	187,880	3.17	2.21
1972	1,690	1,158	208,896	2.95	2.02
1973	1,958	1,330	230,631	3.10	2.10
1974	2,194	1,519	249,945	3.20	2.22
1975	2,283	1,570	254,998	3.27	2.25
1976	2,568	1,824	281,078	3.33	2.37
1977	2,830	2,067	294,737	3.51	2.56
1978	3,133	2,193	302,696	3.78	2.64
1979	3,282	2,367	320,896	3.73	2.69
1980	3,281	2,329	331,728	3.61	2.56
1981	3,363	2,429	325,946	3.77	2.72
1982	3,259	2,460	313,829	3.79	2.86

Source: OPEC

The table showing oil intensity for the entire world shows a fairly abrupt decline after 1973, from 3.31, the peak, to 2.63. This figure is very heavily weighted by the OECD group's experience, since members are still overwhelmingly the biggest oil users in the world. Although other sub-groups show different rates of increase, their weight in world economic activity is too low to have any significant impact on the aggregate oil use pattern.

Table 2.11

ENERGY & OIL INTENSITY: WORLD 1950-1982

	Energy demand	Oil demand	GDP	Energy intensity	Oil intensity
	1,000 b/doe		*$ million*	*boe/$1,000 GDP*	
1950	33,661	10,111	1,831,396	6.71	2.02
1960	56,348	20,427	2,995,818	6.87	2.49
1970	92,906	43,987	5,039,192	6.73	3.19
1971	97,207	46,460	5,258,688	6.75	3.22
1972	101,671	50,215	5,564,471	6.67	3.29
1973	107,928	54,020	5,949,083	6.62	3.31
1974	107,942	53,358	6,090,433	6.47	3.20
1975	107,885	52,642	6,170,081	6.38	3.11
1976	114,379	56,503	6,501,028	6.42	3.17
1977	118,094	58,665	6,775,074	6.36	3.16
1978	122,424	60,622	7,064,485	6.33	3.13
1979	126,701	62,089	7,314,445	6.32	3.10
1980	125,192	59,348	7,470,882	6.12	2.90
1981	123,366	56,531	7,590,454	5.93	2.72
1982	122,300	54,875	7,628,923	5.85	2.63

Source: OPEC

Usage of oil per $1,000 of GDP has thus been taken back to its level of 1962-63. This figure echoes one of the major themes of this book; that demand for refined products peaked ten years ago in most countries, has fallen substantially since then, and henceforth may grow only slowly even once world economic activity gathers pace.

Oil consumption per capita for various countries is shown in Table 2.12. It is clear that the USA is the economy which still has the greatest demand—21.48 barrels per inhabitant in 1982, although this figure is sharply down on the peak of just over 29 barrels in 1978. This 1982 figure is nearly five times the world average and is nearly 15 times the level attained in the LDCs as a group. The OECD group as a whole shows a demand per capita of 14.22 barrels per year in 1982, down from a peak of 18.48 barrels in 1973. Japan (also a member of OECD) shows 18.19 barrels in 1982, a relatively small decline from its 1973 level of 22.00 barrels.

34

Table 2.12

OIL DEMAND PER CAPITA 1950-1982

boe/capita

	World	OPEC	CPEs	LDCs	OECD	USA	Europe	Japan	Rest
1950	1.49	0.36	0.38	0.39	5.20	15.66	1.24	0.14	6.88
1960	2.50	1.08	1.05	0.67	8.41	19.27	3.93	2.13	12.74
1970	4.42	1.45	2.12	1.15	15.95	25.07	11.33	13.08	19.82
1971	4.57	1.55	2.29	1.23	16.35	25.48	11.61	13.82	20.32
1972	4.82	1.54	2.46	1.26	17.54	27.35	12.34	15.46	20.92
1973	5.09	1.72	2.63	1.38	18.48	28.61	12.96	16.88	22.00
1974	4.92	1.91	2.79	1.36	17.44	26.98	12.20	15.51	21.84
1975	4.76	1.93	2.92	1.33	16.59	26.00	11.23	15.16	21.28
1976	5.02	2.18	3.04	1.42	17.64	27.58	12.20	15.76	21.19
1977	5.12	2.41	3.15	1.46	17.99	28.81	11.94	16.42	21.42
1978	5.21	2.50	3.31	1.51	18.18	29.11	12.15	16.26	21.56
1979	5.24	2.62	3.32	1.57	18.33	28.14	12.91	16.32	21.97
1980	4.92	2.52	3.36	1.57	16.69	25.19	11.80	15.03	21.21
1981	4.61	2.56	3.29	1.55	15.19	23.01	10.58	13.79	19.84
1982	4.40	2.52	3.26	1.55	14.22	21.48	10.02	12.67	18.19

Source: OPEC

It would be wrong to look only at the size of the declines since 1973, however; the absolute size of the long-term growth in demand is also important. For since 1950 oil use per capita in the OECD group has grown over two and a half times to reach its 1982 level; in the world as a whole it is nearly three times its 1950 level. This growth implies a great deal of activity in the supply sector, and this is what the next chapters begin to deal with.

American refinery output has remained relatively flat over the last two years. Output of motor gasoline, the most important product in volume terms, has hovered close to 6.5 million barrels/day since early 1981, with the usual slight seasonality in output induced by the peaks of summer demand. Diagram 2.8 shows this. Distillate fuel oil is a smaller-volume product; output of it has held to around 2.5 million barrels/day while residual fuel output has been almost entirely unchanged at close to 1 million barrels/day.

Imports of petroleum follow an erratic path due to domestic demand being met first by home output and only second by imports. Although nuances of quality and price will lead to imports being preferred to domestic output on occasions, the reason for American crude imports falling so dramatically is simply that of a lower level of demand being more completely satisfied from domestic refining capacity.

The largest product imports are those of residual fuel oil. Over the period 1981-83 they collapsed from 900,000 barrels/day to nearly 500,000, before picking up a little. Finished motor gasoline imports rarely exceed 0.2 million barrels/day, a reflection of the capacity available for refining within the US. Distillate fuel oil imports are even smaller; frequently the monthly volume of imports is below 50,000 barrels/day (see diagram 2.9).

The hierarchy of products can be seen in table 2.13, which shows demand in the US for the major refined items in 1981 and 1982. Within total 1982 demand of 5,567.6 million barrels/day, gasoline accounted for 43%. Distillate fuel oil accounted for 18%, jet fuel for 7%, liquefied gases for 10%, and residual fuel oil for 11%. Asphalt accounted for 2% of demand and lubricants for a further 1%.

Table 2.13

DEMAND FOR REFINED PRODUCTS, USA, 1982-83
(million barrels/day)

	1981	1982
Total ★	5,861.1	5,567.6
Gasoline	2,415.6	2,395.6
Kerosene	46.3	46.8
Distillate fuel oil	1,032.5	975.5
Residual fuel oil	762.0	618.4
Jet fuel	367.7	367.7
Lubricants	56.0	50.9
Asphalt	124.0	124.5
Liquefied gases	535.0	563.5

Source: US Dept. of Energy.

Diagram 2.8

REFINERY PRODUCTION BY PRODUCT
(Tons of Barrels per Day)

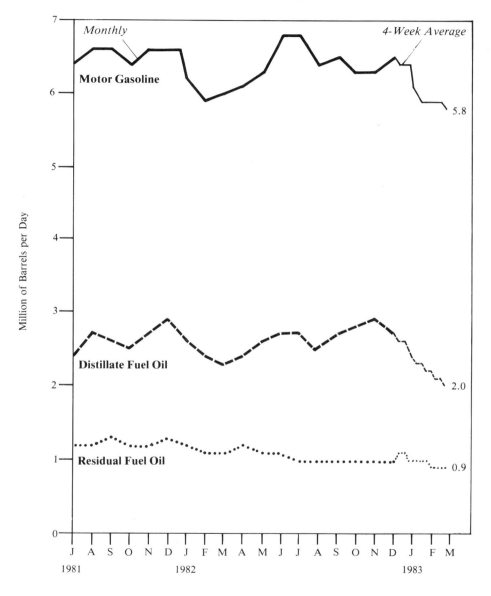

Source: US Dept. of Energy

Diagram 2.9

IMPORTS OF PETROLEUM PRODUCTS BY PRODUCT
(Thousands of Barrels per day)

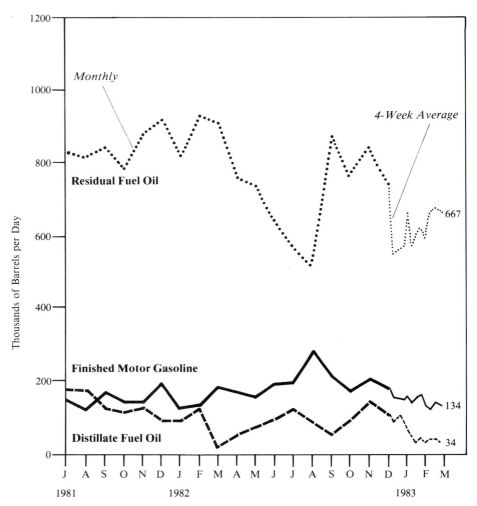

Source: US Dept. of Energy

Table 2.14
WORLD CONSUMPTION OF PETROLEUM PRODUCTS 1972-1982

Country/Area	1972	1977	1982	Yearly Change 1982 over 1972	Yearly Change 1982 over 1977
	Million Tonnes				
USA					
Gasolines	300.8	329.5	295.2	− 0.2%	− 2.2%
Middle Distillates	191.6	224.2	186.6	− 0.3%	− 3.6%
Fuel Oil	133.3	160.8	89.5	− 3.9%	− 11.0%
Others	150.1	151.4	131.7	− 1.3%	− 2.8%
Total	**775.8**	**865.9**	**703.0**	**− 1.0%**	**− 4.1%**
Canada					
Gasolines	24.5	29.8	27.7	+ 1.3%	− 1.4%
Middle Distillates	25.4	26.8	22.7	− 1.1%	− 3.2%
Fuel Oil	16.8	15.8	9.6	− 5.4%	− 9.5%
Others	12.6	13.2	13.0	+ 0.3%	− 0.4%
Total	**79.3**	**85.6**	**73.0**	**− 0.8%**	**− 3.1%**
Western Europe					
Gasolines	125.3	132.8	131.6	+ 0.5%	− 0.2%
Middle Distillates	231.2	244.3	217.1	− 0.6%	− 2.3%
Fuel Oil	256.4	227.1	163.4	− 4.4%	− 6.4%
Others	88.9	93.1	89.0	*	− 0.8%
Total	**701.8**	**697.3**	**601.1**	**− 1.5%**	**− 2.9%**
Japan					
Gasolines	38.3	44.6	38.6	+ 0.1%	− 2.9%
Middle Distillates	41.1	55.7	57.6	+ 3.4%	+ 0.6%
Fuel Oil	124.9	128.9	78.3	− 4.6%	− 9.5%
Others	30.1	31.2	32.5	+ 0.8%	+ 0.9%
Total	**234.4**	**260.4**	**207.0**	**− 1.2%**	**− 4.5%**
Australasia					
Gasolines	10.6	12.9	14.1	+ 2.9%	+ 1.9%
Middle Distillates	7.9	10.8	10.0	+ 2.5%	− 1.5%
Fuel Oil	8.7	9.1	6.6	− 2.8%	− 6.2%
Others	4.5	5.2	4.8	+ 0.5%	− 1.6%
Total	**31.7**	**38.0**	**35.5**	**+ 1.1%**	**− 1.3%**
Rest of World					
Gasolines	64.5	80.3	105.7	+ 5.0%	+ 5.7%
Middle Distillates	104.1	154.0	186.9	+ 6.0%	+ 3.9%
Fuel Oil	138.8	163.3	194.7	+ 3.4%	+ 3.6%
Others	48.8	64.3	79.9	+ 5.1%	+ 4.5%
Total	**356.2**	**461.9**	**567.2**	**+ 4.8%**	**+ 4.2%**
World (excl.USSR, E.Europe & China)					
Gasolines	564.0	629.9	612.9	+ 0.8%	− 0.5%
Middle Distillates	601.3	715.8	680.9	+ 1.3%	− 1.0%
Fuel Oil	678.9	705.0	542.1	− 2.2%	− 5.1%
Others	335.0	358.4	350.9	+ 0.5%	− 0.4%
Total	**2,179.1**	**2,409.1**	**2,186.8**	*****	**− 1.9%**

Note:—"Others" represent refinery gas, LPGS, solvents, petroleum coke, lubricants, bitumen, wax and refinery fuel and loss.
* Negligible change.
Source: BP.

Table 2.14 shows gasoline, middle distillate, fuel oil and other refined products consumed by different country groupings over the period 1972-82 and 1977-82. The figures in the bottom right hand corner point out how, over the years 1972-82, world gasoline demand grew by 0.8% per year, while middle distillates demand grew by 1.3% per year and fuel oil demand contracted by 2.2% per year. Over the more recent period, 1977-82, all three categories have shown negative trends, led by fuel oil with a 5.1% decline.

It is clear that fuel oil demand has fallen sharply in the major developed countries—in the US, for instance, it fell at an annual average rate of 11% over 1977-82, while in Canada it fell almost as quickly, at 9.5%. Japan recorded a similar rate of decrease in demand.

Chapter 3

OIL REFINING

Introduction

THE crisis in the oil refining industry in western countries is at root a fairly straightforward one to understand. The fundamental problem for refiners has been that demand for refined oil products, after growing more than twice as quickly as gross national income for much of the post-war period, has decelerated sharply. Indeed, during the 1979-83 recession demand in some countries for some refined petroleum products began to fall absolutely. Between 1963 and 1976 value added by the oil refining industry grew at 5.8% per year in western countries: this is the type of growth-rate which encourages capacity planners to forecast more and more ambitious levels of demand year after year. Once the 1973 oil price rises (and, in the USA in particular, the physical shortages) had sunk in, however, demand was shifted onto a considerably lower trend path. Even the falls in the real prices of many products in the 1975-78 period failed to trigger much of a reversal of this tendency. The most recent slippage in real prices, following the very weak conditions in the crude oil market of 1982-83, have had the well-reported effect of shifting demand for gasoline very slightly upwards again, chiefly as big cars in the US make a revival, but it is as yet too early to identify the extent to which this is another permanent change.

By 1981 the position in refining was as shown in Table 3.1. It is evident from the table that between 1938 and 1981 worldwide refining capacity had grown by over a factor of ten. A particularly large build-up was seen after 1965, with capacity more than doubling before 1981. The share of Western Europe in total capacity grew appreciably in the post-war period, with the share of the US in the total declining steeply. Eastern European capacity was built up markedly too. The share of the Middle East, a topic to be discussed in this chapter at some length, remains modest. Table 3.2 shows the absolute capacity of refineries, by country, over the period 1972-82, expressed in thousands of barrels per day of throughput. The table shows how steeply capacity has fallen in some countries (for instance, West Germany, where from 3.1 million barrels/day capacity in 1975 fell to 2.5 million barrels/day by 1982) and how little it has fallen in others. In Italy, capacity was still growing until 1979; only thereafter did it begin to fall at all. It is notable that in Spain, refining has been growing strongly over the period covered, with capacity increasing by over 50% between 1972 and 1982, ending up as the sixth biggest volume of installed capacity among European countries. Total Western European capacity

Table 3.1

WORLD REFINING CAPACITY, 1938-1981 (percentages)

Area	1938	1950	1965	1981	
Western Europe	4.4	7.6	23.6	24.8	*20.6*
Middle East	3.8	8.1	5.8	3.9	*3.8*
Africa	0.3	0.2	1.5	2.6	*2.6*
USA	61.8	58.0	29.5	22.3	*21.6*
Latin America	11.0	12.3	11.5	11.0	*11.4*
Far East & Australasia	3.8	2.4	9.6	12.2	*13.1*
USSR, Eastern Europe & China	8.7	6.3	13.0	19.5	*}23.3*
Other	6.2	5.1	5.5	3.7	
World Total	100.0	100.0	100.0	100.0	
Thousand metric tons	363,700	603,100	1,735,000	4,085,000	*8.9*

(handwritten: 1938—)

Sources: Institute of Petroleum (London) *Petroleum Statistics,* 1981 edition; BP (1982).

Notes: (i) The reported capacity of the Middle East fell by 18 million tonnes between 1980 and 1981, reflecting the fall in Iran's serviceable capacity.
    In 1980 the Middle Eastern share in the world total was 4.5 per cent.
  (ii) Figures refer to year-end.

grew by 0.4% over the decade, but peaked in 1976. The year which witnessed the biggest single fall in capacity was down by nearly 10% from the end of the preceding year.

In developing countries, other than OPEC members, Mexico has shown a very rapid build-up of refining capacity, with the total growing nearly threefold over the decade.

For the world as a whole, capacity grew at an annual average rate of 2.9% over the decade. Total capacity peaked in 1981 after nine years of uninterrupted year-on-year growth. It is clear, however, that at close to 80 million barrels/day, refining capacity is very significantly higher than global oil use. In 1982, world oil consumption was averaging 58.5 million barrels/day—leaving a margin of over 20 million barrels/day unused. Of course, a margin of excess capacity is inevitable and desirable, given the

42

Table 3.2
WORLD OIL REFINING CAPACITIES

Year End	1972	1977	1979	1981	1982	YearlyChange 1982 over 1972	1982 over 1977
Country/Area	**Thousand Barrels/Day**						
USA	13,640	16,945	17,920	18,290	17,390	+ 2.5%	+ 0.5%
Canada	1,750	2,220	2,315	2,155	2,080	+ 1.8%	− 1.3%
Total North America	**15,390**	**19,165**	**20,235**	**20,445**	**19,470**	**+ 2.4%**	**+ 0.3%**
Latin America							
Argentina	630	660	660	690	690	+ 0.9%	+ 0.8%
Brazil	795	1,175	1,230	1,530	1,530	+ 6.8%	+ 5.4%
Mexico	590	980	1,395	1,525	1,525	+ 9.9%	+ 9.3%
Netherlands Antilles	800	840	840	780	780	− 0.2%	− 1.5%
Trinidad	440	465	465	455	375	− 1.6%	− 4.3%
Venezuela	1,475	1,445	1,445	1,405	1,360	− 0.8%	− 1.2%
Other Latin America	1,740	2,605	2,765	2,675	2,525	+ 3.8%	− 0.6%
Total Latin America	**6,470**	**8,170**	**8,800**	**9,060**	**8,785**	**+ 3.1%**	**+ 1.5%**
Total Western Hemisphere	**21,860**	**27,335**	**29,035**	**29,505**	**28,255**	**+ 2.6%**	**+ 0.7%**
Western Europe							
Belgium	855	1,055	1,020	965	660	− 2.5%	− 8.9%
France	2,905	3,465	3,420	3,230	2,745	− 0.6%	− 4.6%
Italy	3,630	4,270	4,205	3,990	3,500	− 0.4%	− 3.9%
Netherlands	1,805	1,835	1,815	1,775	1,590	− 1.3%	− 2.8%
Spain	880	1,210	1,475	1,550	1,550	+ 5.8%	+ 5.0%
United Kingdom	2,490	2,655	2,460	2,360	2,135	− 1.5%	− 4.3%
West Germany	2,675	3,065	3,040	2,955	2,520	− 0.6%	− 3.9%
Other Western Europe	2,030	2,925	3,080	3,175	3,335	+ 5.1%	+ 2.7%
Total Western Europe	**17,270**	**20,480**	**20,515**	**20,000**	**18,035**	**+ 0.4%**	**− 2.5%**
Middle East							
Bahrain	250	250	250	250	250	—	—
Iran	605	965	1,045	625	625	+ 0.3%	− 8.3%
Iraq	115	185	265	265	265	+ 8.7%	+ 7.3%
Kuwait	525	575	575	530	545	+ 0.4%	− 1.2%
Neutral Zone	75	80	80	80	80	+ 0.4%	—
Saudi Arabia	425	630	635	660	865	+ 7.3%	+ 6.6%
Southern Yemen	160	145	145	145	180	+ 1.1%	+ 4.5%
Other Middle East	250	415	545	695	760	+ 11.8%	+ 12.8%
Total Middle East	**2,405**	**3,245**	**3,540**	**3,250**	**3,570**	**+ 4.0%**	**+ 1.9%**
Africa	965	1,505	2,020	2,055	2,375	+ 9.4%	+ 9.6%
South Asia	630	785	825	885	1,000	+ 4.7%	+ 4.9%
South East Asia							
Indonesia	415	525	535	515	515	+ 2.1%	− 0.5%
Singapore	550	1,040	1,040	1,020	1,020	+ 6.4%	− 0.4%
Other South East Asia	1,155	1,610	1,850	1,980	1,980	+ 5.5%	+ 4.2%
Total South East Asia	**2,120**	**3,175**	**3,425**	**3,515**	**3,515**	**+ 5.2%**	**+ 2.0%**
Japan	4,360	5,285	5,285	5,675	5,500	+ 2.4%	+ 0.8%
Australasia	705	760	800	815	815	+ 1.5%	+ 1.5%
USSR,E.Europe & China	9,510	13,320	14,650	15,990	16,330	+ 5.6%	+ 4.2%
Total Eastern Hemisphere	**37,965**	**48,555**	**51,060**	**52,185**	**51,140**	**+ 3.0%**	**+ 1.0%**
World (excl. USSR E.Europe & China)	**50,315**	**62,570**	**65,445**	**65,700**	**63,065**	**+ 2.3%**	**+ 0.2%**
World	**59,825**	**75,890**	**80,095**	**81,690**	**79,395**	**+ 2.9%**	**+ 0.9%**

Source: BP

43

need to have a margin in hand for unforeseen demand fluctuations, the need to cope with swings in the mix of inputs and in the pattern of demand as between products, and the fact that as national boundaries coexist with corporate boundaries, excess demand and excess supply will inevitably tend to coexist.

The extent of the deviation of consumption from capacity in the refining business was such that by 1981, substantial excess capacity was beginning to build up. Instead of normal capacity utilization rates of 90% or so (full capacity should really be considered a range, from 90% to 110% or so, with nominal or "nameplate" 100% capacity only a rough guideline) utilization plunged to 60% and below, in Japan, Europe and the USA (see diagram 3.1). These figures reflect the severe contraction in refined products demand after 1979. In 1980, world demand for gasoline fell by 4% over its 1979 level. The following year demand fell a further 3.3%. In the UK, whereas demand for petroleum products in 1981 was 65 million tonnes (the same level as in 1966), capacity to meet demand of 123 million tonnes was in place. Throughout 1982 and 1983, further falls in demand were observed, so that by the first quarter of 1983 demand stood at only 15.2 million tonnes, or 60.8 million tonnes on an annualized but seasonally unadjusted basis. This represented a full 12.0% fall from the same period of 1982, a fall chiefly attributable to very substantial declines in demand for fuel oil and kerosene. To some extent these factors distort the picture, since in 1982 deliveries of fuel oil to power stations were abnormally high, in anticipation of a coal miners' strike. If this factor is taken out, the overall fall in the volume of petroleum products deliveries was probably in the order of 4%.

During the 1980-83 period other countries were seeing similar forces at work. The European Community Energy Council reported in late 1981 that one in four refineries in the EEC could be mothballed without much risk of excess demand appearing before 1990. The International Energy Authority also reported that, on unchanged scrappage policies, the EEC's average level of refinery utilization would be in the order of 40% by 1985.

European refiners have been among the hardest hit. Western Europe is the greatest refining area in the world, with some 44% of the OECD countries' total capacity. In 1981-82 Europe also had about two-thirds of the industry's idle plants. BP was one of the first companies to take the plunge with massive asset writeoffs, with 25% then 40% of its total refining capacity to be abandoned by 1986. To some extent the problems of the European refiners were compounded by their expectation that the European Community would become involved, as in the steel industry, with an effort to arrange capacity closures spread between all the member countries. Further closures are more than likely. Table 3.3 lists the closures

Diagram 3.1

EVOLUTION OF REFINERY UTILIZATION
(per cent)

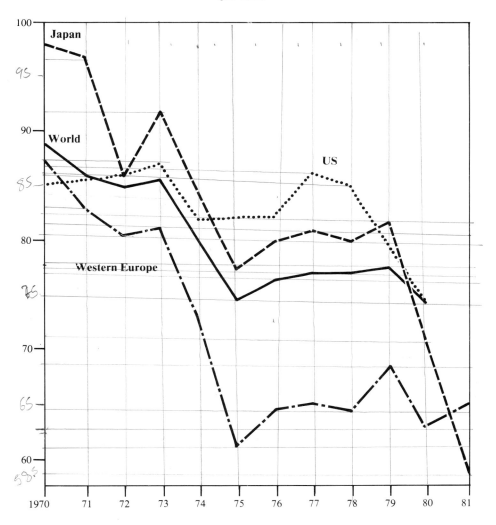

Source: OPEC

Table 3.3

EUROPEAN REFINING CAPACITY LIKELY TO SHUT DOWN
1982-1984
(1,000 barrels/day)

		Present capacity	Per cent shutdown
UK	619	2,482	25
Isle of Grain (BP)	211		
Shell Haven (Shell)	175		
Milford Haven (Exxon)	174		
Ellesmere Port (Burmah)	28		
Belfast (BP)	31		
Italy	483	4,003	12
La Spezia (Agip)	103		
Gaeta, Volpiano (Agip)	200		
Porto Torres (Agip)	100		
Bartonico (Gulf)	80		
France	506	3,291	15
Garbenville (Elf)	139		
Valenciennes (Elf)	85		
Hauconcourt (joint)	104		
Herrlisheim (joint)	92		
Dunkirk (BP)	86		
Netherlands	320	1,707	19
Rotterdam (Chevron)	260		
Rotterdam (BP-part)	60		
Ireland	**56**		
Whitegate (joint)	56		
West Germany	514	2,937	17.5
Dinslaken (BP)	121		
Vohburg (BP)	102		
Speyer (BP-share)	30		
Cologne (Exxon)	115		
Ingolstadt (Shell)	56		
Rauheim (Caltex)	90		
Sweden	45	471	10
Nynashamna (A. Johnson)	45		
Belgium (est. total)	**300**		
Total	**2,843**		

Source: PIW

46

expected before 1984.

Table 3.4 lists the refineries closed in the US during 1982. A total of 57 refineries closed, with a total capacity of 1,146,000 barrels/day. This compares with 1981, when 23 refineries, with a total capacity of 451,000 barrels/day were closed. In a survey of refining the US Department of Energy pointed out that about 40% of the capacity closed was 25 years old at least, suggesting that some firms were selecting their oldest (and probably, least efficient) units to mothball or abandon first. On the other hand, the fact that another 40% of the affected plants were younger than 5 years old implies that other firms were not able to make this choice. Another point worth noting is that over half the refineries closed were very small, with daily throughputs of under 10,000 barrels. 90% of the refineries which closed had capacity of under 50,000 b/day. By far the biggest single closure was that of Dow Chemical's Freeport, Texas, refinery, which had a capacity of 190,000 b/day. Finally, it is interesting to note that most of the larger closures were concentrated in the Midwest and East Coast areas of the US, while most of the smaller and newer refineries which were closed were on the Gulf Coast.

What caused this crisis, and to what extent are these factors still present in 1984? To begin to answer this question one must look first at the structure of the oil refining industry.

Structural Characteristics of Oil Refining

An important characteristic of the refining business is the long lead-time that is involved in the design and construction of new capacity. Lags of 5 to 6 years between beginning design and completing construction of facilities are not unknown. The problem which this raises from a capacity planning point of view, of course, is that one is forced to build to demand forecasts made a considerable time beforehand. The consequence of this was that new capacity was appearing long after planners had spotted the latest trends in demand. Most of the plants caught half way toward completion during this period were finished on the grounds that they still had an attraction on efficiency grounds. Late in 1981 an industry survey established that a further 81 million tonnes, equivalent to about 9% of installed capacity in that year, was due to appear in the USA before 1986.

There are also refining companies outside the US which are taking a more optimistic view of the prospects for demand growth. In the Far East, for instance, low utilization rates (averaging 69% in 1982 after 80% in 1978) in projects, outside Japan, are slated to come onstream adding a total of 47 million tonnes/year of capacity. There are, however, doubts

Table 3.4
US REFINERY SHUTDOWNS, 1982

District/Refinery	Location	Crude Distillation Capacity b/d	Years in Operation
East Coast			
Amoco Oil Co.	Baltimore, Md.	15,000	25+
Ashland Oil Inc.	Buffalo, N.Y.	64,000	25+
Seminole Refining, Inc.	St. Marks, Fla.	15,000	25+
Total		94,000	—
Midwest			
Amoco Oil Co.	Sugar Creek, Mo.	104,000	25+
Ashland Oil Inc.	Findlay, Ohio	20,400	25+
CRA, Inc.	Scottsbluff, Neb.	5,600	25+
CRA, Inc.	Phillipsburg, Kan.	26,400	25+
Dillman Oil Recovery, Inc.	Oblong, Ill.	1,200	4
E-Z Serv Refining, Inc.	Shallow Water, Kan.	9,500	25+
Energy Cooperative, Inc.	East Chicago, Ind.	126,000	25+
Industrial Fuel & Asphalt of Indiana,Inc.	Hammond, Ind.	7,600	25+
Kentucky Oil & Refining Co.	Betsy Lane, Ky.	3,000	25+
Mid-America Refining Co. Inc.	Chanute, Kan.	3,000	25+
Northland Oil & Refining Co.	Dickinson, N.D.	5,000	7
Phillips Petroleum Co.	Kansas City, Kan.	80,000	25+
Texaco, Inc.	West Tulsa, Okla.	50,000	25+
Texas America Petrochemicals Inc.	West Branch, Mich.	11,500	25+
Total		453,200	—
Gulf Coast			
Bayou State Oil Corp.	Hosston, La.	3,000	25+
Bronco Refining Co.	Houston, Texas	2,250	1
Caribou-Four Corners Oil Co.	Kirtland, N.M.	2,400	17
Clinton Manges	Palestine, Texas	6,000	25+
Copano Refining Co.	Ingleside, Texas	11,100	4
Dow Chemical USA.	Freeport, Texas	190,000	1
Eagle Refining Corp.	Jacksboro, Texas.	1,800	1
Giant Industries, Inc.	Farmington, N.M.	13,500	7
Independent Refining Corp.	Pt. Neches, Texas	30,000	4
Independent Refining Corp.	Winnie, Texas	50,000	23
Lake Charles Refining Co.	Lake Charles, La.	28,000	2
Listo Refining Co.	Donna, Texas	3,500	4
Longview Refining Co.	Longview, Texas	14,000	25+
Natchez Refining Co.	Natchez, Miss.	16,000	2
Petraco-Valley Oil & Refining Co.	Brownsville, Texas	12,300	2
Placid Oil Co.	Mont Belvieu, Texas	8,500	2
Quitman Refining Co.	Quitman, Texas	6,600	4
Rio Grande Crude Refining	Brownsville, Texas	9,500	3
Rio Grande Recovery Systems, Inc.	Brownsville, Texas	1,000	2
Schulze Processing, Inc.	Tallulah, La.	1,760	4
Sentry Refining Inc.	Corpus Christi, Texas	25,000	4
Shepard Oil Co.	Jennings, La.	10,000	4
Sooner Refining Co.	Darrow, La.	8,000	2
T & S Refining, Inc.	Jennings, La.	10,500	2
TARCO	Euless, Texas	6,000	20
Tipperary Refining Co.	Wickett, Texas	7,320	4
Vicksburg Refining Co.	Vicksburg, Miss.	7,900	4
Wickett Refining Co.	Wickett, Texas	8,000	25+
Total		493,930	—
Rocky Mountain			
C & H Refinery, Inc.	Lusk, Wyo.	180	25+
Caribou-Four Corners Oil Co.	Woods Cross, Utah	7,200	19
Glacier Park Co.	Osage, Wyo.	10,000	4
Husky Oil Co.	Cody, Wyo.	11,500	25+
Morrison Petroleum Co.	Woods Cross, Utah	6,300	8
Sage Creek Refining Co.	Cowley, Wyo.	1,000	17
Texaco, Inc.	Casper, Wyo.	21,000	25+
Total		57,180	—
West Coast			
Gibson Oil & Refining Co.	Bakersfield, Calif.	4,600	3
Lunday-Thagard Oil Co.	South Gate, Calif.	12,000	14
Sabre Oil & Refining, Inc.	Bakersfield, Calif.	10,000	10
United Independent Oil Co.	Tacoma, Wash.	730	7
West Coast Oil Co.	Oildale, Calif.	21,000	25+
Total		48,330	—
US Total		**1,148,640**	—

Source: National Petroleum News

Table 3.5
MAJOR EXTENSIONS TO EXISTING REFINERIES
Primary distillation capacity in thousand tonnes/year

	Present capacity	Planned total	Probable completion
WESTERN EUROPE			
Turkey			
Izmir	1,000	13,000	1983-4
AFRICA			
Ivory Coast			
Abidjan (Ivory Coast and Upper Volta			
governments-Shell-BP-CFP) ★	2,000	4,000	1982
Nigeria			
Kaduna (NNPC) ★	5,000	6,000	1985
Warri (NNPC) ★	5,000	6,000	1985
Senegal			
Dakar (Soc Africaine de Raffinage)	900	1,350	1983
MIDDLE EAST			
Abu Dhabi			
Umm al-Nar (ADNOC) ★	750	3,750	1983
Egypt			
Mostorod (Suez Oil Processing)	4,675	5,725	1983
Israel			
Ashdod (Oil Refineries) ★	4,000	4,800	
Haifa (Oil Refineries) ★	6,000	8,500	
Kuwait			
Mina Abdullah (KPC) ★	5,000	125,000	1986
FAR EAST			
Australia			
Matraville, NSW (Total)	550	2,300	1985
India			
Bombay (Bharat Petroleum)	6,000	7,000	1984
Cochin (Cochin Refineries)	3,300	4,500	1984
Madras (Madras Refineries)	2,800	5,600	1984
Visakhapatnam (Hindustan Petroleum)	1,500	3,000	1984
Indonesia			
Balikpapan (Pertamina) ★	3,750	13,750	1983
Cilacap (Pertamina) ★	5,000	15,000	1983
New Zealand			
Whangarei (New Zealand Refining)	2,200	3,700	1985
South Korea			
Busan (Kukdong Oil)	500	3,500	1984
Ulsan (Korea Oil Corp)	14,000	21,500	1983
Thailand			
Sriracha (Thai Oil Refining)	3,250	6,000	1984
Sriracha (Esso)	2,300	3,150	
NORTH AMERICA			
Canada			
Edmonton, Alberta (Gulf)	4,000	6,000	1983
Edmonton, Alberta (Imperial)	7,750	10,000	1984
Lloydminster, Alberta/Saskatchewan			
(Husky Oil)	600	1,250	1982
Shellburn, British Columbia (Shell)	1,175	1,725	1984
United States			
Chalmette, Louisiana (Tenneco)	5,000	6,350	1983
Corpus Christi, Texas (Saber Refining)	1,050	3,550	1983
McPherson, Kansas			
(Nat Coop Ref Assoc)	2,700	3,500	1983
St James, Louisiana (La Jet)	1,000	2,500	1982
Warren, Pennsylvania			
(United Refining)	2,000	3,000	1982
CARIBBEAN/SOUTH AMERICA			
Colombia			
Cartagena (Ecopetrol) ★	2,500	3,500	1982
Ecuador			
Esmeraklas (CEPE) ★ *	2,750	3,500	
Mexico			
Madero (Pemex) ★	9,250	16,750	1983
Salina Cruz (Pemex) ★	8,500	16,000	1983
Tula (Pemex) ★	7,500	15,000	1982
Peru			
Talara (Petroleos de Peru) ★	3,250	3,750	1983
Trinidad			
Port Fortin (Trinidad and			
Tobago Oil)	5,000	7,500	1985

★ State owned or state controlled.
* Additional expansion to 4.5 million tonnes/year is being studied.
Source: Petroleum Economist

regarding the extent to which the Indonesian plans for refinery expansion will be completed; the national oil firm Pertamina faced budget cuts during 1983 and is in addition known for budget overruns. In Singapore, capacity will be upgraded more than expanded, doubtless a reflection of the fact that demand in Indonesia and Malaysia will be falling in the 1980s as greater self-sufficiency is attained there. Table 3.5 summarizes the major refinery expansion plans as made public in late 1982.

But capacity scheduling is not the whole story; if it was, the problem facing refiners would be rather simpler than it is. The capacity issue is supplemented by a second—the fact that both demand and supply are shifting between types of input and types of output. For crude oil, the main input to the process is by no means homogenous; while the outputs—the petroleum products themselves—are by no means unchanging either. While it is clear that a variety of products exist, and fulfil various needs, it is crucial to grasp the further point that the pattern of demand as it is composed across product types (the 'demand-mix') is in a state of continuous flux too.

The long-standing tendency has been for demand to shift more and more towards the 'light' end of the barrel, and correspondingly to shift away from the 'heavy' end. This means a more rapid growth of demand in gasoline, jet fuels and other light products, at the expense of the thicker, heavier products such as bitumen.

Widely diverging growth rates for various products have been observed in developed country markets. Why is this?

A number of forces are driving the differential growth rate of the various petroleum products. One way of analysing these forces is by looking at income elasticities of demand, or the proportionate change in demand for a product induced by or associated with a given change in real disposable income. The income elasticity of demand for personal transport is high; it is of the order of 1.5 to 2.0 in most developed countries. Thus a 1% rise in real income feeds into a 2% rise in expenditure on personal transportation. The implication of this is that as incomes grow, demand for cars will grow even faster. If fuel economy were to remain constant, the need for fuel used by an economy to support this transport would also grow disproportionately. In recent years transport has accounted for some 70% of gasoline demand. Thus so long as economic growth proceeds there will be a growing tendency for gasoline to be consumed.

There is also a price effect at work. Gas and coal have been increasingly substituted for heavy fuel oils during the period that oil prices have been

50

rising. Thus the growth in heavy fuel demand which might otherwise have been expected has been undermined by inter-fuel substitution. Governments in many countries have been encouraging power stations and companies to convert so as to be able to use coal rather than oil for steam-generation and electricity-generation. There have, however, been constraints on the extent to which this has been possible, since the infrastructure needed for large-scale coal handling, such as big railway marshalling-yards, has in many cases been sold or destroyed, or in other cases was not built or planned for in the first place. Moreover, industry comprises only 12-15% of demand for steam coal, so that industry's needs alone could not have a massive effect upon coal for oil substitution and thus oil consumption.

There are some 7,500 different products which can be derived from crude oil. They can be grouped into three sets: light products (e.g. gas and gasoline), middle distillates (e.g. kerosene) and heavy products (e.g. fuel oil, residues). In the early days of refining, the light products were of comparatively little interest; gasoline for cars only began to become significant after the First World War. But with the advent of air travel, aviation fuel became an important product. Gas oils and fuels for central heating, fuel oil for power stations and various inputs for petrochemical plants all came to be important. To win sufficient quantities of light products from each barrel of crude input, increasingly complex processes had to be devised to derive more and more light products. Diagram 3.2 shows the breakdown of a notional "typical" barrel of crude in the early 1980s.

Various estimates of the changes to be expected in the proportion of light and heavy products in western countries' demand patterns exist. The International Energy Authority has forecast that refined product demand will fall by 12% between 1979 and 1985, to 1,674 million tonnes for OECD countries. Within this total, the product shares shown in table 3.6 are foreseen. The table suggests that gasoline and middle distillate demand will rise or stay constant, while heavy fuel oil demand will fall. The latter phenomenon is attributed to the build up of coal production and the spread of natural gas use. In contrast with other countries, Japan is expected to witness a general increase in demand for all refined product categories, with middle distillates growing particularly strongly.

Estimates of future European demand by product category made by Esso (see diagram 3.3) point to a gradual increase in motor gasoline's share in total demand to 18% of all petroleum product use. Heavy fuel oil use is forecast to fall, both absolutely and relatively, with only 21% of total demand accounted for by this product by the year 2000.

Diagram 3.2

**PETROLEUM PRODUCTS FROM
AN AVERAGE BARREL OF CRUDE OIL**

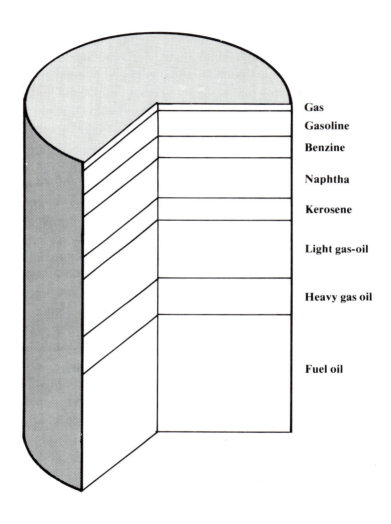

Gas
Gasoline
Benzine
Naphtha
Kerosene
Light gas-oil
Heavy gas oil
Fuel oil

Source: OPEC

Table 3.6

EEC: PRODUCT COMPOSITION OF DEMAND (Percent by weight)

%	1973	1979	1980	1985	1990
Mogas/Naphtha	17.5	20.5	22.1	23.2	25.0
Kerosene/Gasoil	34.2	35.9	35.2	34.5	34.0
Distillates	51.7	56.4	57.3	57.7	59.0
Residual fuel oil	38.0	29.5	28.4	25.5	22.5
Other (incl. refinery fuel)	10.3	14.1	14.3	16.8	18.5
Total	100.0	100.0	100.0	100.0	100.0

Source: EEC

Table 3.7

ANNUAL GROWTH RATES FOR MAJOR OIL PRODUCTS
Percent

	1970-80	1980-90	1990-2000
Naphtha	0.3	0.2	0.9
Motor gasoline	3.6	(0.2)	(0.1)
Turbo/kerosene	1.6	0.7	1.7
Distillate	1.5	(0.7)	(0.1)
Fuel oil	(1.2)	(4.5)	(0.7)
Other	1.0	—	0.7
Total	0.7	(1.5)	—

Source: Esso

Later forecasts by Esso (made at the end of 1982) point to even less energy being consumed in Europe over the period to 2000. Motor gasoline demand is forecast in these later figures to fall very slightly over the 1980-2000 period, chiefly as a consequence of a 30% or so improvement in the fuel economy of the car fleet. Table 3.7 shows these later estimates, and makes clear that a smaller volume of major oil products will be consumed in Europe in the year 2000 than was the case in 1970.

At the same time as demand has been shifting, so too has the supply of the crude used as the primary input to refining. The tendency has been for oil companies to 'upgrade' refineries so that they could more readily use heavy crude oils rather than the light crude for which many had initially been engineered.

Distinguishing light from heavy crudes is not a completely straightforward task. Ambiguities exist in defining the divide between them, and indeed distinguishing heavy crude oil from extra-heavy crudes, tar sands, bitumen and the like is also hazardous.

To distinguish crude oils in this way one of the standards for discrimination which has become widely accepted is the API degree rating, named after the American Petroleum Institute. Water, which is assigned an API rating of 10°, is used as the basis of comparison. One rule of thumb often used in the oil industry is that heavy crudes have a specific gravity less than 20°, while materials with a gravity of 10° or below are called asphalt or bitumen. However, since some crudes have a very low gravity, they are viscous in handling and thus might more appropriately be considered as oils rather than solids. In 1979 crude oils of 34° API or less constituted about 36% of world crude oil supply. By 1990, however, it is expected that they will come to account for 42%-45% of available supply. The rationale for this shift is essentially one of price and of differential conditions of supply. Low quality crudes which are more viscous, costs less than fine, light crudes such as those typically sold from the North Sea, from Nigeria, Saudi Arabia and other sources. The price differential between such types of crudes varies, according quite simply to the state of demand and supply in each market segment. During the first half of 1983 it tended to be unusually small, at around $3 per barrel. As well as there being a straightforward price difference, analysts have for years been expecting most of the major new fields, such as those in Alaska, Mexico and the Orinoco basin of Venezuela, to yield disproportionate volumes of heavy crudes, with the result that the price differential in favour of heavy crude would increase.

Refiners in the US are estimated to have spent around $15 billion in refinery upgrading in recent years. The installations that have to be paid

54

Diagram 3.3

EUROPEAN OIL DEMAND BY PRODUCT

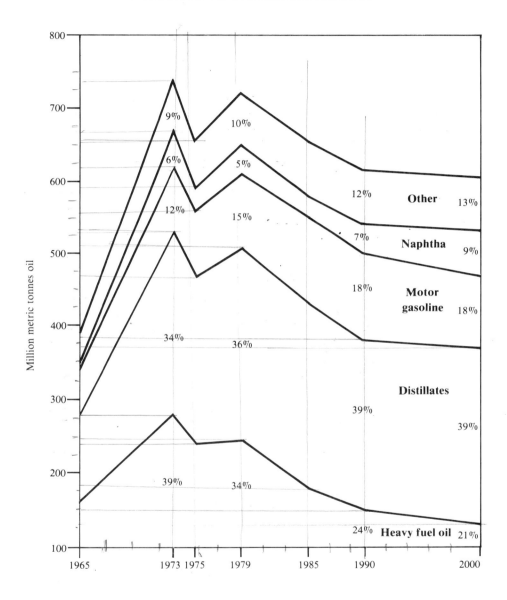

Source: Esso

55

for include considerable networks of pipes and new units to catalyze the sulphur and metal elements of the crudes.

Chevron in 1982 could operate its refineries with a 58% heavy crude diet, but in 1984 expects to be able to use 72% heavy crudes. Texaco's ratio was 30% in 1978 but will be 60% in 1984.

The least efficient are the refineries which still yield up to 50% resid or asphalt from each barrel of input. Since resid and asphalt frequently sell for less than the crude oil volume of the input, there is obviously a keen interest in having a greater proportion of other products, such as gasoline, produced.

In 1983 Exxon Corporation's chief US operating division announced that it was preparing to spend $0.5 billion in order to upgrade its largest refinery so as to cope with lower grade crudes. By 1987 the Bangtown, Texas, refinery will be yielding fuel gas to help run the refinery itself, and will also be yielding a lighter mix of products than at present. The cost of this conversion, other than the capital outlay, lies in the 20% reduction in operating capacity which is a concomitant of regrading. At nearly 500,000 bpd the existing facility is not necessarily always fully utilized, however.

Another approach is that of Valero Energy Corporation, which in 1980 acquired 50% of Saber Energy Inc., a small refinery and gasoline retailing firm, with the intention of converting its small refinery at Corpus Christi, Texas, into a modern low value heavy crude input unit capable of yielding 50,000 bpd of gasoline. In 1980 resid from the Middle East cost $20/barrel and profits were forecast at $16/barrel, or three times the then industry average. But by 1983 when the refinery came onstream, resid cost $26/barrel and light crudes—the more normal choice for refinery input—cost only three dollars more at $29. Moreover, gasoline was selling at only $35/barrel. In other words, too many firms taking the resid route had turned the market against themselves. In Valero's case a likely margin of $8/barrel must now go towards the half billion dollar debt the firm has built up. How then does the resid approach look in 1984?

First, US firms face trouble with anti-pollution standards imposed by the Federal government. Mixing in lighter, but more costly, feed would help in this regard but would of course undermine the cost advantage of taking the low cost inputs in the first place. Second, conditions of short run supply have altered considerably over the past four years. In 1982 and 1983 cash-starved OPEC members were anxious to export more premium-priced light crudes, so that relative abundance turned the price differential in favour of light rather than heavy oils. The $6-8/barrel price differential expected has shrunk to $3-5/barrel. On the other hand, over the long run the world's supply of heavy crudes does outweigh the light

56

crudes by a factor of about 4 to 1, so that eventually a price differential should open up again, favouring users of heavy crudes as refinery inputs.

An aspect of refining in Europe which distinguishes it from the business in North America (and, for different reasons, Japan) is that oil refineries, by virtue of often being located in relatively poor parts of the country, assume some political significance. Plans to shut them down then come up against coordinated objections by trade unions. In the UK, for instance, shop stewards at various refineries interpreted the 1983 EEC estimates of future refinery closures as a threat to British jobs. Seven thousand jobs at British refineries were claimed to be under threat due to the EEC's plan to have Britain shed a further 12.5 million tonnes of refining capacity in a European total contraction of 40 million tonnes. This is equivalent to one large and two small refineries, and, following on recent closures by BP (at the Isle of Grain and Belfast), Burmah Oil (at Ellesmere Port) and Esso (at Milford Haven) it was felt that employment was falling too quickly.

At the same time as such firms as Standard Oil of California and Ashland Oil have been upgrading in this way, other refiners have taken a contrary position. Sun Oil Co., based in Radmor, Pennsylvania, has taken the view that as other firms struggle to upgrade they will, in aggregate, tend to turn the market against themselves by pushing demand for heavy crudes up disproportionately quickly. The likes of Sun may then be able to gain from looser market conditions in light crudes. While so far the evidence does not contradict their position—it is the light crudes which have shown the largest price reductions over the 1982-83 period—the long-term viability of their position is less clear. In mid-1983 it was in fact the lighter crudes which were showing the greatest tendency to firming. One point which has to be factored in is that firms which have gone ahead with upgrading are enjoying productivity gains which just maintaining their old refinery configurations would not have yielded. Thus, Ashland Oil, for instance, is able now to reduce its raw material input costs by 20%-25% since it can now turn 75% of its low-grade crude inputs into gasoline and other light products.

OPEC Refining

Changes in OPEC members' refining capacity due over the next few years are likely to increase the organization's total refining capacity to 8.6 million bpd, as table 3.8 shows. Three points must be made about these estimates, however. First it is not at present possible to offer definitive figures for either Iranian or Iraqi capacity, since the war between them, which began in October 1980, has caused substantial disruption both to physical facilities (particularly in Iraq) and to maintenance and repair.

Table 3.8

OPEC REFINERY CAPACITY, 1982-1988
(1,000 barrels/day)

	Existing 1982	1983	1984	Projects completed 1985	1986	1987	1988	Total capacity 1988
Algeria	438.0	—	—	—	—	—	—	438.0
Ecuador	94.5	—	15.0	—	—	—	—	109.5
Gabon	44.0	—	—	—	—	—	—	44.0
Indonesia	471.0	360.0	—	—	145.0	—	—	976.0
Iran	560.0	—	—	—	—	—	200.0	760.0
Iraq	305.5	150.0	—	—	—	—	—	455.5
Kuwait	594.0	—	170.0	—	206.0	—	—	970.0
S.P. Libyan A.J.	130.0	220.0	—	220.0	—	—	—	570.0
Nigeria	247.0	—	—	—	—	—	—	247.0
Qatar	10.5	50.0	—	—	—	—	—	50.0
Saudi Arabia	1,028.0	495.0	513.0	—	—	300.0	—	2,336.0
UAE	135.0	60.0	60.0	—	—	—	—	255.0
Venezuela	1,314.1	—	—	—	—	28.0	—	1,342.1
Total OPEC	5,371.6	1,335.0	758.0	220.0	351.0	328.0	200.0	8,553.1

Source: OPEC

Thus the figures for Iraqi capacity should be treated as only very rough indicators. Second, the total figure is very sensitive to the building carried out in Saudi Arabia in 1984. As the recent downturn in oil prices has begun to affect the funds available for the Saudi thrust into downstream development, it is increasingly likely that some deceleration in the build-up of Saudi refining capacity will be observed. Third, it should be noted that the planned 1980-81 increases were only in a few instances met. In fact, the only increases in capacity during that period were at the Rabih refinery in Saudi Arabia, where a further 325,000 bpd was installed, and in the UAE, at the Al Ruwais refinery, where a further 120,000 bpd was installed. This accretion to capacity—a total of 445,000 bpd—contrasts with the expected 621,000 bpd increase forecast by the OPEC secretariat in 1981. There is thus need for care in interpreting the figures which relate to current capacity, and for caution in interpreting forecasts of future capacity.

Assuming for the time being that OPEC's own forecasts of refining capacity will work out to be more or less accurate, it is possible to construct a set of forecasts of refinery products supply from those countries, and compare it to forecasts of demand. Table 3.9 shows one such set of demand and supply forecasts. The table indicates that by 1985 the excess of new output over new demand will be of the order of 2.4m bpd. The resultant changes in the organization are a slight tilt in total capacity towards the lighter end of the barrel, with 20% of total capacity devoted to gasoline and naphtha in 1979 but 23% in 1985. Middle distillate capacity will rise too, from 28% to 32% over the same period, while residual fuel oils will contract from 40% to 34%.

Table 3.9

ESTIMATED PETROLEUM PRODUCTS SUPPLY/DEMAND BALANCE
IN OPEC: 1979-1985
(1,000 barrels/day)

	1979	1985	Change
Gasoline/naphtha			
Refinery output	835	1,475	+ 77%
Consumption	585	1,090	+ 86%
Surplus/(deficit)	250	385	+ 54%
Middle distillates			
Refinery output	1,180	2,035	+ 72%
Consumption	965	1,825	+ 89%
Surplus/(deficit)	215	210	− 2%
Residual fuel oils			
Refinery output	1,695	2,170	+ 28%
Consumption	460	700	+ 52%
Surplus/(deficit)	1,235	1,470	+ 19%
Others			
Refinery output	505	635	+ 26%
Consumption	205	285	+ 39%
Surplus/(deficit)	300	350	+ 17%
Total			
Refinery output	4,215	6,315	+ 50%
Consumption	2,215	3,900	+ 76%
Surplus/(deficit)	2,000	2,415	+ 21%

Source: OPEC

Table 3.10 shows the refining capacity and utilization for OPEC members between 1976 and 1981. It is apparent that gross capacity utilization fell markedly over the period, from a high of 78.5% in 1977 to only 64.9% in 1980 and 67.0% in 1981. All of the rise in recorded underutilization came from growth in capacity rather than falls in throughput. Throughput increased from 3.548 million bpd in 1976 to 4.261 million bpd in 1981. By far the greatest refining capacity in the organization is that of Iran (with 19.9% of the OPEC total), Saudi Arabia (with 18.6%) and Venezuela (22.7%).

Table 3.10

REFINERY CAPACITY, THROUGHPUT AND REFINED PRODUCTS'
PRODUCTION IN OPEC MEMBER COUNTRIES
(million barrels/day)

	1976	1977	1978	1979	1980	1981
Capacity	4,556.8	4,822.4	5,224.4	5,319.5	6,237.4	6,357.4
Throughput	3,548.3	3,787.6	3,950.9	4,128.7	4,050.1	4,260.5
Production	3,498.0	3,733.3	3,907.7	4,155.0	4,030.5	4,202.3
OPEC's throughput						
% of capacity	77.9	78.5	75.6	77.6	64.9	67.0
% of crude protection	11.5	12.1	13.3	13.4	15.0	18.7
Total world capacity	74,773.6	77,184.0	78,672.2	80,035.4	82,837.0	82,231.2
OPEC's % of total world capacity	6.1	6.3	6.6	6.7	7.5	7.7

Source: OPEC

What is the reason for this effort to invest further in downstream processing?

First, it should be clear that there is a difference in the extent of different members' ambitions. Venezuela, for instance, has had substantial refining capacity since the 1920s, when US-based firms did a lot of exploring and processing, both on the mainland and offshore. For the other members, the desire has three components. First, domestic demand for refined products (particularly gasoline) has been growing very strongly. This has led to crude oil exports being accompanied by re-imports of refined products, a situation which seems illogical, uneconomic and even humiliating to some member states. Thus there is a straightforward desire for import substitution. Second, adding further value in the oil industry can generate extra revenue, a factor which is of some importance when crude-related revenues are peaking or falling in oil-exporting countries. A country which has been successful in building up some diversification away from crude oil exports is Algeria. By 1983 crude exports were accounting for only 20% of total export earnings. Refined products, gas, and other products made up the balance.

A third impetus behind downstream investments by the OPEC members is the possibility of using oil refining as a growthpole for the economy. Among the attractions of this course of action is the possibility of remote or otherwise inauspicious parts of the country receiving some development stimulus from the oil industry. Although econometric studies have, in the past, shown relatively little spill-over effect from oil refining to the local economy, the appeal persists.

The extent of the fear over OPEC countries' intentions regarding refining capacity has receded to some degree since 1981. In 1979 the OPEC Press Agency was arguing that for the 13 members to move downstream was "a natural and logical development... the gap which exists at present between our natural energy resource potential and our involvement in downstream activities is a truly baffling situation." But, as noted above, the members' ambitions have been curtailed to a degree now; so, moreover, has their ability to find new projects.

A further point, which has only recently come to be studied, is that OPEC members will have to be more circumspect than was initially realized in the way they export refined products, for fear of affecting deleteriously the course of crude prices. Studies at the East-West Center in Hawaii, funded by the OPEC Downstream Project, have pointed out that, by virtue of world markets for refined products being more sensitive to supply changes than are crude markets, those members selling refined products will have to reduce crude exports as they raise product exports, otherwise the crude market faced by the members only able to export unprocessed crude will be glutted. The Center has suggested that if OPEC refined products exports grow to 4 million bpd by 1986, the real spot price of 34° light might be driven down to about $25/barrel. On the other hand, if the build-up in product exports were slower, so that the 1986 volume were not attained until 1990, the crude market would have longer to adjust, and would also stand to gain from any medium-term firming of oil prices as the world economy recovers. The OPEC members have shown that they recognize this problem; at their London meeting in the Spring of 1983 they agreed to avoid 'dumping' of products that would harm exports of unprocessed crude. Table 3.11 shows the calculations offered by the East-West Center regarding the impact of three different levels of products exports on crude oil prices. It shows that a crude oil price of between $23 and $33 would obtain by 1986, the difference being attributable to different rates of build-up of refined products exports.

Table 3.11

SENSITIVITY OF CRUDE OIL AND PRODUCTS PRICES TO
OPEC PRODUCT EXPORT POLICY: THREE SCENARIOS

Exports (in million b/d)	Aggressive			Expected			Conservative		
	1984	1985	1986	1984	1985	1986	1984	1985	1986
Opec Products	3.40	4.35	4.50	2.70	3.45	4.00	1.90	2.30	2.70
Opec Crude Oil	15.45	15.80	15.90	15.75	16.00	16.15	16.05	16.30	16.55
Non-Opec Products	4.80	4.70	4.60	5.10	5.00	4.90	5.40	5.40	5.40
Non-Opec Crude Oil	6.20	6.20	6.20	6.20	6.20	6.20	6.20	6.20	6.20
Avg.Per-Barrel Product Price★	$ 28.22	23.38	23.32	$ 30.23	28.60	27.72	$ 33.07	32.94	32.81
Spot Arabian Light Crude★	$ 26.43	22.16	21.49	$ 27.64	25.71	24.98	$ 31.73	31.65	31.48
Open Net Revenues (billions)*	$168.69	157.83	155.22	$176.11	173.70	170.12	$198.17	204.31	210.18

★ Inflation-adjusted. * Based on spot prices.

Source: Petroleum Intelligence Weekly

Policy Options in the 1980s

Major closures on the scale just outlined obviously suggest severe difficulties for refiners and those oil companies sufficiently integrated to have refining capacity. Indeed, the 1981 estimate of $10 billion in losses for the worldwide refining industry is something which can hardly be expected to continue.

There are three sets of paths open to those who remain in the industry. The following section looks at each in turn.

First, the most straightforwad policy is to keep chasing the demand curve downward by adding further refinery closures until some sort of equilibrium is attained once more. The firms which emerge from this process will be leaner, and lighter on assets, of course, but— ideally—better positioned in respect of both input-mix (the evolving prices and patterns of crude oil inputs having been taken into account) and output-mix (being better positioned to supply a greater proportion of light-end products). This requires a certain amount of nerve in the short-run, since the attraction of holding onto old and loss-making capacity "just in case" can sometimes be overwhelming.

62

A second path is one which could be pursued simultaneously with the first, but which places slightly different emphasis on the role of refining. In this policy, instead of concentrating wholly on becoming a smaller refiner, the firm would look upstream for more and more of its work. Thus, to take Shell Oil as an example, it is clear that exploration and production are the promising areas in which the firm is taking growing interest. Income from exploration and production accounted for 63% of Shell Oil's total earnings in 1981, while these activities were allocated 56% of all capital spending (then equivalent to around $3 billion.)

Becoming a partner with an ambitious OPEC-sourced oil agency once looked as if it might prove to be a third way ahead, at least for some of the smaller refiners. But since 1982, increasingly tight budgets in all the OPEC states have clipped the wings of all but the most well-endowed operators. In 1982, for instance, the government-owned Kuwait Petroleum Corporation (KPC) bought Santa Fé International of California for $2.5 billion. Its intention, as newsmen put it, was to become the eighth sister of the world's integrated oil industry. The firm was then set upon expanding both upstream, into exploration and production, and downstream, into refining and distribution, rather than merely overseeing head office operations and technically owning the crude that was produced in Kuwait. In the event Kuwaiti offers for much of Gulf Oil's refining capacity in Europe were not to bear fruit, but other OPEC interests (Venezuelans in Hawaii, among others) did purchase a small amount of refining capacity. As they experienced difficulty with selling crude oil during 1982 and 1983 (Kuwaiti crude being a relatively low-quality heavy oil) interest turned instead to exploration. KPC announced that it had spent $100 million acquiring leases for oil exploration in 1981, and would plan on spending $500—600 million annually in this way in the future. Allied to this was a desire to add more value to the crude produced at home, by means of building refining and petrochemical plants.

★   ★   ★

Rapid deceleration in demand for crude oil, due to the price rises of 1973 and 1979-80, has precipitated a crisis in oil refining. Too many refineries existed after 1980 to meet current demand, and no authoritative forecasts expected demand to rise to absorb current capacity again, at least until the end of the decade. Refiners then began a severe trimming of capacity. Making the process more difficult, however, were three other factors. First, certain OPEC member states' oil agencies were making increasingly large investments in refining capacity, with the intention of adding more value to the crude oil they produce prior to its export. This only promised to reduce further the effective demand for western countries' refinery capacity. Second, a long-standing shift in demand towards lighter products such as gasoline required that, in addition to volume changes, refiners plan for the need to offer an evolving output-mix. Finally, another tendency, that of lighter crude oils rising in price more sharply than heavy crudes, suggested a need for refiners to be able to cope with greater shifts in inputs while still obtaining a given set of products as output. Together, these three forces were, not surprisingly, causing some difficulty for refiners. A number of ways forward have been identified in the industry and in press discussion. They offer some improvement to refiners' profitability by way of savage capacity cutbacks, concentration on other parts of the oil business, and some form of partnership with OPEC-based agencies. The last of these options is not, during 1983, likely to yield much comfort, since retrenchment is the order of the day in most cash-strapped OPEC states and their oil corporations. Going ahead with capacity changes thus seems to be the most promising prospect in the immediate future.

Chapter Four

PETROL MARKETING IN THE USA AND EUROPE

THIS chapter examines the present state of the petroleum products businesses in the USA and Europe. Greatest emphasis is accorded the US gasoline market, since it is there that six out of every ten barrels of gasoline sold in the developed world are consumed.

The US Gasoline Market: Overview

The relative position of service stations within retailing generally in the US can be judged from the following figures. In terms of aggregate sales in 1982, gasoline service stations rank fifth after food retailers, automotive retailers, general merchandise, eating and drinking places (and above apparel stores). In 1982, all retail stores had a turnover of $1,070 billion. Of this total, some $98 billion (or 9.2%) was accounted for by service stations. The biggest single group, food stores, did two and a half times as much business.

Table 4.1

RETAIL SALES THROUGH GASOLINE STATIONS IN THE USA
(billions of dollars)

	1982	%Change '82-'81	1981	%Change '81-'80	1980	%Change '80-'79	1979	%Change '79-'78
All Retail Stores............	$1,070.2	3.0%	$1,038.8	9.1%	$951.9	6.4%	$894.3	12.0%
Food Group................	249.9	5.2	237.6	9.5	217.0	10.8	195.8	12.2
Automotive Group..........	189.3	4.8	180.7	11.3	162.3	(8.5)	177.3	8.3
General Merchandise........	131.5	3.1	127.5	8.8	117.2	6.8	109.7	10.3
Eating and Drinking Places...	103.6	10.1	94.1	9.7	85.8	7.8	79.6	8.4
Gasoline Service Stations.....	**98.1**	**(3.5)**	**101.7**	**8.7**	**93.6**	**27.9**	**73.2**	**20.2**
Apparel Group.............	94.2	2.9	47.8	7.7	44.4	4.7	42.4	12.2

Source: Survey of Current Business, Department of Commerce, Vol. 63, No. 1, January, 1983.

Looking now to the importance of gasoline and oil in consumers' expenditure, it transpires that in 1982 these items accounted for 5% of all expenditure, and for 12% of expenditure on non-durable goods, which themselves accounted for 39% of all spending. In 1980 and 1981 the share of gasoline and oil within total spending was again 5%.

In the US market two major forces have been at work in the retail gasoline business in the last decade: a steady decline in the number of service station outlets and an effort to enhance the range of products and services sold in each station.

The number of service stations in the US peaked in 1972 at 226,500 units. Since then some 81,000 have vanished, giving an end-1982 total count of 144,690. At the end of 1983 industry sources are forecasting a service station population of the order of 139,000. Between 1972 and 1981 the 28 major oil companies in the country shut 50,882 units and built 5,000 new units. Table 4.2 shows the service station population from 1981 to 1983, with the total broken into company-owned and franchise-owned units. It can be seen that fewer stations closed in 1982 than 1981.

Table 4.2

NUMBER OF SERVICE STATIONS IN USA

	1983	1982	1981	Percent Changes 1982-83	Percent Changes 1981-82
Number of establishments....	139,250	144,690	151,250	− 3.8	− 4.3
Company-owned...........	25,065	26,044	27,225	− 3.8	− 4.3
Franchisee-owned.........	114,185	118,646	124,025	− 3.8	− 4.3
Total sales of products and services ($ million)..........	114,534	106,544	101,665	7.5	4.8
Company-owned...........	20,616	19,178	18,300	7.5	4.8
Franchisee-owned.........	93,918	87,366	83,365	7.5	4.8

Source: Bureau of Industrial Economics, US Department of Commerce:
Franchising in the Economy 1981-1983.

The industry does not find these figures necessarily discouraging, however. First, although the number of outlets as measured here has clearly been falling, the value of sales has continued to rise. As Table 4.1 shows, 1983 saw a 7.5% rise in sales to $114.5 billion. This means that the average turnover per unit has been rising over time; as Table 4.3 shows, in unadjusted terms, sales per outlet grew to $822,200 in 1983. Typically, for every three to five stations closed by an oil company, one new, much larger, outlet is opened. This allows smaller fixed costs to be spread over a much higher volume of throughput, a very valuable change at a time when margins in the industry are very small. Second, the fall in traditional outlet numbers has taken place against a background of growing numbers of outlets not conventionally (i.e.: by the US Census Bureau) defined as

66

service stations. Stores selling gasoline as a side-line are thought to be growing in number, a factor which suggests that there still is ease of entry to the business to a degree. Some 35,000 to 40,000 convenience stores exist in the US, for instance; of these, about 17,500 are thought to sell gasoline.

Table 4.3

SERVICE STATIONS IN THE US, 1972-83:
ESTIMATED SALES AND NUMBER
OF PASSENGER CARS PER STATION

Year	Sales per station $000	No. cars per station
1983	822.2	905
1982	736.1	863
1981	672.4	816
1980	608.0	768
1979	430.0	731
1978	353.0	676
1977	320.0	644
1976	257.0	591
1975	232.0	563
1972	148.8	426

Source: National Petroleum News

Adding these numbers to the totals given previously yields a total of 162,000 outlets in 1982 and a 1983 figure of 157,000. Table 4.4 shows the number of convenience stores selling gasoline in the 1981-83 period. The table indicates that the number of such outlets grew by 6.3% between 1982 and 1983. Total sales in these outlets grew by 9.9% over the same period, to reach $10.2 billion. The two states with the greatest number of outlets are California and Texas, with an estimated 2,000 units each. Florida has an estimated 1,400 units. Four states—Wyoming, Vermont, South Dakota and Alaska—are listed as having no convenience stores.

California possesses the largest number of service stations in any single state, with a total of 11,342. Texas is the next most populated, with a total of 10,050. Every state has lost service stations over the period 1977-83; among the states which have witnessed the biggest drop in service station population is California (down by approximately 3,000 stations, or 21%).

Table 4.4

NUMBER OF CONVENIENCE STORES IN USA

	1983	1982	1981	Percent 1982-83	Changes 1981-82
Total number of establishments.	17,276	16,256	15,524	6.3	4.7
Company-owned...........	10,205	9,741	9,356	4.8	4.1
Franchisee-owned..........	7,071	6,515	6,168	8.5	5.6
Total sales of products and services (Add $000,000).....	10,168	9,252	3,406	9.9	10.1
Company-owned...........	5,847	5,359	4,897	9.1	9.4
Franchisee-owned..........	4,320	3,893	3,509	11.0	11.0

Source: Bureau of Industrial Economics, US Department of Commerce,
Franchising in the Economy 1981-1983.

The second major force which has been at work in the US gasoline business is diversification. Partly as a response to a perceived increase in the risk of being in a business subject to interruptions in the flow in inputs, and with profit margins being subject to many factors not within the operator's own control, service station operators and owners have been increasingly interested in finding new products to add to their accustomed lines. For some outlets' owners, a further impetus doubtless comes from a desire to explore other businesses while retaining a toe-hold in the gasoline industry during their period of experimentation.

The oil majors have almost all now embarked on major renovation schemes of their own, each of them embracing, to some degree, the principle of diversification. Five to six year multimillion dollar programmes have been under way for some years now, their thrust generally being, as Gulf Oil chairman James E. Lee put it, that "we have to learn to sell again."

The effort at ARCO concentrates on having the successful AM-PM Minimarkets spread further through the service station system, and having "MP and G Tune-Up" shops added. A large advertising campaign attempts to lure more drivers away from do-it-yourself tune-ups. The target volume of the new outlets is to be in the 120,000 to 300,000 gallons/month range. (By comparison, small outlets might only sell 40,000 gallons/month.) ARCO's initiative in abandoning its credit card sales entirely in April 1982, and simply cutting prices instead, has yielded it big volume gains in many areas of the country, and the company believes that this diversification scheme will help it consolidate some of this new business.

Texaco embraced the "family of buildings" architectural concept in 1980. It aims to clear away previous clutters of unplanned and uncoordinated fittings and buildings on its sites and present consumers with a more standardized range of services, including car washes, convenience stores, etc. Texaco has gone a long way in rationalizing its range of outlets. From around 11,000 stations, covering all 50 states of the union, in the early 1970s, the company came down to only 3,300 by August 1982. By late 1983 the firm envisaged having only 3,000 units, all of which should be relatively high-volume. From an average of 34,000 gallons/month delivered at each outlet in 1981 Texaco wants to have its average throughput in 1984 at 52,000 gallons/month. Its plans include abandoning sites in 19 states.

Shell's plans include having auto care facilities at 1,900 of its stations, although in recent years it has enjoyed some strength in gasoline-only self-service outlets. The firm now considers any outlet selling below 50,000 gallons/month as a marginal. Shell divested itself of some 3,300 units over the 1975-82 period, during which time it also built 182 high-volume locations.

Divestment of small volume outlets is also the intention at Gulf Oil. Up to 3,400 directly-supplied stations are to be closed, bringing the firm's portfolio closer to 2,000 units. It wishes to see average volumes boosted by 55-60%. The new corporate plan at Gulf aims to cut marketing costs by 50 to 75 cents/barrel, or by $100-150 million, between 1983 and 1987. Manpower in the field marketing function is to be reduced by 35-40%, with stronger head office initiatives supplanting local autonomy. Apart from efforts at cost-cutting, the firm has also decided to go along with a pattern increasingly seen in the early 1980s in the business—withdrawing from certain regions of the country where the retailing presence is weak and concentrating on building market share in a few strongholds. Thus, Gulf's intention is to be strong in the Sunbelt; operations in New York, parts of New England and the Midwest are being reviewed critically to see if there is any point in continuing representation there.

Exxon is pursuing a refurbishment scheme, which will cost $100 million. One thousand outlets are to be improved by 1985. Exxon is also streamlining its marketing operations, by changing 30 distinct retail offices into 22 area offices. Up to 1,000 staff will be made redundant. Amoco is following a drastic pruning policy, with 3,000 outlets cut between 1977 and mid-1983. Conoco lost 3,800 marketing staff in 1982.

The retailers' efforts to cut the number of outlets they control while boosting sales volume at each would be an attractive enough stategy if pursued in isolation. It offers the possibility of cutting unit costs for all firms in the industry. But this is not enough. Most of the firms want to go beyond this and attempt to establish for themselves further attractions in consumers' minds. One way they are trying to do this is by offering automated gasoline dispensing.

Debit cards are increasingly figuring in retailers' minds as a way of helping build sales volume at relatively little cost. People who have not brought cash with them, or prefer not to handle cash at night, may be attracted to gasoline pumps which can be activated by using a form of credit card. Some gasoline firms have had local banks write the necessary software for the dispensers, and the evidence so far available on consumers' reaction points to some increases in volume being obtained.

Experiments in southern states have shown a surprising willingness on the part of consumers to make bigger purchases when using a new credit card system called Auto-card, offered in conjunction with Visa and Mastercard. At one sample of service stations, the average retail transaction involved 45% more fuel when the card was used. In another experiment, previous sales had averaged 8 gallons per purchase; after the introducion of the card they rose to 14 gallons. The time taken to make the purchase fell by 13%, while sales in any adjoining convenience stores—even more surprising—were not reduced by the fact that there was now no need to enter the store to pay for the gasoline. The hardware for these schemes is made by Swedish firm Ericsson's American subsidiary Autotank of Cleveland.

Another system under trial in 1982 and 1983 in California is the Express Cashier, which is a central console on an island of pumps which accepts paper money in $1, $5, $10 and $20 denominations. This device, used by ARCO, has received favourable customer reaction. So far the Express Cashier has been used chiefly in conjunction with conventional pumps. It has been found that the automated pump is particuarly attractive at busy times of the day when queues are building up at the conventional pumps, due to delays in paying cash inside the office. It has proved particularly useful at 24-hour stations.

Having one's gasoline purchase debited directly to one's monthly bank statement, rather than to a credit card account, is a further possibility with some initial success behind it. In Arizona some retailers have installed a system marketed by International Payment System (IPS) of Los Angeles. This system accepts a transactions card, with a special 4-digit gasoline authorization code to be punched onto a keyboard at the pump. The customer's current account is then debited by the amount of the purchase. The gasoline retailer pays about $400—$500 for installation plus a 1.5% service fee to IPS. The purchase is recorded electronically at the service station. Once a day they are all transmitted in a batch to an automated bank clearinghouse. The debits are then allocated to the banks used by each of the customers. For service station owners the appeal is felt to lie in the fact that it reduces the size of the commission paid to the financial institution. Credit card transactions have a 2.5% commission attached to them, although in return they offer a fuller service than the current-account scheme. But at a time of thin margins the difference matters to many retailers. Indeed, the reason for ARCO's decision to discontinue credit cards in April 1982 was the 2.8 cents/gallon cost it incurred by the transaction.

These innovations are likely to have a significant impact on the industry during the 1980s. Some analysts are forecasting that by 1990 debit card and credit card purchases of gasoline will account for the same share of business as self-service has in the late 1970s and early 1980s. In the meantime, however, what is being observed is a series of interim or holding decisions being taken by the oil retailing firms. Phillips Petroleum has cut its bank card processing fee from 3% to 2.5%. In a move to attract more and more credit card customers, Amoco Oil ran a heavy advertising campaign in early 1983. The result was the issue of 100,000 more cards, bringing the firm's total to about 8 million. Shell dealers in the Midwest began a three-month trial of point-of-sale debit cards, in conjunction with AmeriTrust Bank of Cleveland. Ultimately the intention is to have service stations deal with banks' automated teller machine (ATM) cards, debit cards and credit cards. Gulf Oil tested debit card terminals in the Midwest in late 1983 too, in conjunction with Mellon Bank. In reaction to the shift away from credit cards by ARCO, however, local rival USA Petroleum has filed a suit in a federal court alleging illegal price cutting by ARCO.

Table 4.5 shows figures collected in late 1982 relating to the use of credit cards by the customers of the 22 largest oil companies. The table shows that 107 million cards (or nearly one for every two Americans alive) have been issued, and were responsible for 1981 sales worth $26.4 billion. This was equivalent to 5% of 1981 total sales (of the 31 biggest companies, as shown later in Table 4.9) of $528 billion. The table shows also the overwhelming importance of oil companies' own cards within total card sales: 93% of the $28.3 billion sales were made using oil company credit cards.

A major change which has come to service stations in most developed countries in the last decade has been the growth of self-service. In the past, full-time employees would attend to the filling of the car with gasoline and to sundry other services. As real labour costs, both direct and indirect grew, however, and as the incentive for retailers to shave costs so as to preserve their margins increased, forecourt service began to disappear.

Table 4.5

22 LARGEST OIL COMPANY CARD ISSUERS—1981
Oil Company Credit Cards

Oil Companies	Total Cards (millions)	Total Accounts (millions)	Sales Volume ($millions)	Total Credit Sales ($millions)
Texaco	13.00	7.00	$2,600.00	$2,600.00
Amoco	12.50	6.90	3,300.00	3,320.00
Exxon	11.90	6.60	3,500.00	3,500.00
Shell	11.50	6.00	2,700.00	3,100.00
Mobil	11.30	5.80	2,500.00	2,500.00
Gulf	10.00	5.00	2,100.00	2,100.00
Chevron	8.00	4.30	2,400.00	2,800.00
Union	5.90	3.50	1,800.00	2,074.00
Arco	5.90	3.00	1,293.30	1,295.00
Sunoco/DX	4.80	3.00	750.00	895.00
Phillips	3.70	2.20	780.00	975.00
Sohio	2.10	1.20	1,100.00	1,169.00
Citgo	1.60	1.00	270.00	350.00
Conoco	1.30	1.00	310.00	335.00
Marathon	.80	.50	150.00	150.00
Diamaond	.70	.50	140.00	180.00
Fina	.60	.40	162.00	162.00
Total	.50	.30	215.00	360.00
Champlin	.40	.20	67.00	87.00
Kerr-McGee	.40	.20	79.00	97.00
Ashland	.20	.10	57.00	117.00
Husky	.20	.10	128.60	143.70
Totals	107.30	58.80	$26,401.90	$28,309.70

Began honouring bank cards October 1981
Source: The Nilson Report, November 1982

Source: National Petroleum News

1975 saw a major boost to self-service in US stations, with the year beginning with about 10% of stations using self-service but ending with 25% so organized. By the end of 1979 the ratio had climbed to nearly 50%; by the end of 1982, Amoco calculations suggest that self-service sales volume accounted for nearly 66% of all sales. The extent to which self-service is used varies considerably by state, with around 90% of sales thus organized in the states of Utah and Arizona, and most states' proportions falling between 70% and 80%. Maine is the state with the smallest proportion of self-service sales, standing at around 25%. California, the state with the largest number of service stations in early 1983 (11,342 sites in all), uses self-service for around 80% of all gas stations' sales volume. Diagram 4.1 shows how self-service has grown in importance over the period 1975-82.

Sales of accessories at gas stations have tended to grow satisfactorily in recent years. The main products being sold are tyres, batteries, oil and antifreeze. Tune-up services are also offered by certain stations.

Tyre sales have grown through the last ten years—surprisingly, perhaps, in view of the tendency for new and improved tyres to yield increasingly long serviceable mileage. This phenomenon is to some degree being counterbalanced by the rising average age of the US automobile population, so that replacement tyres have been in increasing demand. In 1973 replacement tyre sales in the US overall peaked at 142,000,000. In 1982 the figure was 132,000,000, a rise of 6,000,000 over the 1981 figure. Service stations' share in this market has tended to be less volatile: from 28,000,000 tyres in 1972 and 1973 sales fell away before growing again to 26,400,000 in 1982. The average annual volume of sales per service station, as shown in Table 4.6, has however been rising strongly as the number of outlets has contracted. From 110 tyres per service station in 1968, the average has risen to 183 in 1982. These figures suggest that service stations account for about 20% of the replacement market.

Table 4.6

SALES OF REPLACEMENT TYRES THROUGH SERVICE STATIONS IN USA 1970-1982

	Total Sales	Service Station Sales	Average per Station
	(millions)		
1970	130	26.0	118
1973	142	28.4	130
1975	125	23.7	125
1978	135	27.0	157
1980	107	21.4	135
1981	126	25.2	167
1982	132	26.4	183

Source: Rubber Manufacturers Association

Diagram 4.1

**SELF-SERVICE VERSUS
FULL-SERVICE, 1975-1982 ★**

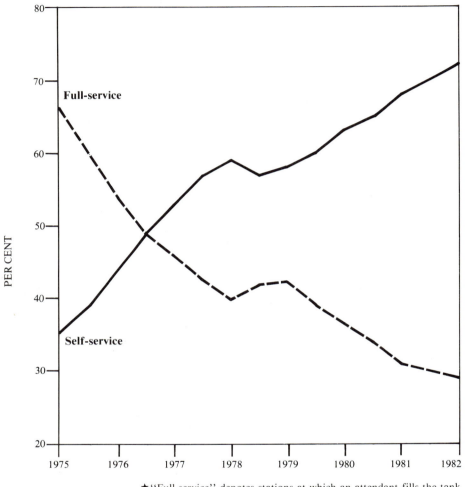

★ "Full service" denotes stations at which an attendant fills the tank

Source: Lundberg Letter, December 1982

74

The share of service stations in the US replacement battery market is considerably lower. Estimates for 1982 point to about 54,200,000 units, or 87% of the entire national market, being replacement sales. Of these, gasoline and oil retailers sold some 2,500,000 units in 1982. More significant as sellers of batteries were department and discount stores, which together sold 13,800,000 units, and battery specialists, who sold 15,750,000 units. The replacement battery market has been growing rapidly since the early 1970s, with deliveries in 1970 of 37,000,000 units growing to 54,000,000 in 1982.

Antifreeze sales (defined as ethylene glycol sales only, since the Chemical Specialties Manufacturers Association no longer surveys the methanol antifreeze market) have been relatively poor since the mid-1970s. The peak year for sales was 1975, when 218 million gallons were sold. In 1981, 179 million gallons were sold, with volume having dropped as low as 137 million gallons in 1978.

Motor oil sales have been declining since the mid-1970s. In 1975 1,014 million quarts were sold, but this fell away to an average level of 900 million quarts for the rest of the decade. In 1982 sales totalled 907 million quarts. Average oil consumption per year by cars has dropped from 10 quarts in 1972 to 8 in 1976 to 7.3 for the 1979-82 period.

Tune-ups are estimated by a survey from the Champion Spark Plug Co. to be predominantly carried out by motorists working on their own accord. In 1982 perhaps 35% of all tune-up work was carried out in that way. In the same year, Champion estimates that repair shops carried out 26% of tune-ups, car dealers 17% and service stations 14%-15%. These shares in the total business are not felt to have changed very significantly over the past four years.

A final factor observed in the industry in recent years has been the gradual pull-out of department stores from gasoline retailing and car servicing. Early in 1983 J. C. Penney, the nation's third largest retail chain, left the business, announcing the closure of its 434 auto centres by the end of the year. Firestone is taking 300 of them. K-Mart, another major chain, cut back its auto service outlets from 1,700 to 1,400 in 1982.

Tying together these elements of the industry in diagrammatic form, diagram 4.2 shows the forces which drive competition between the firms. The diagram is arranged in the manner adopted by Michael Porter in his study *Competitive Strategy*. There appear to be relatively few threats to existing gasoline retailers from the direction of new entrants. If anything, forces point to a slimming in the number of operators in the business, with retail chains such as J. C. Penney losing business. Suppliers are often the retailers themselves, so there is, for a large proportion of the industry, no source of risk from that angle. But for independent retailers, there is an ever-present danger from cuts in the amount of support (or implicit subsidy) offered by the gasoline refiner and/or distributor in times of a "price war". Given the frequently close physical proximity of gas stations to one another in American cities, any failure by the supplier to change the terms of supply so as to allow that retailer to cut his price is likely to be quickly reflected in lower sales volume. Substitutes for gasoline do, of course, exist in the form of public transport use, air travel, walking, car-pooling, and so on. Over the long run, as the statistics referred to earlier in this book have made clear, consumers offer an unmistakable response to higher prices in the form of lower demand. But even a decade after the first oil 'shock' the private car is still overwhelmingly dominant as the American form of transport. This leaves inter-firm competition as the only major source of difficulty for the gasoline retailer.

The US Gasoline Market: Market Shares and Finance

Market shares have moved quite markedly in the US oil business since the early 1970s. In 1981 and 1982 the biggest share of the gasoline market was taken by Amoco, which in previous years had had second and third positions. Exxon remained in third place, after being toppled from first place in 1981. Table 4.7 shows the market shares of the eleven biggest marketers between 1978 and 1982.

Table 4.7
TOP 11 MARKETERS: A FIVE-YEAR OVERVIEW

	1978		1979		1980		1981		1982	
	Rank	% Share	Rank	% Share	Rank	% Share	Rank	% Share	Rank	% Share
Amoco........	2	7.71%	2	7.48%	3	7.40%	1	7.28%	1	7.36%
Shell.........	1	7.70	3	7.30	2	7.44	2	6.89	2	6.78
Exxon	3	7.28	1	7.65	1	7.54	3	6.81	3	6.64
Gulf.........	5	5.83	4	6.36	7	5.39	5	5.69	4	5.85
Mobil........	6	5.55	6	5.78	4	6.01	6	5.38	5	5.63
Texaco.......	4	6.75	5	6.01	5	5.95	4	5.77	6	5.61
Atlantic Rich.	8	3.79	8	3.89	8	3.87	8	3.80	7	4.79
Chevron......	7	4.76	7	5.30	6	5.77	7	5.21	8	4.74
Union........	9	3.28	9	3.49	9	3.50	9	3.31	9	3.25
Phillips	11	3.02	11	2.85	10	2.83	10	2.88	10	2.75
Sun..........	10	3.20	10	3.01	11	2.74	11	2.55	11	2.57

Source: National Petroleum News

Diagram 4.2

COMPETITIVE FORCES IN THE GASOLINE INDUSTRY
AT THE RETAIL LEVEL

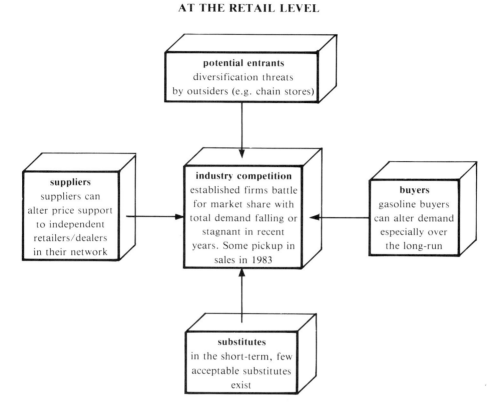

Among the biggest changes seen over the long term are the collapse of Texaco's share, from the highest, at 8.35% in 1971, to 5.63% in 1982; the sudden turnaround in ARCO's share in 1982 (probably to be attributed to its cash sales only policy); and the continuing ebbing away of Sun Oil's market position, reaching a low of 2.55% in 1981 but barely recovering in 1982 to 2.57%.

Taken together, the share of the largest oil companies in total US gasoline sales has been falling for the past decade or so. In 1970, the top 8 distributors held 54.6% of the gasoline market; by 1980, according to a study prepared by the American Petroleum Institute, their share had fallen to 49.3%. Independents have been slowly encroaching on the majors' territory. But the change still leaves the biggest 20 firms over 70% of the gasoline market. In all refined products, a similar process has been at work. Here, independents in 1970 took 17.5% of sales, whereas by 1980 they were taking 28% of sales.

Figures from the Television Bureau of Advertising show (see Table 4.8) that many oil firms spent less on TV marketing in 1982 than they did in 1981. Total expenditure on TV advertising for gasoline, lubricants and related fuels was $145 million in 1982. This represented a 11.1% fall from the 1981 figure of $163 million. 1982 "spot" TV expenditure was $90 million, a rise of 7.3%, while spending on "network" TV was down sharply, by 31% to $54.9 million.

The fact that relatively little, in absolute terms, is spent by oil firms on TV advertising is indicated by the biggest spender Mobil Oil, only ranking as the 75th biggest TV advertiser in US business. The major spenders, the likes of General Motors and Procter & Gamble, allocate sums between $300 and $500 million annually to advertising. Notable in the tables is the near-five-fold jump in spending by ARCO, to $4.4 million—a change in spending that accompanied the strategic shift in the company's pricing policy.

The financial position of the major firms involved in the petroleum business is analysed in Tables 4.9 to 4.12. Table 4.9 shows the 1982 profits made by the 10 biggest oil companies, with the profits deriving from their gasoline business broken out separately. The dismal operating conditions facing most of the world's chemical producers is reflected in these figures; the result is that petroleum profits account in all firms for virtually all of the profits registered. Only in two firms (Standard Oil of Indiana and Atlantic Richfield) did petroleum-derived profits increase between 1981 and 1982; in all the other firms there was a decline in gasoline profitability. Total gasoline-derived profits for 1982 were $24.62 billion, while total

Table 4.8

TV ADVERTISING IN 1982: HOW MUCH THE COMPANIES SPENT

	1982 (Add 000)	1981 (Add 000)	% Change
Spot TV			
Standard Oil (Ind.)	$12,562	$11,631	+ 8%
Exxon	10,804	6,223	+74
Standard Oil (Calif.)	7,756	8,518	− 9
Union Oil of Calif.	6,816	4,291	+59
Gulf	6,756	9,624	−30
Shell	6,243	7,416	−16
Mobil	6,006	8,993	−33
Getty	4,621	3,202	+44
Atlantic Richfield	4,414	746	+492
Sun	3,500	1,799	+ 95
Texaco	3,243	3,746	− 13
Burmah Oil	3,048	2,856	+ 7
Ashland	2,796	2,069	− 35
Pennzoil	2,623	2,522	+ 4
Quaker State	2,466	2,054	+ 20
Standard Oil (Ohio)	2,092	2,621	− 20
Phillips	986	2,092	− 53
Wilco (Kendall)	697	681	+ 2
Amerada Hess	613	—	—
Network TV			
Phillips	$ 9,731	$ 8,780	+ 11%
Shell	9,511	6,145	+ 55
Texaco	7,558	12,016	− 37
Sun	6,461	5,367	+ 20
Quaker State	5,056	3,068	+ 65
Pennzoil	4,560	5,833	− 17
Exxon	4,213	20,577	− 80
Conoco	3,404	7,075	− 52
Ashland	1,700	3,190	− 47
Ford (Motorcraft)	720	1,305	− 45
Witco (Kendall)	609	820	− 26
Gulf	414	205	+102
Getty	320	—	—
The Leaders **(Combined Totals)**			
Shell	$15,754	$13,561	+ 16%
Exxon	15,017	26,800	− 44
Texaco	10,801	15,762	− 31
Sun	9,961	7,166	+ 39
Pennzoil	7,483	8,355	− 10

Source: Television Bureau of Advertising from BAR, April 1983/National Petroleum News

Table 4.9

PROFITS OF TEN MAJOR OIL COMPANIES 1982

Company	Cumulative 1982	% Total	Percent Change From 1981
Exxon			
Total	4,499.0	100.0%	−18.1%
Petroleum	4,407.0	98.0%	−16.1%
Chemical	92.0	2.0%	−61.3%
Mobil			
Total	1,744.0	100.0%	−37.8%
Petroleum	1,723.0	98.0%	35.8%
Chemical	21.0	1.8%	27.4%
Texaco			
Total	1,281.0	100.0%	−44.5%
Petroleum	1,288.0	100.6%	−42.3%
Chemical	—	−0.6%	—
Gulf Oil			
Total	1,906.0	100.0%	−18.4%
Petroleum	2,235.0	117.3%	−5.7%
Chemical	−329.0	−17.3%	—
Standard Oil California			
Total	1,321.0	100.0%	−39.6%
Petroleum	1,337.0	101.2%	−38.0%
Chemical	−16.0	−1.2%	—
Standard Oil Industrial			
Total	1,897.0	100.0%	−0.9%
Petroleum	1,821.0	76.0%	0.4%
Chemical	76.0	4.0%	−24.8%
Atlantic Richfield			
Total	4,053.0	100.0%	7.2%
Petroleum	4,149.0	102.4%	10.2%
Chemical	−96.0	−2.4%	NM
Shell Oil			
Total	1,765.0	100.0%	−8.6%
Petroleum	1,790.0	101.4%	−6.1%
Chemical	−25.0	−1.4%	NM
Standard Oil Ohio			
Total	3,987.1	100.0%	−4.5%
Petroleum	4,010.4	100.6%	−3.0%
Chemical	−23.3	−0.6%	NM
Getty Oil			
Total	1,856.5	100.0	−19.2%
Petroleum	1,855.2	99.8%	−18.7%
Chemical	1.3	0.1%	−90.6%
Total			
Total	24,309.6	100.0%	−16.8%
Petroleum	24,616.6	101.3%	−13.9%
Chemical	−307.0	−1.3%	NM

Source: US Dept. of Energy

losses on chemicals were $307 million. This gasoline profitability was 13.9% down on the 1981 figure. The dominance of Exxon and ARCO in the gasoline business can be seen from the table: in 1982 Exxon and ARCO were responsible for no less than 35% of the top 10 firms' gasoline profits.

Table 4.10 shows the gross sales of the 16 major energy companies based in the US, arranged in two categories. The big eight integrated producers' sales fell by 9.5% between 1981 and 1982, but in 1982 still totalled $357 billion. The other eight integrated firms listed below them experienced a 4.7% sales fall; their sales totalled $101 billion.

Table 4.10

SALES OF MAJOR INTEGRATED ENERGY COMPANIES IN 1982
(million dollars)

Company	1982 Q4	Percent change from 1982 Q3	Percent change from 1981 Q4	1982 Cumulative	Percent change from 1981
Top Eight					
Exxon	26,241.0	3.4	− 11.9	103,649.0	− 9.8
Mobil	16,184.0	4.2	− 10.6	63,820.0	− 6.8
Texaco	11,700.0	− 0.8	− 19.8	48,000.0	− 19.1
Gulf Oil	8,160.0	8.3	2.8	30,658.0	0.6
Standard Oil California	7,887.0	− 9.0	− 35.3	35,943.0	− 22.9
Standard Oil Industrial	7,500.0	− 2.6	− 3.8	29,900.0	− 5.7
Atlantic Richfield	7,095.1	6.5	− 0.6	26,990.6	− 4.3
Occidental Petroleum	5,287.7	19.8	18.3	18,212.2	23.8
Top Eight Total	90,054.7	2.8	− 11.7	357,181.0	− 9.5
Other Integrated					
Shell Oil	5,176.0	1.7	− 0.3	20,214.0	− 7.0
Sun	4,100.0	7.9	− 8.9	16,300.0	− 0.6
Phillips Petroleum	4,063.0	1.5	− 0.7	15,892.0	− 2.4
Standard Oil Ohio	3,216.7	1.3	− 14.8	13,529.2	− 4.3
Getty Oil	3,013.8	− 1.9	− 15.2	12,326.0	− 7.0
Union Oil	2,736.1	− 1.3	− 6.0	10,894.1	− 3.5
Amerada Hess	2,116.4	− 0.4	− 9.5	8,394.1	− 11.1
Kerr-McGee	1,015.0	11.1	− 5.4	3,777.0	− 1.3
Other Integrated Total	25,437.0	2.0	− 7.0	101,331.0	− 4.7

Source: US Dept. of Energy

Table 4.11
GEOGRAPHIC DISTRIBUTION OF PETROLEUM PROFITS
NINE COMPANIES, 1982

Company	Cumulative 1982	Percent of Consolidated	Percent Change from 1981
Exxon			
Worldwide	4,207.0	100.0%	− 16.1%
Domestic	2,120.0	52.6%	− 1.2%
Foreign	2,087.0	47.4%	− 28.3%
Mobil			
Worldwide	1,723.0	100.0%	35.3%
Domestic	878.0	51.0%	25.2%
Foreign	845.0	49.0%	44.1%
Texaco			
Worldwide	1,289.0	100.0%	− 42.3%
Domestic	475.0	36.9%	− 46.1%
Foreign	814.0	63.1%	− 39.9%
Gulf Oil			
Worldwide	2,235.0	100.0%	− 5.7%
Domestic	1,381.0	61.8%	0.3%
Foreign	854.0	38.2%	− 14.1%
Standard Oil California			
Worldwide	1,337.0	100.0%	− 36.0%
Domestic	963.0	72.0%	21.7%
Foreign	374.0	28.0%	− 59.6%
Standard Oil Industrial			
Worldwide	1,821.0	100.0%	0.4%
Domestic	1,207.0	66.3%	1.4%
Foreign	614.0	33.7%	− 1.6%
Shell Oil			
Worldwide	1,290.0	100.0%	− 6.1%
Domestic	1,290.0	100.0%	− 6.1%
Foreign	0.0	0.0%	
Standard Oil Ohio			
Worldwide	4,010.4	100.0%	− 3.0%
Domestic	4,010.4	100.0%	− 3.0%
Foreign	0.0	0.0%	
Getty Oil			
Worldwide	1,855.2	100.0%	− 18.7%
Domestic	1,017.1	54.8%	− 17.0%
Foreign	838.1	45.2%	− 20.7%
Total			
Worldwide	20,467.6	100.0%	− 17.6%
Domestic	14,041.5	68.6%	− 9.2%
Foreign	6,426.1	31.4%	− 31.4%

Source: US Dept. of Energy

Finally, table 4.11 shows the extent to which the profits of the major oil companies were derived from operations within the US. For the top 9 companies, on average 68.6% of petroleum profits were earned from domestic American activities, and 31.4% from overseas operations. Texaco, with 36.9% of its earnings derived in the US market, was the firm most heavily reliant upon overseas earnings. Between 1981 and 1982, in aggregate term, foreign profitability declined by 31.4% whereas domestic profits fell by only 9.2%.

Preliminary first quarter 1983 figures point to revenues of $117.4 billion and profits of $4.2 billion being earned by the 26 biggest oil corporations. The profit figure shows a 14.2% drop on the equivalent period of 1982. Depressed demand, feeding into a relatively fixed stock of assets and causing problems of underutilization, was again the major difficulty facing the industry. Nonetheless, six companies did manage to improve their performance, three of them (Exxon, Mobil and Standard Oil of California) by virtue of the cut in the price of Saudi market crude which, through their membership of Aramco, they purchased. Refining operations were a major drag on profitability, with some firms finding that profitability from that activity was restored in 1982 but lost again in 1983. Shell Oil's refining, transport and marketing loss in the quarter for instance, was $40 million. Standard Oil of Indiana lost $56 million on the same activities, while Gulf Oil lost $96 million.

Earnings reported for the first three quarters of 1983 show that seven of the nine major oil companies enjoyed higher profits than in the preceding quarter (see table 4.12). Among the bigger improvements were those of Mobil with a 42% gain (chiefly from foreign earnings), and Gulf Oil, whose profits were up by 74%. Profits at Mobil were up despite revenues having fallen by 7.8% to $14.2 billion. Mobil's mass retailing subsidiary Montgomery Ward & Co showed a $2 million profit as opposed to a $38 million loss the year before. Net income for the first three quarters of the year was down by 9.9% to $42.9 billion. Texaco also reported a fall in revenue, this time of 10% to $10.6 billion, and a 9.2% gain in profits. For the nine month period in 1983 net income fell 16% to $30.6 billion. Sun Oil's sales fell by 4.6% to $3.77 billion but its third quarter net income was 10% up on the quarter. Cost cutting was stressed at the firm; among the factors identified as having helped cut costs was lower headquarters staff expenditures. Phillips Petroleum recorded a 3% drop in revenue in the third quarter and a 15% rise in earnings. Over the first three quarters of the year profits were down by 3.5% while revenue was down by 4.3% to $11.3 billion. Gulf Oil's first three quarters' net earnings were up 1% while revenue fell by 4.9% to $21.4 billion. Shell Oil's chemical products division earned $14 million in the third quarter of 1982. Earnings from the oil products division fell by 20% to $111 million. Amerada-Hess earnings fell by 6.2% although, unusually in that quarter, revenue rose by 11% to reach $2.35 billion. For the first three quarters taken together, profits rose 19% and revenue was 2.9% down at $6.1 billion.

Tenneco's third quarter earnings were 20% down on 1982 levels. Revenue for the period fell by 5% to $3.43 billion. For the first three quarters revenue was down by 4% to $10.6 billion.

Table 4.12

NET INCOME OF MAJOR OIL COMPANIES:
NINE MONTHS 1982, 1983

$million	1983	1982	% change
Mobil........	1,056	815	+ 30
Gulf.........	681	674	+ 1
Phillips	474	491	− 3.5
Sun..........	392	386	+ 1.6
Texaco.......	977	975	+ 0.2
Shell........	1,084	1,166	− 7
Occidental....	320.7	153.6	+ 10.9
Amerada.....	148.3	125	+ 19
Tenneco......	475	531	− 11

Source: Wall Street Journal

The bold—and, as it has transpired, justified—action of ARCO's top management in striking out in a bold strategic manner as described above raises the interesting question of where next integrated oil companies can look to steal ahead of their competitors, as well as the question of where small independents can hope to gain in the interstices of the larger competitors. There are, after all, relatively few options immediately apparent: cutting costs; squeezing margins; diversifying at the retail level; reordering the priorities in the vertically integrated chain of operations to emphasize the upstream end more; introduce price structures, for leaded, unleaded, premium and other brands which consumers will find attractive; and price cash and non-cash sales differently. Having exploited one, with the move to cash-only sales, and already being active in exploration and drilling, and being a leader in convenience stores and other forms of retail-level diversification, ARCO now faces a harder task in devising future strategic moves. By no means all the initiative, however, has to come from within the oil industry itself. By opening discussions with Shell Oil and Gulf, J. C. Penney, the chain store retailer, has been able to use certain excess capacity in its huge trans-continental communications network (second in scale only to AT&T, perhaps) to assist the gas companies in performing credit card checking. Previously, Shell Oil spent some 40-50 cents each time it had to make a credit card check from a gas station. Now, using Penney's network, Shell enjoys a 2-second turnaround time and online access to credit card firms' files. The exact cost reduction is not known, but is clearly sufficient to make it attractive to Shell.

This illustration is of greater significance than showing the sort of new tie-ups which could evolve between oil companies and others outside the traditional confines of the business. For it suggests that there is scope for the imaginative firm to attempt to redefine the confines and nature of its own business. Here, J. C. Penney has become not merely a retailer but also a financial services firm (a move bolstered by its acquisition of a small bank on the East Coast). It is in a position to take over certain financial tasks in their entirety from other firms. And in turn it suggests that other firms can borrow from this example, and add and divest functions or divisions that, for various reasons, they feel they can do particularly well or peform unusually badly.

One implication of this is that oil companies may be inclined to reorganize themsleves so as to highlight management control and operating performance strengths and weaknesses more quickly and more clearly. Thus, vertically integrated firms could be tempted to operate as autonomous divisions, with the boundaries of control reset so that each division is not merely a cost centre for the next division to which it passes on its products: by reference to the open market, internal transfer prices could be allowed to float freely and divisions no longer 'fed' work purely because they are there.

Time will tell how far oil company managements are willing to have the elements of their firms examined critically in this way. But it would be surprising if, after three or four years of cost-cutting and convulsive change elsewhere in American industry, and in an environment in which oil firms have had to react very deftly to changes in their own market, they were to prove unwilling or unable to scrutinize their internal operations with a view to proposing a few more upheavals.

The European Petrol Market: Overview

In Europe a number of forces are acting upon gasoline dealers and marketers. In general what the industry foresees is gloomy: too much capacity exists, too narrow margins, and too much competition from other companies to make service station diversification the boon it has been to many firms in the U.S.

First, many firms are having difficulty sustaining their existing gasoline distribution networks. In the UK in mid-1983, it was estimated that the oil companies together were losing $45 million each month in providing subsidies for their dealers. Shell and BP were each losing $6 million per month. Standard Oil of California attempted to sell its Chevron chain of service stations in the UK and the rest of Europe. It owns 250 stations in the UK and over 4,000 elsewhere in Euroep. After some hesitancy in the midst of negotiations, US Gulf Oil sold its major downstream operations in three European countries, Belgium, Luxembourg and the Netherlands, to Kuwait Petroleum Corporation (KPC). KPC now possesses 750 service stations, nine oil terminals, a refinery and a lube plant near Rotterdam. It is anticipated that the brand name Gulf will continue to be used for some years, after which KPC may introduce its own name brands. How far KPC will restructure Gulf's old operations is not yet clear. If it were to emulate recent US practices it would close outlying stations and consolidate its grip on a well-defined geographical area. But being a newcomer to the business, KPC may wait for a while before deciding where its strengths lie.

What makes the closure of some of Europe's service station network likely is the fact that dealer profit margins there are considerably below US levels. Comparable figures referring to 1981 show that on a barrel of gasoline, the oil company's profit and cost margin on top of the cost of raw materials and taxation is 20% in the US and 23% in Japan but only 8% in Western Europe. In West Germany the margin is 9% and in France 5%. In the UK it has been put below even that.

Tight margins could be tolerated if there was a way to mitigate their effect through attracting extra business volume via convenience stores and the like. However, there is evidence to suggest that, in the UK at least, repair work is drifting away from garages and towards speciality stores which offer guaranteed service at fixed prices. Four such chains are Kwik-Fit Europe, PH Standard, Lucas and Auto Safety Centres. Kwik-Fit Europe has 235 centres and in 1982 had a turnover of £44 million ($66 million). By 1986 it expects to have 500 centres. As service intervals for cars fall to 12,000 miles, there is a smaller volume of servicing business, offset only by the gradual increase in the number of cars on the road (the 'parc') and (as in the US) a slight recent ageing of the existing car fleet. Much of this business is highly profitable for service stations and, to the extent that it continues to ebb away the narrowness of the profit margins on petrol, retailing will more quickly encourage exit from the business.

Among the several instances of oil companies consolidating their retail chains is the plan by Chevron Oil to pull out of refining and downstream activities in Italy. In West Germany, BP is cutting its 3,200 unit network of service stations so as to curb downstream losses in its German subsidiary, Deutsche BP. In 1982 the division lost DM 900 million. In France, where the position is complicated by tight state intervention, both via ownership of the main oil groups and via price control, the trend is towards cut-backs in the extent of the gasoline distribution network.

For much the same reasons as were discussed in the context of the big US-based integrated oil firms, most European oil agencies have been faced with huge losses since the 1981-83 recession began to contract demand for refined petroleum products. Excess capacity in oil refineries and in associated infrastructure, such as supply depots, warehousing, distribution, corporate staffs and so on have conspired to push many companies firmly into the red. Mention has already been made of BP's German losses; elsewhere on the Continent losses have also been intolerably large. The impact of these on corporate planning efforts has chiefly been to prompt companies to look for substantial cutbacks in their operating capacity. To a greater or lesser extent all the main firms have come to the view that the sooner they cut back the sooner their profit and loss statements will look healthy—albeit at the cost of severely written-down balance sheets. This thinking has naturally had an effect on these firms' involvement in all stages of processing.

In 1982 the total loss of the West German oil refining industry was DM 5.5 billion. Since 1974 cumulative losses in the industry have reached DM 13 billion. In France, the publicly-owned oil majors have been shielded to some extent by state intervention to set floor prices at the retail stage. Nonetheless, Elf lost FF 3 billion in 1982 on its refining operation alone, and that only accounted for 22% of French refining capacity in that year. The particular difficulty facing French refiners has been the swift fall of the franc against the dollar, which, since oil is priced in dollars, means that importers' domestic currency costs have soared.

A large set of factors has affected those responsible for taking a long-term view of refined products demand. Two in particular should be noted.

First is the expectation—borne out through most of 1983, as it happened—that the European recovery from recession would be slower than in the USA. As the economy in America grew extremely rapidly through the year but as European economies, with the somewhat surprising exception of the UK, remained relatively stagnant, planners in oil companies draw conclusions about the long-term health of much of mainland Europe. As long as the US dollar remained high, they reasoned, the real domestic cost of imported oil and oil products to European consumers will be high and rising. Thus, in West Germany, where the currency had sunk from an average level against the dollar of DM 2.50 at the start of the year, to end the year as low as DM 2.76, consumers faced, even before any tax changes, the equivalent of a 10% price rise on oil products. Similarly with the UK, where all-time lows were set by sterling against the dollar during 1983. But even aside from the impact of the dollar upon domestic European oil product prices, there were other grounds for expecting that economic growth might return more slowly to Europe than to the USA.

Many of the peripheral economies—Sweden, Denmark and Portugal are examples—were still beset with chronic budget deficit problems. The difficulty which these pose is that they require, in the short-term, relatively high interest rates for funding. Resources have to be induced into the country to fund the deficit after domestic taxes and borrowing have been utilised while in the longer-term they require either an absolute fall in living standards so that the size of the accumulated debt may be reduced, or a diminished rate of increase in living standards so that more taxes can be devoted to servicing the debt. In any case, disposable income, after taxes and interest payments, is reduced and correspondingly less is available for purchasing such things as automobiles and gasoline. Elsewhere in Europe, in Spain, Italy, France and the UK—far bigger economies of much greater concern to refiners and distributors—much the same story was being played out. Even as exports rise in 1984, to take up the slack as US firms found themselves increasingly priced out of overseas markets by the ever-rising dollar, these economies will in all probability experience a slow and uncertain recovery.

A second point is that, precisely because of the above budget problems, taxes on petroleum products will continue to be considered fair game as targets for governments needing money in the years to come. As is well-known, taxation on oil in European countries is significantly higher than in the US (and, oddly enough, than in many developing countries). This is not at all likely to change—indeed, in the light of European governments' continuing budget crises, it is likely if anything that the real burden of taxation will increase.

Before looking at the European markets in more detail, Tables 4.13 to 4.16 summarise some of the broader European trends.

Table 4.13 shows trends in petrol consumption by country between 1970 and 1982. Consumption rose in all countries between 1970 and 1980 to a substantial degree due to the increase in the size of the European car parc, and despite the rising costs of fuel. Since 1980, consumption has levelled out or fallen in most countries.

Table 4.13

MOTOR GASOLINE CONSUMPTION IN WESTERN EUROPE
1970-1982, BY COUNTRY
(Thousand tonnes)

	1970	1978	1979	1980	1981	1982
Austria	1,574	2,358	2,414	2,443	2,408	2,387
Belgium	2,207	3,108	3,132	2,948	2,719	2,669
Cyprus		96	98	94	93	99
Denmark	1,497	1,792	1,578	1,574	1,441	1,369
Finland	1,026	1,338	1,407	1,336	1,345	1,374
France	12,281	17,587	17,701	17,746	18,115	18,132
Greece	535	1,363	1,375	1,387	1,447	1,513
Iceland	51	91	88	89	92	
Ireland	617	976	981	1,018	1,020	986
Italy	9,200	11,090	11,960	12,090	12,000	11,950
Luxembourg		224	279	286	311	
Malta	37	41	43	39	40	
Netherlands	3,035	3,954	3,979	3,904	3,685	3,633
Norway	925	1,340	1,412	1,395	1,380	1,405
Portugal	474	754	749	751	784	810
Spain	2,771	5,454	5,743	5,644	5,655	5,513
Sweden	2,761	3,660	3,636	3,516	3,463	3,480
Switzerland	2,113	2,608	2,594	2,744	2,851	2,889
Turkey	935	2,214	2,018	1,953	1,927	1,965
United Kingdom	14,234	18,348	18,686	19,145	18,718	19,250
West Germany	15,492	23,015	23,307	23,721	22,269	22,730

Source: National Agencies

The reduction in the number of motor gasoline outlets has occurred in major European countries as it has in the USA. Table 4.14 shows the drop which has occurred in the four major European countries; compared with the 27% decrease in the USA between 1973 and 1979, outlets fell by 36% in West Germany and a third in the UK. This has continued in Europe as it has in the USA; in the UK, for example, the number of sites fell by a further 9% between 1979 and 1983.

Estimates by Esso, shown in Table 4.15, demonstrate that average fuel sales per outlet have risen with the number of outlets closing. In most countries the average sales of fuel doubled between 1974 and 1981.

Table 4.14

WESTERN EUROPE
MOTOR GASOLINE DISTRIBUTION: REDUCTION IN SALES
OUTLETS IN VARIOUS COUNTRIES, 1968-1979
Number of sales outlets

	1968	1973	1979	1979/73 %
France[1]	47,500	50,000	46,000	− 8
West Germany	46,700	42,000	27,000	− 36
United Kingdom	38,500	3,500	23,500	− 33
Italy	35,000	41,000	38,000	− 7
United States	226,000	216,000	158,000	− 27

Source: Union des Chambres, Syndicales de l'Industrie du Pétrole
[1]Statistics from the Ministry of Industry

Table 4.15

WESTERN EUROPE
AVERAGE SALES OF MOTOR FUEL (GASOLINE & DIESEL)
PER OUTLET IN VARIOUS COUNTRIES, 1974-1981
(Fuel sold in thousand litres)

	1974		1981	
	Number of sales outlets	Ave.fuel sales per outlet	Number of sales outlets	Ave. fuel sales per outlet
Austria	5,400	675	4,600	1,090
Belgium	11,130	320	8,900	740
France	49,000	485	43,200	710
Italy	39,200	470	38,800	610
Netherlands	11,600	440	6,500	1,370
Switzerland	6,080	450	5,200	900
United Kingdom	32,720	680	24,500	1,080
West Germany	35,520	660	24,500	1,310

Source: Esso estimates

Within the distribution structure in Europe there has been a steady increase in sales through "post payment" self-service stations. As Table 4.16 shows, these have increased in all countries to account for a dominant share of petrol sales in many countries, most notably, in Scandinavia and West Germany.

90

Table 4.16

WESTERN EUROPE
"POST-PAYMENT" SELF-SERVICE STATIONS AS PART OF TOTAL
MOTOR FUEL DISTRIBUTION NETWORK IN VARIOUS COUNTRIES
1978-1981
(Percentages)

	1978	1979	1980	1981
(a) Post-payment self-service stations as a proportion of the total number of sales outlets				
Austria	4	7	11	12
Belgium	6	10	11	13
Denmark	27	50	55	60
Finland	53	59	64	69
France	4	6	7	9
Ireland	1	1	2	2
Italy	0.5	0.5	0.6	0.6
Netherlands	9	10	12	18
Norway	27	31	35	44
Sweden	85	88	89	90
Switzerland	7	9	22	26
United Kingdom	18	20	23	27
West Germany	34	38	45	56
(b) Volume of motor fuel sold by post-payment self-service stations as a proportion of the total volume sold by all sales outlets				
Austria	6	12	16	19
Belgium	17	24	33	37
Denmark	60	78	81	88
Finland	60	70	73	80
France	15	13	17	20
Ireland	5	7	10	13
Italy	7	1	1	1
Netherlands	30	34	39	45
Norway	60	64	68	75
Sweden	75	75	84	85
Switzerland	24	25	44	48
United Kingdom	36	40	45	52
West Germany	58	63	67	72

Source: Esso estimates

The European Petrol Market by country: Austria

Consumption of gasoline in Austria increased steadily over the last period from 1970 to 1980 due to the growth in the country's automobile population. A further factor underpinning the strength of the Austrian market was that by virtue of the so-called "hard schilling" policy whereby the Austrian currency was allied to that of West Germany, Austrian import prices of oil fell appreciably during the latter part of the 1970s.

Nevertheless, petrol consumption has levelled out during the 1970s, falling marginally away from the peak year of 1980. In 1981, 68% of sales were of super grade petrol compared with 63% in 1970.

Table 4.17
AUSTRIA
CONSUMPTION OF NORMAL AND SUPER MOTOR GASOLINE,
1970-1981
('000 tonnes)

	Normal	Super	Total
1970	589.5	984.3	1,573.8
1978	606.5	1,751.9	2,358.4
1979	679.7	1,734.5	2,414.2
1980	791.1	1,645.2	2,436.3
1981	767.3	1,640.2	2,407.5
1982	n/a	n/a	2,390.0

Source: BMH

Table 4.18
AUSTRIA
NUMBER OF FILLING STATIONS, 1981-1982, BY COMPANY
(End of December)

	1981	1982
Elan	687	672
Shell	659	625
Martha	588	557
Mobil	527	508
Esso	398	405
Aral	319	299
BP	313	293
Total	173	170
Agip	170	156
Other companies	800	791
TOTAL	4,634	4,476
Of which self-service stations	587	776

Source: Fachverband der Erdölindustrie Österreichs

The number of gasoline stations has been falling in Austria as in other European countries. At the end of 1976, there were 5,250 stations compared with 4,476 at the end of 1982. Of these, 17% were self-service. The main operators are shown in Table 4.18.

Belgium and Luxembourg

Consumption of gasoline for automobiles in Belgium has fallen steeply since 1980. From 2.57 million tonnes in 1973 gasoline consumption rose to 2.95 million tonnes in 1980. It then fell to 2.67 million tonnes in 1982. A rather different pattern held with petroleum products overall, with total demand peaking in 1973 at 25.7 million tonnes and falling to 18.18 million tonnes by 1982. In Luxembourg, gasoline consumption has grown rapidly, from 103,000 tonnes in 1965 to 311,000 tonnes in 1979.

The number of service stations in Belgium fell between 1973 and 1982, from 2,618 to 1,987. Counting garages and other places where gasoline was sold, the total number of points of sale declined from 11,634 in 1973 to 7,575. The number of people employed in the gasoline sector (other than in refining) fell too, from 4,905 in 1973 to 4,227 in 1982.

Table 4.19

BELGIUM
MOTOR GASOLINE—SALES OUTLETS, 1974-1983
(Number of outlets as at January 1)

Type of outlet	1974	1981	1982	1983
Garages	6,255	4,489	4,137	3,683
Filling stations	1,449	1,131	1,250	1,198
Service stations	2,618	2,147	2,048	1,987
Other outlets	1,312	870	823	707
Total	11,634	8,637	8,258	7,575

Source: Fédération Petrolière Belge

Denmark

Consumption of gasoline in Denmark rose by 5% between 1970 and 1980 but fell back by 13% between 1980 and 1982. This was partly due to the impact of passenger cars run on diesel and liquid petroleum gas, which accounted for 3.5% of cars in use in 1982 compared to 1.4% in 1979. Since 1979, the boom in sales has come from LPG rather than diesel.

Table 4.20

DENMARK
PASSENGER CAR PARC, 1979-1982, BY TYPE OF FUEL USED
(Units at January 1)

	Gasoline	Diesel	LPG	Total
1979	1,387,703	15,306	4,721	1,407,730
1980	1,397,531	18,871	7,041	1,423,443
1981	1,353,848	20,747	14,952	1,389,547
1982	1,317,930	22,807	26,130	1,366,867

Source: Statistiske Efterretninger

Finland

Although total sales of petroleum products in Finland have been falling since 1980 (from 11.3 million tonnes in 1980 to 9.9 million tonnes in 1982) sales of motor gasoline have been rising. From 1.34 million tonnes in 1980, they grew to 1.37 million tonnes in 1982. Figures for the first half of 1983, expressed in annualised terms, show consumption of 1.34 million tonnes. Underlying this growth in demand has been a rise in the Finnish car parc, from 1,392,827 cars in 1980 to 1,532,697 cars in 1982. The average distance driven by each car in Finland has, however, tended to fall in recent years, from 18,500 km in 1980 to 17,800 km in 1982.

Table 4.21

FINLAND
SALES OF PETROL AND DIESEL OIL TO CONSUMERS
1960-1982

Tonnes	Ordinary petrol	High octane petrol	Diesel oil
1960	308,979	29,003	420,201
1965	366,488	274,373	580,614
1970	466,431	559,126	741,925
1975	561,009	780,381	880,472
1976	549,470	783,980	882,877
1977	556,705	776,336	899,764
1978	575,579	762,676	923,681
1979	620,210	787,055	1,046,874
1980	641,447	694,757	1,098,816
1981	661,911	683,336	1,122,361
1982	1,373,633		n.a.

Source: Finnish Petroleum Federation

The number of service stations in Finland has not shown the fall that other countries have witnessed. In part this will be due to the fact that in such a physically large country with a small population, there will always be a need for a large network of dispersed stations. The total number of service stations stood at 1,967 at the end of 1980, 1,979 at the end of 1981 and 1,964 at the end of 1982. Of those in existence in 1982, 1,230 were self-service and a further 201 were partially self-service. The growth of cash machines at self-service locations has been brisk. In 1980 there were 806 such machines installed; in 1982 the figure was 858.

Table 4.22

FINLAND
NUMBER OF MOTOR FUEL RETAIL SALES OUTLETS, 1980-1982, BY TYPE
(At December 31)

	Total service and filling stations	Service and Filling Stations		Other outlets	Total outlets
		Of which self-service only	Of which partly self-service		
1980	1,967	928	321	n.a.	n.a.
1981	1,979	1,128	208	n.a.	n.a.
1982	1,964	1,230	201	311	2,275

Source: Finnish Petroleum Federation

Table 4.23

FINLAND
MOTOR GASOLINE SALES: NUMBER OF SERVICE AND FILLING STATIONS
BY COMPANY, JANUARY 1, 1983

	Service Stations	Filling Stations
Esso	310	50
E-Öljyt	154	22
Kesoil	219	50
Shell	379	73
Suomen BP	2	1
Teboil	284	49
Union	293	93
TOTAL	1,641	338

Source: Finnish Petroleum Federation

France

In France consumption of petroleum products in general has been falling since the early 1970s. Demand in 1973 stood at 111,809 thousand tonnes, but by 1979 had fallen to 105,752 thousand tonnes. In 1982 it stood at 81,600 thousand tonnes. Motor gasoline demand has not, however, fallen during this period. Between 1973 and 1982 it grew from 15,772 to 18,132 thousand tonnes. This occurred despite a slight fall (13,000 km in 1978 to 12,400 km in 1982) in the average distance driven in each car. The main factor underlying growth in volume of demand was a large (14.5%) growth in the car parc in France, from 17.72 to 20.3 million cars over the 1978-82 period. By 1982, diesel cars accounted for a 6.3% market share.

Table 4.24

FRANCE
TOTAL PASSENGER CAR PARC AND PROPORTION OF DIESEL CARS
1970-1983
(Thousand units on January 1)

	1970	1979	1980	1981	1982	1983
Total parc	11,860	17,720	18,440	19,130	19,750	20,300
Of which diesel	108	610	730	890	1,077	1,288
Percentage	0.9	3.4	4.0	4.7	5.5	6.3

Source: Chambre Syndicate des Constructeurs d'Automobiles

Table 4.25

FRANCE
ESTIMATED TOTAL FUEL CONSUMPTION OF PASSENGER CARS
1978-1981, BY TYPE OF FUEL
(Thousand cubic metres)

	Normal Gasoline	Super Gasoline	Diesel
1978	3,300	15,380	1,450
1979	3,315	15,525	1,700
1980	3,280	15,625	2,050
1981	3,169	16,068	1,963

Source: L'Argus de l'Automobile

The French gasoline market is dominated by seven retailers, whose market shares in 1982 were as follows: Total (19.5%), Elf (17.6%), Shell (12.5%), Esso (11.9%), BP (8.5%), and Mobil (6.1%). Other firms, such as chains run by supermarkets, account for 21% of the market. At the beginning of 1983 these companies possessed the following number of service stations: Total (8,300), Elf (9,400), Shell (2,810), Esso (4,500), BP (2,500) and Mobil (1,900). These stations were either owned outright by the oil companies (such stations account for some 7,000 of a total of 37,000 stations, excluding supermarkets' outlets) or are run as 'brand retailers.' These are the great majority; they sign supply agreements with oil companies but remain independently-run. There are 26,500 of these.

Table 4.26

FRANCE
MOTOR GASOLINE RETAIL SALES: NUMBER OF OUTLETS BY
CATEGORY AND MARKET SHARES PER CATEGORY, MARCH 1983

	Number of outlets	Market share (%)
Stations owned by oil companies	7,000	35
Brand retailers	26,500	35
Free retailers	3,500	10
Supermarkets	1,500	20
TOTAL	38,500	100

Source: Union des Chambres Syndicates de l'Industrie du Pétrole

The number of service stations has continued to fall in France, from 47,500 in 1968 to 46,000 in 1979; the peak of 50,000 was reached in 1973. This is rather less than the extent of other countries' falls and has left the French stations selling on average only about 70% of the volume achieved in typical stations in West Germany or the UK.

The important factor underlying this relatively small volume of sales per outlet is price control. The French government maintains a minimum price level; retailers who do not observe it are threatened with having their own supply cut off. An instance of this occurred over the summer of 1983, when the Leclerc and Carrefour supermarket chains both began price discounting in excess of the 10 centimes per litre allowed by the government. (This discount was off the then prevailing price of 4.91 to 5.05 FF/litre.) The energy minister personally condemned these infringements and nearly 60 legal actions were opened. Since EEC rules prohibit minimum price fixing by monopolies there may however be a limit to the time during which the French government is able to carry out this policy.

Ireland

Data for 1978 shows that 417 service stations existed. In the same year there were a further 312 repair and service shops. Consumption of motor gasoline in Ireland has risen only slowly since the early 1970s, with low real GNP restricting demand. During the 1970s a series of substantial tax increases, affecting both automobile and gasoline prices, were enacted. These have also continued into the 1980s as the country's budget crisis deepens. After growing to 617,000 tonnes in 1970, gasoline consumption reached 796,000 tonnes in 1973 and subsequently changed only a little until 1979, when it began to approach 1 million tonnes. 1981 marked the peak level of demand, at 1.072 million tonnes. In 1982 demand fell back to 986,000 tonnes. In 1982, there were 716,452 cars in use, of which 14,374—around 2%—were diesel.

Table 4.27
IRELAND
PASSENGER CAR PARC, 1982, BY HORSE POWER
AND TYPE OF FUEL USED
(Number of cars on September 30)

Horsepower	Gasoline	Diesel	Total
Up to 16	687,622	11,344	702,105
Over 16	11,017	3,030	14,347
TOTAL	698,639	14,374	716,452

Source: The Society of the Irish Motor Industry

98

Italy

The number of passenger cars in use in Italy rose sharply betwen 1974 and 1983, and with them, fuel consumption. As in other countries, however, petrol consumption decreased between 1980 and 1982, partly due to the rise in cars run on diesel, LPG, and methane. These jointly accounted for over 10% of cars in use in 1983.

Table 4.28

ITALY
PASSENGER CAR PARC, 1974-1983,
BY TYPE OF FUEL USED
(Thousand units at January 1)

	1974	1980	1981	1982	1983[1]
Gasoline	13,055	15,687	16,235	16,800	6,910
Diesel	60	363	480	702	1,030
LPG[2]	370	640	665	678	740
Methane	15	110	120	120	120
TOTAL	13,500	16,800	17,500	18,300	18,800

Sources: Automobile Club d'Italia and Unione Petrolifera
[1]Provisional figures
[2]Estimates

Table 4.29 compares the use and performance of diesel cars with gasoline-driven cars between 1973 and 1982. Although the number of gasoline cars in use rose by 34% between 1973 and 1982, the average distances travelled fell sharply. As the annual fuel consumption decreased, so did total gasoline consumption.

The parc of diesel-driven cars rose from a mere 55,000 in 1973 to 850,000 in 1982. The distances travelled, which are considerably higher than the gasoline-driven models, have also fallen sharply but total fuel consumption has risen rapidly due to the increased cars in use.

Table 4.29

ITALY
FUEL CONSUMPTION AND DISTANCE TRAVELLED BY
GASOLINE—AND DIESEL—DRIVEN PASSENGER CARS, 1973-1982

	1973	1974	1980	1981	1982[1]
Gasoline-driven cars					
Parc at mid-year (thousand units)	12,310	15,135	15,600	16,172	16,490
Distance travelled yearly (kms/car)	12,300	10,450	10,400	10,000	9,800
Fuel consumption rate (kms/litre)	11.8	11.7	11.8	11.9	12.0
Annual fuel consumption (litres/car)	1,040	892	877	838	817
Annual fuel consumption (kgs/car)	763	655	644	615	600
Total consumption of all gasoline-driven cars (thousand tonnes)	9,930	9,910	10,040	9,950	9,890
Diesel-driven cars					
Parc at mid-year (thousand units)	55	315	415	580	850
Distance travelled yearly (kms/car)	40,000	38,000	36,000	34,000	32,000
Fuel consumption rate (kms/litre)	11.0	12.9	13.0	13.1	13.2
Annual fuel consumption (litres/car)	3,650	2,950	2,770	2,595	2,425
Annual fuel consumption (kgs/car)	3,000	2,450	2,300	2,150	2,010
Total consumption of all diesel-driven cars (thousand tonnes)	165	770	950	1,250	1,700

Source: UP estimates
[1]Provisional figures

There were 38,000 outlets for motor gasoline in 1982 compared with 39,200 in 1970.

Table 4.30

ITALY
MOTOR GASOLINE-NUMBER[1] OF SALES OUTLETS,
1970-1982, BY TYPE

	1970		1980		1982	
	Number	%	Number	%	Number	%
Motorway	286	0.7	440	1.1	440	1.2
Service stations	7,278	18.6	7,400	19.2	6,900	18.2
Filling stations	8,500	21.7	10,200	26.2	9,160	24.1
Kiosks	14,200	36.2	14,200	36.6	16,500	43.4
Single pumps	8,936	22.8	6,560	16.9	5,000	13.2
TOTAL	39,200	100.0	38,800	100.0	38,000	100.0

Sources: Esso, Unione Petrolifera
[1]Estimates—for 1970 and 1980 by Esso, for 1982 by UP

100

Netherlands

The pattern of Dutch internal consumption mirrors other countries; a steady rise to 1980, but sharp decreases since. In terms of passenger cars, in use, out of 4.63 million on the road in 1982, 15% were either diesel or gas-driven. The main increase was in Lpg models.

Table 4.31

NETHERLANDS
PASSENGER CAR PARC 1978-1982
BY TYPE OF FUEL
('000s)

	Petrol	Diesel	Gas	Total
1978	3,772	104	180	4,056
1979	3,966	128	218	4,312
1980	4,057	169	289	4,515
1981	4,013	200	381	4,594
1982	3,925	224	480	4,630

Source: Centraal Bureau voor de Statistiek

Table 4.32

NETHERLANDS
DELIVERIES OF MOTOR FUEL TO THE INTERNAL MARKET,
1976-1982
('000 tonnes)

	Motor Gasoline			Diesel[1]	Lpg[1]
	Super	Normal	Total		
1976	3,164	495	3,659	1,766	211
1977	3,266	534	3,800	1,809	269
1978	3,368	586	3,954	2,253	355
1979	3,363	616	3,979	2,024	457
1980	3,225	679	3,904	2,121	642
1981	2,937	748	3,685	2,242	721
1982	2,838	795	3,633	2,106	858

Source: Centraal Bureau voor de Statistiek
Note: [1]Includes commercial usage

At the end of 1982, there were 9,155 stations in operation in the Netherlands selling motor gasoline. Shell accounted for 16% of the stations, but is estimated to have a 25% market share.

Table 4.33

NETHERLANDS
MOTOR GASOLINE: SALES OUTLETS AND
MARKET SHARES OF PRINCIPAL COMPANIES (END 1982)

	Number of stations	Of which self-service	Market share (percentage)
Shell	1,500	405	25
Esso	740	210	11
Chevron	1,030	120	9
BP	660	110	8
Mobil	365	135	6
Fina	390	50	4
Total	340	40	3
Texaco	270	30	3
Aral	220	70	3
Elf	340	10	2
Other[1]	3,300	100	24
Total	9,155	1,280	100

Source: RAI

[1]Includes stations owned by, or linked to, smaller supplying companies and also many "independent" stations and single-pump installations.

Norway

Gasoline consumption in Norway has continued to rise against the general European trend. Again, as with Finland, the opportunities to decrease internal fuel consumption in the country are geographically limited. Consumption rose from 925,000 tonnes in 1970 to 1.4 million tonnes in 1980, showing a further small rise to 1982.

There are 2,531 outlets selling petrol in Norway.

Table 4.34

NORWAY
NUMBER OF MOTOR GASOLINE SALES OUTLETS, 1973-1982,
BY TYPE
(At December 31)

	Service Stations	Other outlets	Total
1973	2,179	1,474	3,653
1981	1,949	682	2,631
1982	1,913	618	2,531[1]

Source: Norsk Petroleumsinstitutt

[1] Of these 2,531 outlets, 1,1236 are company-owned and 1,405 are dealer-owned. 1,200 outlets are equipped for self-service.

Portugal

The relative under-motorization of Portugal—in 1981 there were barely 900,000 cars registered, a density of only one car per 11 people— and the extremely high real price of gasoline coupled with low real earnings keep gasoline consumption in Portugal low. In 1982 consumption was 810,000 tonnes or only 4% of the level of demand in West Germany in that year. Despite the unfavourable factors just referred to, gasoline consumption has nonetheless been growing strongly, from 474,000 tonnes in 1970 to 751,000 tonnes in 1980. The 1974 and 1979 price rises barely seem to have interrupted the growth of demand at all. As for total petroleum products demand, their volume has also gained, from 3.43 million tonnes in 1970 to 7.6 million tonnes in 1980 and 8.4 million tonnes in 1982. Once again, the years 1974 and 1979 showed hardly any interruption in the upward climb. A reason for this surprisingly buoyant growth may have been that the Portuguese economy was growing strongly itself at that time.

Figures for 1980 point to 1,900 service stations existing in Portugal. Many are thought to be barely economic and survive only because their gasoline business is augmented with repair work, etc. An estimated 46% of the outlets sell less than 600,000 litres annually. There is little self-service in the Portuguese distribution network. In 1980 about 75% of gasoline sold was super grade.

Table 4.35

PORTUGAL
SALES OF NEW PASSENGER CARS, 1978-1982
BY TYPE OF FUEL USED
(Units)

	Gasoline	Diesel
1978	38,100	6,576
1979	40,002	5,496
1980	45,303	5,276
1981	63,729	6,909

Source: ACAP

Table 4.36

PORTUGAL
MOTOR GASOLINE CONSUMPTION, 1973-1980, BY SUPER/NORMAL
(Tonnes)

	Super	Normal
1973	531,768	180,341
1979	563,608	184,989
1980	571,272	180,053

Source: Associacao Nacional de Revendedores de Combustíveis

Spain

Consumption of petroleum products in Spain more than doubled between 1970 and 1982, with consumption in the latter year at 29.5 million tonnes. Unlike Portugal, its neighbour, Spain has seen consumption decline very markedly since its peak of the mid-1970s. Between 1981 and 1982 alone, petroleum demand fell by 13.7% with gasoline demand falling by 0.5% and fuel oil by 24%. The year in which gasoline demand was highest was 1979, with 5.74 million tonnes used. The late 1970s were awkward years for the Spanish economy with low GNP growth after a period of rapid expansion. A series of falls in the peseta's value against the dollar also made oil products more expensive, pushing down demand.

Diesel-driven cars have again made their impact in Spain. In 1982, 12% of the total cars in use in Spain were operating on diesel, compared with under 2% in 1978.

104

Table 4.37

SPAIN
NEW REGISTRATIONS OF PASSENGER CARS
BY TYPE OF FUEL USED
1978-1982
(Units)

	Gasoline Number	%	Diesel Number	%
1978	643,681	98.42	10,352	1.58
1979	598,368	96.41	22,284	3.59
1980	540,673	94.17	33,476	5.83
1981	460,811	91.12	44,905	8.88
1982	472,160	88.13	63,573	11.87

Source: Ministerio del Interior

Table 4.38

SPAIN
MOTOR FUEL SALES OUTLETS, 1981-1982,
BY OWNERSHIP AND TYPE
(Number of outlets on December 31)

	1981 Monopoly[1]	Private	Total	1982 Monopoly[1]	Private	Total
Service stations	69	3,422	3,491	86	3,416	3,502
Supply units[2]	297	—	297	307	—	307
Isolated pumps	555	57	612	547	54	601
Indoor garage units[2]	—	114	114	—	111	111
Indoor garage pumps	—	145	145	—	141	141
Other indoor facilities	—	120	120	—	122	122
TOTAL INSTALLATIONS	921	3,858	4,779	940	3,844	4,784

Source: CAMPSA

[1] Installations owned by the State Oils Monopoly and managed or leased to third parties by CAMPSA (Compania Arrendataria del Monopolio de Petroleos). CAMPSA exists to administer, under the higher management of the Ministry of Finance, the State Oils Monopoly first established in 1927.

[2] Groups of more than two different products.

105

The number of gasoline distribution stations in Spain at the beginning of 1983 was 4,784. This represented an increase of 5 units over the year before level. More narrowly-defined service stations numbered 3,502 in 1982, up from 3,491 in 1981. A further 908 small units, with one or two pumps, also existed, and another 111 indoor garage units were recorded. Campsa, the public agency which is the state oil monopoly's retail presence, operates 86 of these stations; the rest are run as private concessions.

Sweden

Consumption of gasoline in Sweden rose by 27% between 1970 and 1980 before levelling off in a typical European fashion between 1980 and 1982. Diesel cars accounted for 4% of the cars in used in 1982.

Table 4.39

SWEDEN
PASSENGER CARS BY TYPE OF FUEL USED
(Units in use at end of year)

	Petrol	Diesel	Other
1974	2,551,467	87,406	12
1975	2,666,160	94,092	12
1976	2,779,054	102,252	5
1977	2,756,706	100,429	6
1978	2,755,190	100,984	4
1979	2,761,979	106,316	7
1980	2,769,591	113,358	7
1981	2,774,457	118,729	56
1982	2,811,453	124,305	227

Source: Statistiska Centralbyran

In 1983, there were 4,431 outlets selling petrol in Sweden. The numbers of outlets fell dramatically between 1968 and 1982, virtually halving in strength, although the numbers of self-service outlets increased.

The largest petrol retailer is Shell with 850 outlets and a 19% market share. Others of significance include Esso with 500 outlets, Texaco with 550 outlets and BP with 655 outlets. The market shares are shown in Table 4.41

106

Table 4.40

SWEDEN
MOTOR FUEL: NUMBER OF SALES OUTLETS BY TYPE 1968-1983
(As at January 1st)

	1968	1973	1980	1981	1982	1983
Service stations	4,520	4,409	3,478	3,437	3,322	3,276
Filling stations	763	444	359	355	336	351
Self-service[1]	17[2]	186[3]	209	214	235	264
Single pumps	3,644	1,874	637	585	550	540
TOTAL	8,927	6,727	4,683	4,591	4,443	4,431

Source: Svenska Petroleum Institutet
[1] With cash-operated machines
[2] 1974
[3] 1976

Table 4.41

SWEDEN
MOTOR GASOLINE: 1982 SALES VOLUMES[1] AND
1981/82 MARKET SHARES OF PRINCIPAL SUPPLIERS
('000s of cubic metres)

	Volume 1982	% 1982	% 1981
Esso	579	12.3	13.0
Shell	920	19.5	19.4
Texaco	355	7.5	7.8
BP	678	14.4	14.0
Gulf	498	10.6	10.8
Nynäs	—	—	0.4
OK	989	21.0	20.3
Mobil	214	4.5	4.5
Fina	109	2.3	2.2
ARA	105	2.2	1.4
SP	25	0.5	0.5
Others	240	5.2	5.7
TOTAL	4,712	100.0	100.0

Source: Svenska Petroleum Institutet
[1] Excluding bunkers
[2] This represents a rise of 0.8 per cent over the 1981 volume of 4,675 ('000s m³)

Switzerland

Petroleum product sales stood at 11.2 million tonnes in 1982, a 5.5% fall from their level of a year earlier. Over the past decade, the volume of products sold has diminished by no less than 27.5%. Within this total motor gasoline demand has grown, however, with heavy oils taking by far the biggest part of the decline. Gasoline sales rose by 5.8% and 3.9% in 1980 and 1981, but in 1982 they fell by 1.2%. In part this reflected the economic downturn that was then starting to affect Switzerland. Sales then reached 2.89 million tonnes.

Table 4.42

SWITZERLAND
CONSUMPTION OF NORMAL AND SUPER MOTOR GASOLINE,
1971-1982
(Tonnes)

	Normal	Super
1971	413,344	1,945,305
1978	450,751	2,157,637
1979	439,124	2,155,139
1980	452,097	2,292,162
1981	430,470	2,420,395
1982	496,438	2,392,281

Source: Union Petrolière, Zürich

In January 1983 there were 4,977 company-owned service stations in Switzerland. The 1982 total was 5,153. Of the 4,977 stations, 2,333, or 47%, are self-service. Over the decade up to 1983, the fall in the number of service stations was 25.6%. A marked shift has been taking place, towards self-service in Swiss stations; in 1982 only 36% of all stations were classified as self-service. It is notable that self-service outlets account for 59.4% of the value of sales, suggesting that volume per unit for them is rather higher than for the full service stations they are replacing.

108

United Kingdom

Unlike most European countries, petrol consumption in the UK has remained remarkably steady over recent years. Between 1970 and 1980, consumption rose by a substantial 35%, and then despite a dip in 1981 rose further in 1982 and 1983.

Table 4.43

UK CONSUMPTION OF PETROL AND DERV FUEL (Volume)

Thousand tonnes	1978	1979	1980	1981	1982	1983 (est)
Petrol	18,350	18,680	19,150	18,720	19,250	19,700
Derv	5,870	6,060	5,850	5,550	5,730	6,200
TOTAL	24,220	24,740	25,000	24,270	24,980	25,900

Source: Institute of Petroleum/Euromonitor

The number of petrol sites in operation has fallen in the UK as it has elsewhere; there were 24,108 outlets in 1982 compared with 28,300 in 1978. In the earlier year, only 18% were self-service; this has since risen to 30%.

Table 4.44

UK PETROL SITES 1978-1982

Number	1978	1979	1980	1981	1982
Company-owned	8,632	8,166	8,011	7,796	7,563
Self-service	4,966	5,340	6,022	6,712	7,145
Attended	23,329	21,140	19,505	18,048	16,927
TOTAL	28,295	26,480	25,527	24,760	24,108

Source: Institute of Petroleum

Although the absolute number of company owned stations has declined, they took a slightly higher share of the total in 1982 than 1978. Clearly however, they, and vertical integration in general, are not seen as a universal panacea for declining market share. Self-service outlets, bigger with fewer staff costs, also seem to be a route to higher profitability.

Consumption of petrol per car varies from year to year, tending to reflect changes in relative petrol prices and the general level of economic activity. The worst recent year was 1981, when consumption fell to 329 gallons per car, compared to a ten-year high of 357 gallons in 1978. Changes in model mix ought, if the much publicised trend towards lower fuel consumption cars is to be believed, to result in a downtrend in this figure. The effects are as yet minimal and offset by higher mileages.

Some 24,000 outlets sell petrol in the UK. Esso, supplying 3,478 of them, emerges as the best represented petrol brand.

Table 4.45

NUMBER OF SITES PER COMPANY 1982

	Self-Service	Company-Owned	Retailing Derv	Motorway	Total
Esso	1,065	1,186	1,689	42	3,478
Shell	1,332	1,199	892	24	2,972
BP	850	777	893	26	2,335
Texaco	625	932	522	18	1,779
National	463	389	607		1,368
Jet/Globe	235	268	222		1,077
Mobil	661	610	436	28	1,045
Burmah Group	148	203	246		990
Fina	195	242	367		878
Total	403	605	289	7	821
Ultramar/Summit	14	34	133		603
Elf	349	237	194	1	591
Anglo		1	133		569
Pace	25	4	115		515
ICI	83				470
Gulf	198	181	139		446
Others	499	695	1,123	2*	4,171
TOTAL	7,145	7,563	8,000	148★	24,108

Source: Institute of Petroleum
★including shared sites
*AMOCO

110

An important change gathering pace is the diminishing importance of the company-owned chains and the concomitant growth of independents. There is fairly free movement in and out of the business, so that in any year, eight or ten new brands appear and about the same number might vanish. In 1982, for instance, eight new names began to trade, with a total of 184 outlets between them. A number of sites tend to be made available each year through the contraction of the large integrated chains. In 1982 Esso relinquished no less than 360 sites (over 10% of its network) while Shell gave up 317 sites, BP 181, Texaco 105, Fina 47, Mobil 40 and so on. National abandoned by far the largest number—436. The importance of the top five retailing chains has been falling over time. In 1982 the top five accounted for 54% of all British sites but at the start of 1983 they accounted for under 50%. As the discussion of the US market made clear, there are several factors at work here, among them the desire to regroup stations of strength (reflecting ease of access to refineries, volume of turnover, etc) while relinquishing weaker sites.

The UK is thus exhibiting the same trends as in the USA towards petrol retailing; placed in a less profitable line of business, the aim has been to achieve higher throughput per station; small uneconomic units have been closed and large, self-service outlets built. In efforts to enhance profit potential, service stations are installing shops selling tobacco, confectionery and leisure goods in addition to motor accessories; some are linking to garden centres and even setting up video libraries. With increases in passenger car travel, the service station is being tipped by many commentators as the "corner shop" of the future.

West Germany

Consumption of motor gasoline in West Germany is the highest in Western Europe. In 1982 total demand was 22.73 million tonnes. Demand had already reach a peak of 15.5 million tonnes in 1970, and from then on demand grew only slowly. The actual maximum level of demand was reached in 1980, when 23.7 million tonnes were consumed. Premium sales accounted for 53.8% of demand in 1982, down from 55% in 1981.

The number of service stations in West Germany has been declining, in common with other European countries. At the beginning of 1980 the outlets were as shown in the table below. The peak number was achieved in 1969. In that year there were 46,684 outlets, but by 1979 the number had fallen by 18,141—a decline of 39%. Special arrangements govern motorway outlets. These are run by the state-owned Gesellschaft für Nebenbetriebe der Bundesautobahnen (GfN) which in 1982 operated 267 outlets.

Table 4.46

WEST GERMANY
CONSUMPTION OF NORMAL AND SUPER MOTOR GASOLINE
(INTERNAL MARKET), 1978-1982
(Thousand tonnes)

	Normal	Super	Total
1978	9,465	13,549	23,014
1979	9,992	13,316	23,308
1980	10,682	13,039	23,721
1981	10,287	11,982	22,269
1982	—	—	22,730

Source: Mineralölwirtschaftsverband

112

Table 4.47

**WEST GERMANY
SERVICE STATIONS[1], BEGINNING OF 1983,
BY COMPANY**

	Number of stations	Of which self-service stations[2]	Self-service sales as % of total
Aral	4,993	3,880	91.5
Esso	2,630	1,582	83
Shell	2,429	1,653	92
Texaco	2,338	1,604	84.4
BP	2,171	1,314	85
Avia	1,287	761	86
Fina	721	421	73.8
Fanal	694	464	87
Elf	581	524	97
Agip	539	516	93
Chevron	461	298	81
Jet	387	387	100
Total	275	200	87
Deltin	225	149	90.0
Westfalen	175	125	89.5
Pam	151	26	n.a.
Rückwarth/Emhagol	105	48	77.4
Baywa	104	—	—
Montan-Union	98	11	n.a.
Eller Montan	82	80	96
Union Kraftstoff (UK)	79	68	90.1
HWB (Bohlmann)	62	62	100
Transit/Startol	61	61	100
Mabanaft Group	57	25	63
Tramin	56	52	90
SVG	50	35	n.a.
Free service-stations[3]	1,140	285	30
Others	approx. 1,000	approx. 600	n.a.
TOTAL	22,951	15,231	approx. 85

Source: Erdöl-informationsdienst

[1] Excluding motorway service stations

[2] Self-service for motor gasoline

[3] Bundesverband Freier Tankstellen

Table 4.48

**WEST GERMANY
NUMBER OF MOTORWAY SERVICE-STATIONS
JANUARY 1, 1982, BY COMPANY[1]**

Aral	62	UK	2
Shell	35	Westfalen	2
Texaco	29	Autol	1
Esso	27	Baywa	1
BP	19	Bomin	1
Avia	15	Club	1
Fanal	8	Deltin	1
Jet	8	Deutsche Ölimport	1
Elf	7	Efa	1
Fina	7	Elo	1
Agip	6	Fôrster	1
Chevron	4	Hammer	1
SVG	3	Kessel	1
Total	3	Kuttenkeuler	1
Bavaria	2	Montan-Union	1
Eller Montan	2	Pam	1
Framin	2	Rückwarth	1
HGK	2	Supol	1
Rhein-Main-Kraft	2	Tillmann	1
Tramin	2	Topp	1
		TOTAL	**267**

Source: Erdöl-informationsdienst

[1] Motorway service stations are leased out to the oil companies by the state-owned Gesellschaft für Nebenbetriebe der Bundesautobahnen mbH (GfN). The above table shows the breakdown of the outlets by brands. In some cases, however, the name of the operating company of a group of suppliers is listed.

114

Chapter Five

OIL AND PETROLEUM PRODUCTS POLICY

ATTITUDES towards oil have been considered essential parts of government policy for decades. In part due to a desire to raise tax revenue, in part due to a desire to dull the demand for an expensive item which had to be imported with foreign exchange that was scarce after the Second World War, and in part due to security considerations, oil taxation has tended to be both heavy and complex in most countries. In the US, one expert judged as long ago as 1948 that "our laws controlling the petroleum industry are a fantastic and inordinately complicated patchwork of state and federal regulation. Our system of oil law is wasteful and expensive."[1]

Policy in the USA

The fact that the US was a net oil exporter until 1949 postponed much consideration of the security aspects of a military power being dependent on other countries for fuel with which to run its armed forces' equipment. Quotas on oil imports, which were for the most part voluntary in the 1950s were made compulsory by the mandatory Oil Import Program of March 1959. This scheme lasted until April 1973. Its intention was to protect domestic oil by keeping out imports, other than from Mexico and Canada. This "drain America first" policy had the unintended consequence in the words of one author of "safeguarding American supply by using it up."[2] By guaranteeing 90% of the American market to indigenous producers the policy was instrumental in depleting the US oil reserve, faster than would have been the case in less tightly regulated circumstances.

More recently, attention in the US has turned to direct taxation of oil products, and in particular gasoline, as a means of reducing consumption and thus imports. In this manner oil policy in the US has come increasingly to resemble policy in Japan and Western Europe, but the position in most less developed countries is rather different.

---

[1] E. Rostow, *A National Policy For the Oil Industry,* Yale University Press, 1948.

[2] J. E. Spero, "Energy Self Sufficiency and National Security," in R. H. Connery and R. S. Gilmour, eds. *The National Energy Problem,* Lexington, 1974.

Taxation of gasoline at the retail level is virtually universal. In the US there are, as Table 5.1 shows, considerable differences in states' taxation on gasoline. In addition to the 9 cents/gallon federal tax (which was raised from 4 cents/gallon on April 1, 1983) states levy taxes ranging from 5 cents in Texas to 14 cents in Washington D.C. and New Hampshire. (The Hawaii taxes in Honolulu county exceed this; these figures include both state and county taxes.) Adding the federal and state taxes together, it transpires that even in the highest tax states of Washington D.C. and New Hampshire taxes were equivalent to a mere 23% of the average retail price of gasoline.

Until the federal tax increase of 1983, in the lower-tax states the tax bite on each gallon was a low as 7% of the average retail price.

The amount of tax collected in each state from gasoline duty totalled $9,973 million in 1981, up by 4.1% from $9,577 million the year before, as Table 5.2 indicates. By far the biggest amount—$777 million—was collected by the state of California.

116

Table 5.1

FEDERAL AND STATE GASOLINE TAXES, 1983, USA[1]
(Cents per Gallon)

	Gasoline		Gasoline
Federal[2]	9.0	Michigan	13.0
		Minnesota	13.0
		Mississippi	9.0
Alabama	11.0	Missouri	7.0
Alaska	8.0	Montana	9.0
Arizona	10.0	Nebraska	13.9
Arkansas	9.5	Nevada	10.3
California	9.0	New Hampshire	14.0
Colorado	9.0	New Jersey	8.0
Connecticut	11.0	New Mexico	10.0
Delaware	11.0	New York	8.0
District of Columbia .	14.0	North Carolina	12.0
Florida	8.0	North Dakota	8.0
Georgia[3]	7.5	Ohio	11.7
Hawaii[4]		Oklahoma	6.6
Hawaii County	13.5	Oregon	8.0
Honolulu County . .	15.0	Pennsylvania	11.0
Kauai County	12.5	Rhode Island	11.0
Maui County	14.5	South Carolina	13.0
Idaho	12.5	South Dakota	13.0
Illinois	7.5	Tennessee	8.0
Indiana	11.1	Texas	5.0
Iowa	13.0	Utah	11.0
Kansas	8.0	Vermont	11.0
Kentucky	10.0	Virginia	11.0
Louisiana	8.0	Washington	12.0
Maine	9.0	West Virginia	10.5
Maryland	11.0	Wisconsin	13.0
Massachusetts	9.8	Wyoming	8.0

[1] The table sets rates of general application, exclusive of local taxes. The State rates are effective as of January 1, 1983 and will be updated as appropriate.

[2] Federal tax changed as of April 1, 1983.

[3] Georgia. An additional tax is levied at the rate of 3% of the retail price.

[4] Hawaii. Rates are combined state and county rates.

Source: Commerce Clearinghouse State Tax Guide.

Table 5.2

STATE GASOLINE TAX COLLECTIONS—1981

	Gasoline Tax Rates per gallon 12/3/81	Net Taxable Gallonage in 1961($000)	Total State Motor Fuel Tax Revenues ($000) 1980	1981	% Change
Alabama	11.0	2,126,341	$200,331	$193,783	−3.3%
Alaska	8.0	180,049	14,976	14,872	−0.7
Arizona	8.0	1,446,963	123,067	118,089	−4.0
Arkansas	9.5	1,303,034	132,036	128,207	−2.9
California	7.0	11,330,390	816,349	777,169	−4.8
Colorado	9.0	1,528,776	112,261	124,493	10.9
Connecticut	11.0	1,327,964	158,174	152,270	−3.7
Delaware	11.0	305,283	28,335	30,053	6.1
Dist. of Columbia	11.0	175,291	17,024	21,260	24.9
Florida	8.0	4,838,540	413,825	392,028	−5.3
Georgia	17.5	3,086,487	248,565	236,787	−4.7
Hawaii	8.5	345,733	28,010	29,915	6.8
Idaho	11.5	504,578	48,776	56,414	15.7
Illinois	7.5	5,102,930	374,361	373,987	−0.1
Indiana	10.5	3,054,646	269,608	302,506	12.2
Iowa	13.0	1,714,616	150,830	183,188	21.5
Kansas	8.0	1,443,548	115,045	115,281	0.2
Kentucky	9.0	1,867,177	189,076	182,522	−3.5
Louisiana	8.0	2,211,334	185,158	177,998	−3.9
Maine	9.0	523,013	48,906	46,808	−4.3
Maryland	9.0	1,957,794	186,100	178,176	−4.3
Massachusetts	9.8	2,291,485	218,000	252,898	16.0
Michigan	11.0	4,108,255	461,866	431,845	−6.5
Minnesota	13.0	2,139,238	218,757	268,083	22.5
Mississippi	9.0	1,317,197	128,969	125,856	−2.4
Missouri	7.0	2,763,325	197,913	195,580	−1.2
Montana	9.0	523,232	48,776	49,559	1.6
Nebraska	13.6	950,508	102,771	145,566	41.6
Nevada	10.5	533,054	33,701	45,134	33.9
New Hampshire	14.0	407,339	45,194	56,135	21.5
New Jersey	8.0	3,348,705	287,354	279,901	−2.6
New Mexico	9.0	843,962	68,052	73,289	7.7
New York	8.0	5,569,766	473,307	462,496	−2.2
North Carolina	12.5	3,086,132	296,124	336,181	13.5
North Dakota	8.0	455,518	30,661	34,514	12.6
Ohio	10.3	5,368,283	384,787	471,839	22.6
Oklahoma	6.5	2,013,945	127,197	129,509	1.8
Oregon	8.0	1,471,538	88,941	86,364	−2.9
Pennsylvania	11.0	5,070,441	584,412	658,200	12.6
Rhode Island	12.0	380,232	37,511	43,718	16.5
South Carolina	13.0	1,674,770	179,555	199,746	11.0
South Dakota	13.0	463,754	48,225	67,128	39.2
Tennessee	9.0	2,623,250	219,134	243,640	11.2
Texas	5.0	8,735,415	472,132	459,392	−2.7
Utah	11.0	788,209	74,057	80,200	8.3
Vermont	11.0	249,987	22,020	25,476	15.7
Virginia	11.0	2,717,465	285,517	276,674	−3.1
Washington	13.5	1,970,632	246,651	253,461	2.8
West Virginia	10.5	903,712	99,315	94,250	−5.1
Wisconsin	13.0	2,314,361	197,532	253,683	28.4
Wyoming	8.0	434,071	37,096	37,022	−0.2
Total US		111,904,998	$9,577,740	$9,973,545	4.1%

Source: Federal Highway Administration

The overwhelming impetus to policy-makers has been the rise in the price of crude oil, and the effects that that has had on the macroeconomic climate facing governments all over the world. Table 5.3 looks at one measure of the oil price increases: the value of fuel imports (oil, coal, gas) as a percentage of the total value of imports. It can be seen that for the OECD countries as a whole, fuels accounted for only 10% of their import bill in 1970 but grew to account for nearly 28% of their total import bill in 1981, before falling slightly the following year, to 26%. For the EEC countries taken separately, fuels accounted for 10.7% of total imports in 1970, 24.9% in 1981 and 23.6% in 1982. For the US, fuels made up 7.2% of the import oil in 1970, then peaked at 33% in 1980 before falling to 26.6% in 1982. Japan is the country which has had the biggest proportion of its total imports taken up by fuels: in 1981 they accounted for no less than 51.5% of the total.

Two factors have tended to make this proportion fairly unvarying over the last three years. One is that despite the recession cutting domestic demand for fuels, this has been reflected in a sluggish growth of world trade. After growing very rapidly in the 1960s, and slightly less briskly in the 1970s, the volume of world trade remained broadly unchanged between 1980 and 1982. This meant that, in aggregate, exports could not be boosted to defray the higher bills being paid for oil imports. A second factor which has affected countries outside the US has been the fall in their currencies against the dollar. This has boosted the domestic currency price of oil (since oil is traditionally priced in US dollars) and has offset any gains they have managed to make by way of increased exports. Should the dollar return by the mid-1980s to a range considered more 'normal' than its 1982-83 trading ranges, the domestic cost of crude oil to most importing nations will fall and the importance of crude and other fuels in total imports should decline correspondingly.

Another way of looking at the impact which fuel price rises over the last decade have had on oil-importing countries is to look, as shown in Table 5.4, at the ratio of these countries' manufactured goods export prices to the price of fuel imports. This can be expressed as an index with 1975 set equal to 100. The table shows a fairly steep decline in the purchasing power of manufactured goods exports. In the case of Austria, for instance, the index has fallen from 205 in 1970 to 100 in 1975 to 56 in the last quarter of 1982. This means that roughly four times the volume of manufactured goods has to be exported in 1982, compared to 1970, to acquire the same amount of imported fuel. Changes of this magnitude, which can be seen as a form of terms of trade change, are so severe as to precipitate policy action by governments.

Table 5.3

SOME INDICATORS ON FUEL IMPORTS—OECD COUNTRIES

Region, Country or Area	1970	1975	1978	1979	1980	1981	1982
as percent of total value of imports							
Total OECD.....	10.0	21.4	19.2	22.1	25.9	27.9	26.1
Europe	10.2	18.7	16.3	18.8	21.8	24.9	23.8
EEC...........	10.7	19.5	16.5	18.7	21.7	24.9	23.6
Australia.......	4.9	8.8	8.3	10.6	13.8	13.6	—
Austria	8.3	12.6	10.7	12.4	15.5	18.7	—
Belgium-Lux ...	9.1	14.1	12.4	14.2	17.4	20.4	—
Canada	5.2	11.0	8.2	9.4	12.4	12.3	—
Denmark.......	10.4	18.5	15.5	19.6	22.4	24.1	—
Finland	11.5	19.0	22.2	26.5	29.0	30.7	—
France.........	12.0	22.6	19.5	21.5	26.6	28.8	27.0
Germany, F.R..	8.3	17.7	16.2	19.7	22.5	24.3	—
Greece.........	6.9	22.1	18.7	21.1	23.4	22.0	—
Ireland	8.1	14.1	11.0	12.1	14.8	14.7	—
Israel..........	5.0	15.4	13.9	18.2	26.5	—	—
Italy...........	14.0	26.7	23.6	23.7	27.9	34.5	—
Japan	20.7	44.3	39.8	41.0	50.0	51.5	—
Netherlands	10.9	17.8	15.7	20.2	23.7	26.6	—
New Zealand...	6.7	14.4	13.4	16.1	21.0	19.5	—
Norway........	7.7	9.8	11.8	15.2	17.4	14.4	—
Portugal	9.2	15.3	17.0	20.9	24.1	24.4	—
Spain..........	13.3	25.8	28.4	30.2	38.7	42.5	—
Sweden	10.6	17.4	16.6	22.0	24.2	24.1	—
Switzerland.....	5.4	10.3	8.1	11.8	11.2	12.2	—
UK............	10.5	17.8	11.8	11.9	13.5	—	—
USA	7.2	25.6	24.5	29.3	33.0	30.9	26.6
Yugoslavia	4.8	12.3	14.3	18.9	23.6	24.0	—

Source: United Nations

Table 5.4

RATIO OF MANUFACTURED GOODS EXPORT PRICES
TO FUEL IMPORT PRICES

1975 = 100	1970	1975	1978	1979	1980	1981	1982	1982 I	II	III	IV
Total OECD..	267	100	106	87	62	53	56	54	58	56	55
Europe......	239	100	105	86	63	57	52	51	53	53	52
EEC.........	245	100	105	87	63	50	52	50	52	52	52
Austria......	205	100	98	83	59	—	—	—	—	52	56
Belgium-Lux.	224	100	95	87	63	49	—	48	49	47	—
Canada	381	100	96	86	64	56	—	52	69	66	67
Denmark	207	100	105	79	61	51	—	51	54	—	—
Finland	156	100	98	71	65	57	—	58	61	60	—
France	246	100	104	92	63	50	48	49	50	47	47
Germany,F.R.	222	100	108	77	56	42	—	41	44	46	45
Greece	304	100	97	80	54	—	—	—	—	—	—
Italy	279	100	101	91	65	52	—	53	56	—	—
Japan	301	100	113	91	59	57	—	58	67	57	55
Netherlands .	249	100	107	83	60	49	—	52	53	—	—
Norway	178	100	96	81	64	62	—	62	69	61	56
Sweden	229	100	104	73	66	—	—	—	—	—	—
Switzerland ..	192	100	108	65	61	53	—	57	58	56	54
UK.........	278	100	118	99	77	—	—	62	63	61	59
USA........	349	100	106	86	59	60	68	66	69	68	70

Source: United Nations

Policy decisions can affect those involved in the petroleum products business in three ways. First, policy measures introduced to affect the volume of crude oil imports to a country will tend to do so by raising the price, to consumers, of products derived from that crude oil. Second, policy measures can be taken directly at the retail or wholesale level, to raise revenue by imposing higher taxes. Policy may impinge on the business in a third way if there are regulations governing the use of domestic refining capacity. And finally, there can be a panoply of ad hoc policy measures which can influence the industry. Examples would include the emergency imposition of physical fuel rationing, or restricting the use of cars on certain days.

A wide range of institutional measures to deal with future energy problems have been widely discussed in the last decade. Table 5.5 shows a full list, compiled by William Hogan. It includes proposals to tinker with existing arrangements—such as altering the rules whereby the IEA deals

Table 5.5

PROPOSED INSTITUTIONAL MEASURES TO DEAL WITH ENERGY SHORTAGE
An Agenda for Cooperation

I   International framework
    Reforming the IEA
        Sharing proportional to consumption
        Subcrisis cooperation
        Pricing according to high-ten rule
    Summit nations
        Focus on wealth transfers
        Oil-import-value-share targets
        Provide lead for IEA
II   Demonstrating commitment
    Share all oil supplies
    Remove anti-trust restrictions
    Control trading companies in spot market
III   Emergency preparedness
    Demand restrictions
        Tax/tariff
        Administrative controls
        Fuel switching investments
        Allocation programmes
        Emergency management teams
    Supply expansion
        Fill strategic oil reserves
        Expand storage capacity
        Develop management plans
        Natural gas and coal stockpiles
    Macroeconomic management
        Tax policies
        'Prebates'
        Recycling
IV   Long run adjustments
    Import restrictions
    Excess capacity construction
    Supply diversification
    Guaranteed returns
    Avoid unilateral oil deals
V   Military and diplomatic options
    Protect oil fields
    Buy time for energy options

Source: William Hogan

with an energy shortage—as well as more substantial proposals to alter taxes, impose allocation controls, and to use strategic oil reserves. (At mid-1983 the US Strategic Petroleum Reserve possessed 350 million barrels of oil.) Long term matters such as diversifying supply sources and thinking out in advance the nature and extent of the military options—if any—which could be deployed to protect oil importing countries' essential interests are also listed. The list serves to give a flavour of the variety of ways in which oil and oil product policy can be pursued.

Policy decisions affect drilling for oil, at the highest end of the system. For instance, licensing changes in Western Europe in the early 1980s have led to a major resurgence of drilling in the area. Forty offshore wells in Europe were operating in mid-1983, a level not seen since 1976, and the second highest rate ever seen. The overall rate of drilling, including appraisal and development wells, stood at a record level in 1983. What this illustrates is that despite the 1982-83 fall in crude oil prices, production can still rise if the policy stance is suitable. In other parts of the world, public policy affects drilling through other means. In India, for instance, the investment allocation to crude oil drilling was increased from $4.8 billion to $8 billion for the sixth five-year plan of 1980-85. The two nationalized oil companies have been prospecting offshore for crude with which to reduce the volume of crude imports. In 1983 crude imports should be of the order of 0.3 million barrels/day, with petroleum products imports just under half that amount again. The Ministry of Energy and Petrochemical Industries in Algeria also has been making conditions for exploration more attractive for foreigners. Conditions for gas prospecting have also been enhanced.

Policy in Less Developed Countries

Most developing countries' governments have a keen interest in suppressing the demand for oil products—and gasoline in particular—in their countries. There are five reasons for this.

First, by their very nature most LDCs are foreign exchange-scarce. They persistently run deficits on the current account of their balance of payments. In 1982, non-oil exporting LDCs ran an aggregate current account deficit of $70 billion, which in 1983 grew to an estimated $89 billion.

Second, oil and refined oil products play a not inconsiderable part in increasing these deficits. The facts that, according to the World Bank, developed countries' income-elasticity of demand for oil over the period 1971-81 was of the order of 1.0 while in LDCs it was closer to 1.3, coupled with the LDCs' higher annual average growth of GDP (2.7% per year over 1970-80 compared to 2.5% in developed countries) means that oil demand rose more quickly in LDCs than elsewhere. An example of government efforts to introduce some measures for conservation is provided by Egypt. The 1982-83 to 1986-87 five year plan envisages using less oil for electricity generation. With hydropower and natural gas being used more extensively.

Third, to the extent that the governments of LDCs are interested in the equity or distributional aspects of economic development, they will tend to try to tax products used with luxury goods, such as private cars, more heavily than staples such as foodstuffs or clothing. Thus, the embryonic car industries in LDCs are likely to be faced with heavy taxes on gasoline for some time to come.

Fourth, the fact that many cities in LDCs have become extremely congested (Bangkok and Lagos would be good examples) prompts further action to limit private car usage. The main policy levels which can be manipulated to achieve this—taxes on cars and taxes on gasoline—are thus of obvious importance to planners and governments.

Fifth, many governments are anxious to act on the recommendations of many experts and international institutions over the years to increase their outputs of non-oil energy sources such as coal and traditional energy sources like wood.

Policy is not always consistent or well-developed, however. In Zaire, where the government struggled to reduce the refined products import bill from $25 million in 1982 to $15 million in 1983, retail gasoline prices were boosted from 5.5 to 12.5 kwacha per litre. Elsewhere in black Africa, on the other hand, gasoline prices were reduced, for example by 8% in Zambia in May 1983. These were announced as being made possible by the lower price of crude on world markets.

An analysis of gasoline, kerosene and fuel oil prices in 53 countries by the World Bank revealed that, over the 1972-80 period, governments tended to use the typically large element of taxation in retail prices to cushion the effect of higher crude prices. Although the international trading prices of these three products rose by some 350% over the 1972-80 period, domestic prices, in real terms, in these 53 countries rose by only 71%. In oil-exporting developing countries, real prices actually fell by 30%.

The extent to which countries have been able to reduce their reliance upon oil products and substitute other sources of energy varies considerably.

124

Table 5.6 shows, in aggregated form, all countries' patterns of energy use. It is clear that, for the world as a whole, the growth of refined products usage has been effectively checked. Other sources of energy, such as water power, have increasingly been taking their place.

Table 5.6

CHANGES IN WORLD ENERGY USAGE, 1972-82 and 1977-82, ANNUAL AVERAGE RATES, %.

	1972-82	1977-82
Oil products	—	−1.9
Natural gas	2.3	2.4
Coal	2.3	2.2
Water power	3.2	3.5
Nuclear	18.9	10.4
Primary energy use, total	2.0	1.0

Source: BP

The outlook for policy towards oil and oil products is difficult to foresee, but it is likely that the following will be among the policies pursued.

First, in the EEC some greater form of multilateral discussion and decision-making regarding refinery over-capacity is likely. This need not take the form of a formal policy, as was seen in the steel industry. A more probable model is the petrochemicals industry; in that case various producer groups got together to discuss problems and, occasionally, to allocate capacity cutbacks between one another and even to swap capacity ownership. The fact that the problem of over-capacity is unlikely to go away points to the strong likelihood of some official forum being welcomed.

Second, taxation on refined oil products may well be increased in some European countries at least. The reasoning for this is that while fighting increases in the consumer price index is an imperative to all governments, and higher gasoline taxes feed straight into the index, to the extent that the dollar falls against European currencies over the 1984-86 period, there will be scope for increasing taxes while benefiting from the lower import price. To consumers the retail price will be unchanged, but the mix of raw materials, cost to tax within the retail price will have changed. Since increases in government revenue are likely to be keenly sought after, this seems to be a probable policy.

Third, in less developed countries greater efforts at increasing taxation on refined products (or at least towards reducing subsidies) are likely, given the increasing awareness in many countries of the role which correctly formulated public policy can play in assisting growth-rates.

Overall, policy seems likely to be largely unchanged: gasoline taxation will continue to be substantial in most countries.

Chapter Six

THE AVAILABILITY OF CRUDE—
HOW MUCH FOR REFINERS?

THIS chapter looks at one important aspect of the future facing the petroleum products industry; how much crude oil, and of what sort, is likely to be available to it. It examines the factors which are likely to have a bearing on the availability of the raw material. Chapter Seven then goes on to examine the various influences on demand for refined products which are likely to be at work over the next decade. Chapter Eight pulls together the threads which have been running through the book to look at the business of producing, refining and marketing oil towards the end of the twentieth century.

OPEC in the Oil Markets of the 1980s

During the 1960s OPEC's share of world trade in crude oil was very substantial, only falling below 90% after 1966. It has never fallen below 80%. But if that share were to be substantially eroded the leverage of OPEC over the world oil market would stand to be diminished too. This erosion would most probably come about through non-OPEC countries increasing their net oil exports, particularly if, at the same time, OPEC members' net exports fell as a result of rising domestic oil needs meeting output ceilings. The growth in availability of non-oil energy sources would similarly affect OPEC's control of energy markets, but this topic is not pursued in this study, since it deals less specifically with demand for OPEC oil than with world demand for energy generally.

As had been predicted, the rise in the price of oil during the last decade has eventually stimulated a substantially greater level of oil search activity. This is particularly true of the end of the decade, by when it had become clearer that the slippage in the real price of oil that had taken place for much of the period since 1973 could not be relied upon to persist or reappear after further rises in the official 'marker' crude price. By 1979, indeed, there were very few countries of the world which had not embarked upon some form of oil exploration. In 1982 no fewer than 73 countries were producers of crude oil, defined as yielding 2,000 tonnes per year, contributing to total world output of 230,500 thousand tonnes. Judgements on the likely degree of success which these lowest efforts will meet with differ largely, as one would expect. The latest thinking of the World Bank is that "the potential of the vast majority of the import-dependent countries has been explored at best superficially; enough to

127

determine that large, easily exploitable reserves are not present, but not enough to establish whether there are smaller deposits that could make an important contribution to their own energy supplies''.[1]

After a record amount of oil drilling in 1980, 1981 witnessed another record. Oil commentators foresaw a 10% rise in the number of wells drilled on top of the 1980 world total of 6,976, which was itself a 16% advance upon the 1979 total. In the USA, in particular with the decontrol of gas prices following the Natural Gas Policy Act and of oil prices in early 1981 the active drilling count soared. The September 1980 rig count (at 3,138) was the highest since 1955, while for 1981 an average count of 3,525 rigs (a 12% rise) was recorded. But it is elsewhere in the world that the most spectacular rises occurred. Africa, the Far East and Central and South America each showed a near doubling of rig activity between 1977 and 1983.

Which countries, not presently thought of as oil exporters, might then emerge from this bout of exploration as significant elements in the world oil market? Among the countries which witnessed the largest proportionate rise in drilling activity in 1981 over 1980 were Bolivia (+43%), Brazil (31%), Chile (22%), Ecuador (20%), Peru (15%), Angola (50%), Cameroons (23%), Egypt (15%), India (21%), Malaysia (50%), Pakistan (23%), Thailand (25%) and Australia (28%). Drilling does not, obviously, imply eventual production, but prospects in some of these countries merit discussion.

In Malaysia 117 wells were drilled in 1981 in an effort to raise output from its 1980 level of 280,000 bpd. A number of productive wells have been reported in Ivory Coast, offshore, and augur well for the future. Although they are presently facing great problems, authorities in Turkey hope that with the help of a $62 million World Bank loan for water injection recovery into their biggest oilfield, they can raise output from 45,000 to 90,000 bpd, nearly one-third of demand. Drilling in Peru is also expected to build upon the current output level of 200,000 bpd of which some 76,000 bpd is exported. Exploration in Central America is, as always, hampered by political risk, and in the case of Trinidad, by tax policy too, but elsewhere in the Americas, Argentina remains "lightly explored"[2] and is to experience new drilling efforts, whereas elsewhere on the continent ten years of drilling in Brazil has as yet yielded rather disappointing results. The renewed efforts in Venezuela to exploit the heavy crudes there will result in enhanced recovery from the mid-1980s.

128

A New Market Structure to 1990?

The significance of the sort of enhanced oil discovery in the 1980s as discussed above has three dimensions. First, the energy security and balance of trade—among other factors—of those countries which evolve from net oil importers to net exporters will obviously undergo substantial change. Second, conditions in the oil exploration business clearly also stand to be altered by the augmented level of drilling activity; hiring rates for rigs and stock market evaluations of oil tool firms have been improved, for instance. But there is also a third point: that general oil market conditions, rather than merely local conditions, stand to be altered to the extent that new sources of supply, which become significant when aggregated, appear. What is at stake here is the possibility of changes in the structure of the oils market brought about by an increase in the number of participants in the business and their share in total trade.

Conventional economic theory suggests that the smaller the number of participants in the market, the greater the ease of collusion by those participants in fixing prices. In the limiting case of market dominance—monopoly—there is only one supplier. In cases of a few suppliers—oligopoly—similar price-setting can follow but naturally the longevity of any such agreement depends upon, among other factors, the acquiescence of all the participants. Covert price-cutting by one or more of the participants in the hope of increasing market share, and thereby total revenue, will threaten the existence of the group.

Expressing these notions more formally leads to the economic theory of cartels. There are three conditions which must be fulfilled to create a successful cartel. To achieve the objective of a cartel, which is simply to enjoy sales at a price higher than that which would obtain under conditions of competition between whatever number of firms found it profitable to operate in the industry, it is essential that there should be an agreed objective (usually to achieve a certain selling price); a system for sharing output cut-backs when supply threatens to exceed demand at their chosen price; and—to facilitate the above two—only a few participating sellers. The more the number of participants grows, the harder it will be to maintain market discipline; the easier it will be for participants to be lured from the task of maximising joint revenue for the cartel as a whole and towards maximising personal or private revenue at the others' expense.

The significance of all this for oil markets is that the more countries that enter the market as sellers, the greater the potential for price-cutting and for the tenets of *sauve qui peut* to replace concern for collective interests.

The necessity of resorting to price-cutting as an entree to the market, on the part of new suppliers, arises because these countries' oil agencies do not yet have oil distribution systems of their own. To induce the established oil and oil products distributors to begin taking their oil in preference to, or in addition to, output from other sources, price inducements have in the past been offered. Penrose documents this pattern in the 1960s thus:

> "The new companies cut prices to sell their oil, for many lacked adequate integrated outlets and, having no stake in the older areas, were unconcerned about the effect of their price cutting on prices elsewhere".[3]

More recently, the fact that the new OPEC national agencies face distribution networks owned by others has been acknowledged by the deputy secretary-general of OPEC. Dr. F. J. Al-Chalabi recently wrote of the OPEC members having become "pure sellers of crude oil, . . . without having any link with subsequent oil operations". This interruption in the hitherto prevailing patterns of vertical integration means that the "past pattern of 'one buyer' . . . has been replaced by a pattern of several buyers (for services, materials, rigs, pumps, etc.) entering the market separately, and even in competition among themselves".[4]

So far relatively few cases of OPEC members buying their way into established crude and/or oil products distribution networks have occurred. (Chapter Three noted the case of FPC taking over some Gulf capacity in Europe.) Saudi Arabian and Qatari agencies have taken shareholding interests in downstream producers in Italy and France respectively (as well as hired capacity from oil majors for refining), while the National Iranian Oil Company in 1975 acquired a minority stakeholding in one South African oil refinery. Outside OPEC, Mexico's Pemex agency in 1979 took a small stake in a Spanish refinery. Although such cases do not as yet amount to significant shares in oil products sales in any developed country, time will in all likelihood bring more penetration. As Frankel observed in his definitive 1976 study, "it only takes two years to build a refinery but ten years to develop a market position".[5]

Having looked at the possibility of non-OPEC oil sales rising in the 1980s, another crucial factor is the level and consistency of net oil exports from OPEC members themselves. The few indications that are available of internal oil needs in OPEC states point to sharp rises in the coming decade. One such forecast, from the OPEC economics secretariat itself, has presented this rise in domestic demand in terms of necessarily lower net exports, assuming from the outset that total output will not exceed the

130

peak levels of 1977 of 31.3 million barrels/day. But such a figure need not be seen as immutable: it can be altered by policy choices as outlined below.

OPEC governments' intentions as to optimal production volumes have become increasingly opaque. For not only has the desired physical volume of installed capacity been altered on many occasions; but the desired level of "surge" or short-term peak capability has also been adjusted.

Demand within OPEC

To a greater or lesser extent all the 13 OPEC member states are engaged upon the expansion of their industrial capacities. In this they share an objective of virtually all developing countries. From accounting for 9.0% of world manufacturing value added in 1977, it is predicted that on the basis even of the trends established over 1960-77, the developing countries in aggregate will account for some 14% by the year 2000.[6] Within that total of course, the OPEC countries will account for relatively little. Latin American developing countries account for around one-half of the total developing country involvement in manufacturing. Moreover, since countries tend to find their manufacturing sector taking an accelerating share of GDP during the per capita income level of $265-$1,075, few OPEC countries can expect to observe significant additions to their manufacturing capacity for decades to come. Nonetheless, both in the construction of such manufacturing and processing capacity, and in its operation, it is clear that greater amounts of energy will be needed than hitherto. Indeed, much of the industrialization being undertaken or contemplated in the OPEC states is of a particularly energy-intensive nature. Aluminium smelting and petrochemicals production are two instances.

Thus the internal demand for petroleum and petroleum products by OPEC members themselves is set to rise throughout the 1980s. The OPEC secretariat has estimated the rise in demand as being from 2.7 m bpd in 1981 to 6 m bpd or more by 1990, and 16.8 m bpd by 2000. A substantial, though unqualified, extra volume of associated gas is also expected to be used.[7]

Table 6.1 shows how internal demand for refined products by OPEC members has grown since 1969. It is apparent that the annual average growth rate, at 12.5%, is considerably in excess of the rate in other groups of countries over that period. During the years 1967-77, world consumption of refined oil products grew at an annual average rate of 5.5%. For South America the growth-rate was 4.6%; for Central America and the Caribbean, 4.4%; Africa, 6.3%; and Asia (including Australia and Japan), 2.4%.

Table 6.1

OPEC CONSUMPTION OF REFINED PRODUCTS, 1969-1981
(1,000 bpd)

	1969	1970	1975	1980	1981	% change 1980/1	Average annual growth rate % 1969/81
Algeria	32.3	39.5	70.1	108.1	115.8	7.1	11.2
Ecuador	21.6	23.9	38.4	70.7	79.0	11.7	11.4
Gabon	14.1	17.0	10.8	27.8	29.8	7.2	6.4
Indonesia	106.7	108.7	210.0	397.4	442.4	11.3	12.6
I. R. Iran	165.9	182.4	352.1	400.0	400.0	0.0	7.6
Iraq	66.4	61.9	98.0	191.2	191.2	0.0	9.2
Kuwait	12.5	14.4	23.9	44.0	46.9	6.6	11.7
S. P. Libyan A. J.	20.9	16.4	51.2	85.4	90.9	6.4	13.0
Nigeria	23.6	28.4	68.1	171.0	235.8	37.9	21.1
Qatar	2.4	1.6	4.2	9.0	10.0	11.1	12.6
Saudi Arabia	39.0	42.5	117.3	474.6	576.8	21.5	25.2
UAE	2.0	2.2	25.5	76.2	83.2	9.2	36.4
Venezuela	140.7	148.2	211.3	343.2	364.1	6.1	8.3
Total OPEC	648.1	687.1	1,280.9	2,398.6	2,665.9	11.4	12.5

Source: OPEC

The result of this rapid internal growth in OPEC countries has been a rise in excess of 400% in domestic refined products demand, to 2.666 m bpd equivalent by 1981. This has grown irrespective of crude oil output and exports, so that in years of below average output (such as 1975) the proportion taken up by domestic needs has grown considerably faster than the year-on-year growth rate would imply. By 1981 domestic demand had risen to the equivalent of 11.9% of total OPEC crude oil output.

Care must be taken in interpreting such estimates, however. As the section discussing the concept of the energy-GDP ratios shows, the measurement of total energy use is necessarily imprecise. This is particularly so for developing countries, where non-commercial energy accounts for a substantial—although, once again, intrinsically unquantifiable-portion of all energy use. This is particularly true in rural areas and at low per capita income levels. Particularly low income countries such as Mali, Ethiopia, Haiti and Upper Volta have recently been estimated to use in total some ten times the amount of commercial energy they consume.

The import of these observations for forecasters of OPEC members' internal energy demand lies in the fact that as recorded energy comes to take an increasing share of total energy used, so will energy demand rise disproportionately quickly as GDP rises. Growing shortages of such traditional fuels as firewood will buttress this tendency, in that the "firewood crisis" and similar shortages will precipitate onto commercial energy markets consumers who previously satisfied their needs in non-commercial ways.[8]

Other recent estimates by the World Bank refer to firewood, charcoal, crop residues and other wastes often being responsible for more or less all energy usage in rural areas of developing countries.[9] Africa is judged to be most dependent upon such fuel sources, with Latin American countries the least so. Altogether, "between 0.5 and 1 billion (people) use agricultural and animal wastes to fuel their cooking fires".[10]

It is necessary to refer to this level of domestic demand as an equivalent, however, because OPEC exports are not simply the surplus after home needs have been satisfied. This is because domestic refined products are acquired from overseas as well as domestic sources, depending on the nature of the products and the extent of demand. Just like all countries, OPEC members are subject to the complexities of the oil business with its 130 or more crude oil types traded, and the multiplicity of refined products on the market. A constraint on OPEC members' ability to satisfy their own internal needs is their refining bottlenecks. OPEC refining capacity has grown considerably less quickly than domestic demand for refined products: over 1968-1981, the annual average growth rate of 5.6% was rather less than half that demand, at 11.4%. While some members are able to refine nearly half their crude output, the majority can manage considerably less.

The OPEC Output Response

The previous section has alluded to the potential importance of new oil-exporting sources upon the structure of the oil market in the coming years. This section now looks at the factors which will determine the ability and the willingness of the OPEC states to confront this new configuration. It begins by briefly reviewing the current physical output capabilities of OPEC members, then assesses which factors will be most significant in

133

determining the extent to which present and prospective capacity will be utilized by each member, bearing in mind the differing—and sometimes conflicting—objectives which are dependent upon the level of oil output.

Table 6.2 shows how exports of crude oil grew, during the period 1969-81, from 18.8 million barrels/day in 1969 to a peak of 27.6 million barrels/day in 1977, then fell to 18.4 million barrels/day in 1981.

Table 6.2

OPEC CRUDE OIL EXPORTS, 1969-1981
(1,000 b/d) AND MEMBERS' SHARES OF EXPORT VARIATION

	1969	1975	1980	1981	% change 1980/81	% max. deviation between peak and minimum exports
Algeria	883.0	877.5	715.5	521.8	− 27.1	98
Ecuador	1.2	145.8	110.2	105.5	− 4.3	58
Gabon	84.2	205.3	151.2	124.9	− 14.4	72
Indonesia	516.8	994.7	1,034.6	1,050.4	1.5	62
I. R. Iran	2,847.4	4,671.0	796.7	815.9	2.5	574
Iraq	1,445.4	2,058.8	2,459.0	697.4	− 71.6	370
Kuwait	2,441.7	1,803.4	1,296.5	813.8	− 37.2	225
S. P. Libyan A. J.	3,069.5	1,431.1	1,691.4	959.2	− 43.3	131
Nigeria	542.2	1,713.3	1,960.2	1,228.4	− 37.3	80
Qatar	353.0	428.3	465.7	391.0	− 16.0	46
Saudi Arabia	2,794.7	6,601.1	9,223.2	9,017.9	− 2.3	40
UAE	624.9	1,661.4	1,697.3	1,439.0	− 15.2	38
Venezuela	2,476.0	1,472.2	1,287.2	1,266.0	− 1.7	71
Total OPEC	18,808.0	24,063.9	22,888.7	18,431.2	− 19.7	50

Source: Calculated from OPEC data

Judging what capacity utilization rates these output figures correspond to is hazardous, for two reasons. First, only outsiders' estimates of installed capacity are available: there are no universally agreed figures to which reference can be made. Thus, estimates of Saudi Arabian capacity during 1980 ranges from 10.5 m bpd to 12 m bpd or even more. This disagreement stems in part from the fact that, as in manufacturing industry, 100% capacity utilization is a notional level, best interpreted as a band or range of possibilities rather than a point estimate. This is because for brief periods output can be raised sharply (so-called surge capacity) but not sustained. Technical constraints, constraints on storage, well maintenance, etc., intrude on surge capacity. A commonly used concept of capacity is 90 day sustainable output, which is obviously, by definition, a more consistent level of output than surge output. But even with this measure one must make assumptions about the extent and quality of maintenance procedures being carried out.

Table 6.3 shows the estimated productive capacity of each member in 1980, as assessed by the Central Intelligence Agency in their International Energy Statistical Review. But the provisos mentioned above naturally apply to these estimates as much as to any others.

Table 6.3

OPEC MEMBERS' SUSTAINABLE PRODUCTION CAPACITY 1980
(thousand bpd)

Saudi Arabia	12,500
Iran	7,000
Iraq	4,000
Kuwait	2,900
Venezuela	2,600
Libya	2,500
Nigeria	2,500
Abu Dhabi	2,150
Indonesia	1,800
Algeria	1,200
Qatar	650
Dubai	390
Ecuador	250
Gabon	250

Source: Central Intelligence Agency, *International Energy Statistical Review,* various issues

Saudi Arabia as Swing Producer

Since the Saudi Arabian oilfields dominate OPEC in volume terms it is worthwhile studying the Saudis' interest in oil production in some detail. It has been mentioned that Petromin, the Saudi oil agency, is proposing to increase sustainable output to up to as much as 14-18 m bpd by the late 1980s—a figure which would further buttress the country as the dominant producer within OPEC. The two immediate questions which arise from this observation are: first, will there be greater willingness, on the part of the Saudis, to use this enlarged production capacity to affect prices more directly? And second, to what extent, and under what circumstances, might the Saudis actually be willing to do so?

The question of the Saudis' interests in increasing their normal output of crude oil is, clearly, a complex one and one to which only tentative answers may be given. This is due not only to the secrecy which surrounds Saudi Arabian decision-making at its highest levels, but also because the evidence points to a considerable degree of flux in that decision-making itself. Priorities between objectives (particularly, as is often the case, when they are at least partially conflicting) appear to shift between public pronouncements from the Saudi authorities; while the authorities' perceptions of the other OPEC members' intentions and interests also appear to change over time. It is easiest to confront the question if it is broken into two parts.

First, one might ask how strong is the Saudis' desire to begin acting consistently as a "swing producer" within OPEC. The swing producer idea is one which has exercised the minds of OPEC-watchers for years. In its simplest form, it is the contention that one country, or (alternatively) a small sub-group of countries, acts to enforce its own price or output objectives for the group of members as a whole by using its large share of total OPEC capacity. By virtue of its dominant share of OPEC output Saudi Arabia is conventionally identified as the swing producer of OPEC. Similarly, in most interpretations of OPEC behaviour the Saudis are perceived as always wanting to impose their will on the other members by using their margin of slack, or spare output capacity, to alter output where necessary. By the same token, the Saudis' alleged position as "low absorbers" (i.e. a country with an excess of funds over its ability to invest them domestically) gives them the power to cut back output, once again to adjust aggregate OPEC output to the level that optimises the Saudis' own objectives.

Certainly, there is no real technical difficulty in stopping or slowing oil output from a well: the reservoirs are not necessarily damaged by this. But it is not clear that, in the past, the Saudis have in practice consistently been acting as swing producer to the rest of the organisation. If one construes the swing producer to be that country which most drastically scaled down its production plans over the long-term, then the Saudis certainly suit the title, given that in the early 1970s they were proposing to build up sustainable oil output towards 16-20 m bpd. Now that a move to install even 12 m bpd is of greater significance, it is apparent how substantially the Saudis have moved to curtail proposed output. But one need not interpret that policy option as indicating a desire to act as swing producer; there is another, day-to-day, form of swing producing. If, using this alternative definition, one views the swing producer as that country which most consistently accounts for disproportionately large output adjustments within the OPEC total then Saudi Arabia is in this case also a clear candidate. Saudi Arabia consistently accounted for around one-third of the year-on-year change in the total OPEC exports until 1979. And even after 1979, when Iran accounted for a bigger proportion of the total OPEC export turnaround (eg, in 1980, 42% as against 10% for Saudi Arabia) this was due to the intense upheaval and revolution which naturally affected the oilfields for a period. The next most variable exporting country between 1972 and 1979, Iran, in most years accounted for around one-half as much of the export turnaround between years as Saudi Arabia. The country responsible for the third greatest export variability over this period was Iraq. Because of the considerable variation in the members' average level of exports, Saudi Arabia and, to a lesser extent, Iran could account for disproportionately large shares of total export variability without themselves changing their own export volumes so considerably from year to year. Thus, despite its dominance in the total of export variance from year to year, Saudi Arabia's maximum deviation of exports between peak and trough years over 1973-81 was 40%. The country with the greatest peak-to-trough deviation was Iraq (77% before 1979 and 370% over 1973-81) followed by Algeria (98%). The OPEC total variation was 50%, between peak exports of 27.64 mb/day in 1977 and trough exports of 18.43 mb/day in 1981. A few independent commentators have recently come to recognise the constrained sense in which the Saudis can be said to work as the swing producer within the organisation. Among them Al-Nasrawi has recently confirmed that "the behaviour of Saudi Arabia does not reflect an understanding among OPEC members that the leadership role does in fact belong to Saudi Arabia... Saudi Arabia did not have the controlling voice in OPEC's pricing councils".[11]

By the late 1980s, then, it is clear that the Saudis will be better placed to deploy—if they so wish— a wider margin of discretionary capacity than they can presently wield. But so far it has not been possible to decide exactly under what circumstances this capacity would be brought into play. This implies some hesitancy on the part of the Saudis in working their oilfields at full capacity, and the reasons for this are now examined.

The best place to begin is to judge what consequences would follow from the clear assumption of the swing producer role. Three sets of consequences—one economic, two political—follow.

First, economic consequences follow from the time-preference of different OPEC members where some have significantly greater reserves than others. Table 6.4 shows the extent of each member's oil reserves at end-1981. Other things being equal, the bigger a member's reserves, the greater the possibility of maximizing "life time" oil revenue being achieved by not precipitating too rapid a shift by consumers into oil-saving technologies. Such a country will presumably therefore argue for moderate oil price increases rather than large and volatile increases. Given the Saudis' large reserves (as Table 6.6 indicates, the country possesses 38% of all OPEC members' known reserves) their predisposition will be towards slower price rises.

Table 6.4

OPEC CRUDE OIL RESERVES AS OF YEAR END 1981
(million barrels)

Algeria	8,080
Ecuador	850
Gabon	480
Indonesia	9,800
I. R. Iran	57,000
Iraq	29,700
Kuwait	67,730
S. P. Libyan A. J.	22,600
Nigeria	16,500
Qatar	3,434
Saudi Arabia	167,850
UAE	32,176
Venezuela	19,888
Total OPEC	436,088

Source: OPEC

Related is the point that too great a degree of instability in oil market prices has an effect upon inflation rates, and thus currency movements and interest rates in industrialized countries. To the extent that OPEC members—again, to a widely differentiated extent—retain financial investments in industrialized countries, it follows that their real rate of return and real capital values stand to be jeopardised. At the end of 1979, the Bank of England estimated the accumulated OPEC foreign assets to stand at $240 bn. Of this total, some 80% was held in developed countries' financial centres, with over half the total held as bank deposits.

A set of diplomatic consequences follows from the assumption of the swing producer role. These would follow in the Saudis' case through their becoming unambiguously the most important influence on oil prices, with all the economic and political (not to mention emotional) liabilities that involves. The issues discussed below, while not necessarily resulting from such a position, are certainly conceivable outcomes. The point of issue then becomes the Saudis' judgement of the likelihood of these outcomes occurring. Should officials believe their coming about is likely, this will be sufficient to deter their pursuing the price-leadership course. The three issues involved are as follows.

First, there is the danger of cross-border incursions into the Kingdom by guerrillas from North Yemen. This is, to the Saudis, just one of several regional dangers they face, given the continuing interest and involvement of the USSR in supplying arms to the government of North Yemen. As the attraction of a military success over the Saudis grows, with the Saudis acting increasingly obviously in the interests of the West, Marxists in both the Yemens might find small raids into Saudi territory irresistible.

Second is the fear of loss of control as the Israeli conflict gathers momentum. Having openly supported the PLO in the past, if now a more radical group should seize the anti-Israeli initiative, will the Saudis feel compelled to go further along the path with them? The danger here is of the momentum of ideas and events in the Israeli conflict outpacing the evolution of the Saudis' own thinking. Yet the more the Saudis assume the mantles of regional power-broker, coupled to oil-price leader, the more they will be expected to make gestures and pursue an anti-Israel line they may not truly believe to be the most constructive.

The third problem which the Saudi government stands to find arising is that of rifts in the upper echelons of the hierarchy of Saudi society if decisions on oil policy are felt too consistently to be favouring Western countries' interests. For there is a constant tension in the attitude of the Saudi élite and its lower echelons towards the interest of external powers. While, on the one hand, there is recognition of the significant cultural, military and economic consistency of interests between Saudi Arabia and the West, there is also a desire to try to shape a "third option"—something independent of US or Soviet super-power alignment. Writing in a slightly different context (that of Soviet-Arab relations in Nasser's period) Heikal put the Arab empathy with the West thus:

> "The West was the world they knew. They recognised the worth of liberal democracy, and would have been happy to see it flourish in their own countries... All the formative influences—the books they had read, the history they had learned, the films they had seen, came from the West."[12]

This third option would be shaped, initally at least, from the institutions already set up to foster non-alignment. Apart from the non-aligned states organisation itself, the Islamic Conference and, more recently, the Council for Gulf Cooperation are likely channels for this ambition.

January 1981 saw the Saudi Arabian government act as hosts to some 40 nations' representatives at the Islamic summit at Taif. There is no doubt that the Saudis lavished a great deal of money and effort on the meeting even if they were not unambiguously happy about the rather strident tone of the resolutions which came out of it. The "Mecca Declaration", issued at the end of the conference, insisted that henceforth the PLO should play an active part in discussions of the Israeli question.

The six members of the Gulf Council (Saudi Arabia, Kuwait, Bahrain, Oman, United Arab Emirates and Qatar) met on May 26-27 1981 (just after the regular OPEC members' conference) to establish a framework for security collusion in future. This interest in the creation of more autonomous organisations, which look to the super-powers if at all, and only then for arms supplies, must be seen as part of a growing disenchantment by most Middle Eastern OPEC members with Soviet and American courting of their area.

What does the foregoing indicate about the course of Saudi-US relations in general? Much was written on this subject during the 1970s; in particular, it has become almost a conventional wisdom that the USA and Saudi Arabia entered into a relationship of mutual support after 1973, whereby the Saudis exerted their influence in the oil market to suppress further price rises, while the USA used its unique influence with Israel to hasten the arrival of a negotiated end to the Middle East hostilities.

It should be borne in mind that this relationship, like all relationships, has its boundaries, and that sometimes other priorities may intrude. For the 1980s the importance of this relationship should not be overstated.

First, the relationship has a military component. It has frequently been argued that the Saudis are inextricably bound to the US because they are dependent upon US exports of military equipment, as well as trainers and resident advisers. But there are three problems to taking this proposition as a framework for Saudi actions in the 1980s. First, arms can be bought from many places other than the USA. France, Germany, Spain, Italy, China and the USSR, to mention only some countries, have enterprises capable of supplying defence equipment. While some of these sources are clearly not capable of delivering the most elaborate technologies, at least France apart from the US could deliver jet fighters and bombers comparable to US equipment. Indeed, some diversification in Middle

140

Eastern armaments ordering away from the US has been apparent since the early 1970s.

In addition to the first point, the Saudis have in practice found an ambivalence within various US administrations regarding arms supplies to the Gulf, just as there has been a reciprocating ambivalence regarding the desirability of a US presence there. The Saudis have, in the past, been humiliated by the American prevarication over supplying them with fully-equipped aircraft. Once an order has been made, to have the attractiveness of that order scrutinized so publicly by the US authorities not only calls into question the ordering country's standing in the eyes of Washington, but also effectively signals to other interested parties that for the USA, supplying military effectiveness to the remaining "pillar" of American's erstwhile "twin pillars strategy" in the Middle East is by no means a policy imperative. Seen in this light the Saudis' announcement of February 1981 that they were contemplating arms purchases from the USSR (and that in March, dealing with a possible purchase of German Leopard II tanks) is to be interpreted as a signal that Saudi military development is by no means dependent upon American acquiescence.

A further question which will continue to circumscribe the US-Saudi relationship, at least insofar as it is based upon shared military interests, concerns the type of equipment most keenly sought by the Saudis. The US, clearly, is capable of supplying technologies of the utmost sophistication. But if some of the hardware needed is of a less sophisticated variety, then naturally the choice of supplier widens considerably. Given that there is at least as much danger lying in parochial small-scale insurgency as in full-scale military initiatives against the Kingdom's rulers, it follows that, to some extent at least, the most appropriate military mix will contain the type of goods that could be obtained from many suppliers other than merely the USA. The example of the Shah of Iran being toppled despite the considerable extent and quality of the hardware available to his various security forces has almost certainly underlined for other Middle Eastern rulers the fact that the operation and maintenance of hardware is of at least equal importance as its intrinsic sophistication.

It will continue to be crucial for the Saudi authorities to reappraise the state of their relations with the USA. Chiefly this is because for the Saudis that relationship carries certain costs, and, logically, the extent and nature of those costs must continually be weighed against whatever benefits are directly attributable to the US link. The problem for the Saudis has been posed by Penrose in these words: "Saudi Arabia has to balance its friendship with the United States against the hostility its price policy has aroused among its Arab and other OPEC colleagues".[13] The price policy to which reference is made in this quotation is the continuous pressure applied on the other members of OPEC to restrain price rises as quid pro quo for US support and efforts at cajoling the Israelis towards some form of accommodation with the rest of the Arab world. A similar interpretation of the Saudi-US relationship has been drawn by the International Institute for Strategic Studies, which argued that:

> "The special relationship between Saudi Arabia and the US was central to the 1975-78 oil order... the former used her influence and leverage within OPEC... the latter undertook to protect Saudi Arabia and her present regime against external threats..."[14]

These interpretations clearly contain an element of truth. But there is more to the tensions between the two countries than either of the commentators quoted above might allow. In essence the growing problem for the Saudi government is that any relationship it retains with the US government is increasingly a focus for anti-Saudi sentiment. An article, reportedly written by a group of well-educated concerned Saudi professionals, and published in the Washington-based "Armed Forces Journal" late in 1980 points to these dangers. At one point the article claims that "the Saudis now view the United States as the most serious threat to their own security".[15]

In essence the embarrassment the Saudis increasingly feel over their relationship with the USA is that other Arab countries can interpret the Saudis as standing ready to act as trojan horse for a US-organised field force in the event of instability in the region. There has been a great deal of talk, from Henry Kissinger among others, about groups of Western countries mounting "rescue" missions in Saudi Arabia. The latest instance was the Reagan-Thatcher announcement along these lines late in February 1981. Yet nothing will faster alienate Saudi officials from their US link than such talk, given that the bulk of it is premised upon "scenarios" in which the House of Saud is collapsing, and in which the Western countries exercise as of right an option to protect raw material supplies, irrespective of the effect such outside intervention might have upon neighbouring countries' actions. Such tensions have recently been recognised by Valerie Yorke, who has written that "American peace diplomacy has already dangerously polarized the Arab world... The US... might inadvertently bring about destabilizing repercussions inside the Arabian peninsula".[16] As long ago as 1977, Rustow also identified what he believed to be the "highly ambivalent" nature of the Saudi-US relationship,[17] and in early 1981 the Iraqi foreign minister, responding to the idea of a Western security task force in the Middle East, argued that "we think it would mean an increased attempt by the Soviet Union to emphasise its presence in the area". Rustow's voice was one of the few at that time, however. More conventional was the view of the Washington-Riyadh axis typified by the lead article in *The Economist* of June 18 1977, which ran: "Saudi Arabia needs American technology... and weapons... it is the Saudi-American connection... that helps to keep most of the Arab world on a pro-Western rightward-leaning path".

This, then, is the policy conundrum that Saudi policy-makers will face in the 1980s with respect to their US relations. The likelihood of exacerbating regional problems while attempting to cajole a distant, only dimly comprehended and frequently wayward foreign policy partner, whose own policy apparatus is severely constrained by pro-Israeli lobbyists, is going to continue to provide problems. This in turn confirms the attraction for the Saudis of making intensified efforts at shaping new options for themselves.

The above analysis points, then, to a dilemma for Saudi Arabia. The figures for exports and export variability during 1973-79 indicate the predominant share of export cutbacks and surges within the OPEC total taken by Saudi Arabia, yet it has also been shown that operating with too high a profile in output or export manipulation brings with it certain political difficulties. The question which then arises for the 1980s is how the two issues can be traded off one against the other. The following section tries to judge how this problem will be resolved.

For Saudi Arabia, and possibly Venezuela, two options appear to exist in the coming years. The first option, briefly stated, is to attempt to persist with, and expand upon, the largely *ad hoc* system of bilateral dealing between Western and Middle East governments. The second option is the introduction of some form of long-term Strategy Agreement.

Middle Eastern Security Problems

A final set of security problems facing the Middle Eastern OPEC states concerns the USSR and the Eastern bloc. Two issues are involved in this. First, it has long been a strain of much strategic analysis in the West that the USSR has territorial designs upon a number of Middle Eastern countries. Second, it is argued that the impulse to act on these ambitions will become increasingly irresistible in the 1980s, as the Soviet Union meets failure in attempts to exploit its own oil and gas resources, particularly those in Siberia. To the extent that the USSR ceases to be a net oil exporter and has to countenance net oil imports at some stage in the 1980s, the invasion option will therefore become, some argue, almost an imperative. Linked to this is a related issue: the rate of growth of the Soviet economy

143

in the 1980s and the attendant energy needs of that growth rate.

The growth rate of GDP of the USSR in the 1980s appears likely to be below its 1970s performance. But the extent of the growth is of less significance here than the nature of the growth—in particular, it is important to judge how far the share of industry in GDP will remain at its present very high level in Eastern bloc countries generally, and the USSR in particular. For this high share implies a relatively large input of energy is needed for a given rise in GDP; in more advanced economies where the share of GDP has begun to fall, or has at least stabilised, growth can be less energy-intensive.

Perhaps no area of research into the Soviet economic outlook has received so much attention in the West as the energy sector. The catalyst to much of this work was the CIA report entitled *Prospects for Soviet Oil Production*, published in 1977, forecast the USSR becoming a net importer of crude oil—possibly to the extent of around 3.5-4.5 m bpd—by 1985. This is in contrast to net crude exports from the USSR of 80 m tonnes annually during the second half of the 1970s.

Since the publication of that report a great deal of research has contested the CIA results, arguing that at least as many leading indicators of Soviet oil production point to rising output as do to stagnating output. Most spectacular of such estimates are those from the Swedish specialists Petrostudies, who in 1980 forecast rising Soviet oil output throughout the 1980s and net exports remaining close to current levels. Pointing to the rising importance of oil in the Soviet Union's total foreign currency earnings other than from gold (crude oil's share rose from 28% in 1978 to 34% in 1979) they argue that the foreign currency imperative will result in still greater resources being devoted to oil exploitation. Similarly sanguine predictions are those of Dr. David Wilson, who argues in a 1980 forecast that oil output would rise steadily throughout this decade, due chiefly to success in overcoming the territorial problems and skilled labour constraints entailed by working in the Western Siberian region. He foresees output rising from 606 m tonnes in 1980 to 700 mn by 1985 and 750 mn by 1990, with exports (160 m tonnes in 1980) peaking in 1985 to 170 mn tons but tapering off to 120 m tonnes in 1990.[18]

By contrast, the same year saw two sets of forecasts pointing in exactly the opposite direction. In March a Gosplan department head, V. Filanovsky, publicly berated the insufficiency of the Soviet exploration effort in an article in the Soviet magazine *Planovoye Khozyaistuc*. Similarly, the CIA, revising both their 1977 Soviet report and their August 1979 update (entitled *The World Oil Market In The Years Ahead*) gathered a panel of experts from the CIA, Pentagon and Department of Energy. These testified to the Senate Foreign Relations Committee in

144

March 1980 that oil output from the USSR would certainly fall, leading to net imports by 1985.[19]

The technical arguments centre chiefly on the Soviets' ability to produce and maintain in sufficient quantities their own electric pumps for high-pressure water injection techniques of oil recovery and deep well drilling accessories.

Analysts in the West have taken these questions a stage further, confronting not only the question of Soviet ability to export oil, but also their willingness to do so. The background to this question is that the USSR is presently responsible for underwriting a considerable portion of the other COMECON states' energy needs. Since, however, these states pay for these imports not in hard currency but in so-called convertible roubles, crude oil exports from the USSR to COMECON imply an opportunity cost to the Soviets in terms of hard currency earnings forgone. While it is apparently not possible to point to the existence of any coherent Soviet guidelines in their allocation of crude as between COMECON demand and hard currency demand, nor do clear statistics exist beyond 1976, specialists have estimated how the calculus might be constructed. E. A. Hewett has, for instance, suggested a series of trade-offs facing Soviet planners, choosing between the rising real costs of the subsidy to COMECON, the effect of foreign currency oil purchases by COMECON nations forced onto the free market, and having to sell their manufactures in the West for whatever price they can command, and the possibility of political instability arising from significantly lower growthrates as a result.[20]

Recent weaker market conditions have led Soviet oil managers to begin price cutting to gain market share. In March 1983 the price of Soviet crude was cut to $1.50/barrel less than the Saudi market price of $29, to $27.50/barrel. Output of 12.3 mn barrels/day in the first half of 1983 means that the USSR has recently been the fifth biggest crude oil exporter to West Europe, and the US. Bigger export volumes come only from Saudi Arabia, Iran, Venezuela and Mexico. During the early 1980s Soviet oil exports have grown quickly, from slightly under 1 mn barrels/day in 1980, to 1.1 mn barrels/day in 1981 and 1.5 in 1982. In 1983 the total may reach 2 million barrels/day. In 1982 earnings from this source totalled around $15.5 billion. This export growth may reflect a decline in the amount of oil made available for domestic consumption. Certainly, Eastern European fellow members of COMECON received smaller oil deliveries in 1982: estimates of a 12% drop between 1981 and 1982 have been made in the West.

Later in 1983, however, as the oil market began to tighten slightly, the

USSR raised its oil price by 50 cents/barrel, shortly after Egypt had talked of raising its price. This increase followed a formal approach by OPEC officials to Soviet oil managers. Algeria was mandated by its OPEC members to request Soviet cooperation in the oil market. The UK and Mexico were also included at this time in a series of informal talks.

Having looked briefly at the question of the Soviet Union's own oil resources, the next section confronts the wider question of the Soviet Union's strategic interets in the Middle East.

Essentially the research into the USSR's strategic interests in the Middle East can be seen as falling into two categories. The first consists of those who believe on the one hand that as part of a coherent global strategy the USSR is interested as a matter of course in gains in the Middle East. The second school, while not denying that there could be some circumstances under which the USSR would take military action in the Middle East, emphasises the lack of any discernable worldwide strategy in its policies. Among the most eloquent of spokesmen for the former position is William Pfaff, who has written of a new impetus to Soviet involvement in the Middle East, which is now, with increasing frequency, military in character (through proxy armies such as Cuba's) rather than, as hitherto, merely ideological and technical assistance.[21]

Views which stress the hesitant, reactive and defensive nature of much Soviet involvement in Middle Eastern affairs, as well as its parochial scale have been referred to by one author as subscribing to the "muddle" theory. In essence this emphasises the enormity of the Soviet Union's domestic difficulties regarding labour shortages, harvest failures, ethnic splits, etc., as inevitably constraining the scope of foreign intiatives. Pointing to a number of opportunities for greater involvement (for instance the upheavals in Iraq in 1958) having been lost rather than capitalised upon, this school concludes that Soviet policy "still follows rather than directs" events.[22] Failures by the USSR to comprehend the true nature of the Israel-Palestine conflict, and to incline towards an inappropriate class interpretration rather than as a protracted social conflict with untidy implications for nation state realignments have been indicated.[23] So too have the USSR's evident failures in the past to entrench themselves as desirable allies for Third World nations: Campbell has observed that "the record of the past two decades shows that a country that invites the Soviets in does not necessarily accept their control and can, when it chooses, invite them out."[24] Similarly, Heikel has pointed to the inability to expand into Africa beyond the bridgehead provided by Gamal Nasser[25] and Binyon to Soviet communism as "an ideology that now appears exhausted, self-serving and less and less attractive to the Third World."[26] An overview of strategic problems in the 1980s by four institutes

146

similarly pointed out how "from an economic point of view the Soviet cupboard is bare and unattractive."[27] Many times the Soviets have been wrong-footed, according to these commentators. The confusion over the Iran-Iraq war after late 1980, not being able unequivocally to support one side over the other, was frequently cited as an illustration.

Developments since the fall of the Shah probably confirm the chaos theory more than any other. And that interpretation is probably the one increasingly favoured by OPEC Middle Eastern states as well. The chief reason for this has been, in complete consistency with the theory, the USSR's evident discomfiture over the instability in Iran and the war with Iraq. Despite its treaty of friendship with Iraq, dating from 1972, and the encouragement from King Hussein of Jordan in providing arms for the Iraqi's in late 1980, the Soviet Union also had to look to its treaty with Syria, an ally of Iran. In the event the Soviet prevarication has confirmed the basic unpreparedness of any coherent scheme.

Government Trade in Oils

The possibility of OPEC members' share of world traded crude in the 1980s shrinking to the point where market entry by new sources, with the disruptive consequences hypothesized above, becomes a significant factor, will very probably give rise to a reaction. In the face of a shrinking share of world trade in crude oil and the corresponding diminution of their pretensions to international influence already predicted, some OPEC states may attempt to wield their crude exports increasingly as an instrument of foreign policy.

One of the most significant structural and political changes to come over the world oil market in the 1970s—that of the increased involvement of governments—will have pervasive effects in the 1980s. In 1973, the multinational oil majors still enjoyed direct access to some 75% of the non-Communist world's crude exports, or some 30 million b/day. Despite their having had their autonomy cut back by the unilateral price fixing forced upon them by the host country governments at the Teheran-Tripoli disussions, the multinationals could at least still control the bulk of physical supply. Growing realization among the OPEC governments that the concession system in which they participated was yielding little in the way of indigenous spin-off or ancillary industrial activity, or, more crucially, nothing in the way of production monitoring and control, prompted a series of measures to preempt certain of the companies' activities. The process of growing control in the 1960s and early 1970s is well known by now. By 1976 the multinationals had access to only 50% of traded crude oil; by 1983 this proportion had further shrunk to around

147

40%. The host country governments had correspondingly been chipping away at the multinationals' authority over production policy so that by the beginning of the 1980s the position is as shown in Table 6.5. In future this trend is set to continue as before: "the time is drawing near when OPEC's national oil companies will handle nearly all of their export sales of crude oil and refined products".[28]

Table 6.5

GOVERNMENT TRADE IN OILS, BY EXPORTING NATION

	mb/day	% of country's exports, 1979
Iraq	2.1	60
Iran	1.1	40
Nigeria	0.4	20
Libya	0.5	30
Kuwait	0.2	10
Saudi Arabia	0.8	9
Others	0.7	(3)
Total	5.8	average 20%

Source: J. M. Mohnfeld, "Changing Patterns of Trade", *Petroleum Economist,* August 1980, pp.329-332
Note: Since the precise extent (and nature) of government involvement in trade in crude is extremely difficult to judge, the figures reproduced above are necessarily only indicators of broad magnitudes.

Oil-exporting countries export more of their relatively abundant heavy crudes first, leaving the increasingly more valuable and less easily substitutable light crudes in the ground. The optimal output-mix from the point of view of the oil-exporting country governments is thus one related to reserve patterns rather than to demand patterns.

The practical result of this as far as oil clients are concerned is that they may increasingly be required to accept shipments of more heavy crude than they want as a condition of receiving the volume of light crudes they need. For refiners, this raises the need to reorientate refineries to cope with greater than anticipated volumes of heavy crudes. This process of 'debottlenecking' has been underway for some years, but has been accompanied also by the mothballing of some of the parts of refining capacity that could not readily be switched from sweet low-sulphur crudes

148

to the more prevalent sour crudes with more sulphur.

The second result which could follow from the greater host state control of oil exports concerns the frequency of price changes. Notice of 30 days is conventional in the industry but observers expect shorter periods of notice—possibly 5 days—to become commonplace.

Related to this is another manifestation of less leeway in the government/company relationship. Rather than allowing actual liftings to oscillate around the agreed volume by a degree, to match shipments more precisely with regional, seasonal or other client needs, it is thought probably that these 'contractual tolerances' will be eliminated or tightened up to a degree. The result of such a move would clearly be a less responsible and flexible supply side.

The same fear is raised by the increasing use of quid pro quos in the business. In these cases the exporting government exacts certain conditions, which may or may not be related to the oil sector, from the intending client. Recent instances involve both OPEC members and non-members. Saudi Arabia has, for instance, in the last year suggested that each million dollars of industrial investment undertaken will be 'rewarded' by guaranteed access for the firm to 500 barrels of oil per day. Such 'incentive' oil schemes, undertaken with Western multinational contractors' help, will in the case of Saudi Arabia be yielding 0.75 million barrels/day in entitlements by the mid-1980s.

An even more overtly political element to trade in oils follows from the enhanced ability given by GTO for the exporter to dictate to which countries the oil can be re-sold or on-traded (sometimes referred to as "positive destination controls"). Certain nations (for example, South Africa) may be proscribed as oil clients through the decisions of certain exporting countries or through the decisions of multilateral agencies. Any trader buying with the intention of reselling elsewhere is thus in a riskier position than before, and complex sequences of shuffling between companies and company divisions may be necessary to disguise the true destinations of cargoes. (A great deal of careful scrutiny of oil supplies obtained by South Africa, for instance, by the Shipping Research Bureau of Amsterdam, tries to ascertain which firms' ships make the deliveries.)

The problems this could give rise to would be more rapidly realised, moreover, if the duration of the typical oil supply agreement is shortened. A number of OPEC members have already curtailed contract lengths to one year, and have indicated that actual deliveries can in practice deviate from agreed delivery volumes at short notice and under a number of circumstances other than straightforward *force majeure*.

149

Another consequence of greater GTO is that the flexibility of the market will almost certainly be impaired. The multinationals' control of output and of output-mix, coupled with their sophisticated distribution network, allayed dangers of prolonged supply interruption or failure to balance refinery runs. By contrast, the output-mix and volume decisions are to an increasing degree to be taken other than by the distributors: they will necessarily also be less concerned with efficiently eliminating mismatches of supply and demand at the parochial level due to their preoccupation with lifting heavy crudes in greater than desired quantities.

Furthermore, an orthodox response to conditions of greater risk in crude supplies would be greater resort to greater stockpiling, and, on occasions, greater use of the spot market. Stockpiling carries two costs, however. The first is, as with all strategic stocks of materials, an opportunity cost equivalent to the return which could have been obtained by applying the money needed to retain the stock or inventory in some other use. The second cost is that building up inventories is of course itself a form of demand. In periods of tight market conditions, the need to stockpile therefore makes all the more likely a demand-led precipitation of higher spot, if not term, prices. The effect of stockpiling upon both spot and official OPEC oil prices was an important part of OPEC criticism of developed country governments' purchasing behaviour in the last quarter of 1980, when the Iran-Iraq war had initially removed around 3.9 million barrels/day from world supply. The International Energy Agency members were urged not to attempt to maintain their stocks at the prevailing very high levels (in excess of 100 days in early October) but to run them down to fulfil a greater proportion of demand from previous rather than current purchases. Among OPEC circles there was talk of "irresponsible stockpiling policies", which were exceeding a "normal" level of 30 days supply, reaching 90 days worth.[29] The irresponsibility of this action by the importing countries was attributed to a failure to understand the intrinsically responsible behaviour from OPEC members which would have been prompted by any sign of oil shortage developing. As an article highly critical of the industrialized countries' stockpiling behaviour suggested, "the consuming world has received ample proof that in all delicate market situations OPEC countries have taken the right decisions". The only clear consequence has been huge windfall profits for the oil companies, in the view of the OPEC analysis. Both companies and developed country governments, OPEC now argues, are, in their stockpiling behaviour, "quickly gathering a new and unhealthy sophistication.[30]

The hostile reaction, on the part of OPEC, towards stockpiling behaviour in the last two years or so is itself the third problem exacerbated by GTO. It raises another issue over which conflicts of interest can easily

150

arise. In the terms of the early sections of this report, however, it may increasingly be the case that new entrants to the oils market will make part of their selling appeal, initially at least, freedom from the political dimensions of supply that are expected increasingly to characterize supplies from OPEC sources.

The need for stocks to assume permanently a higher proportion of total demand for oil obviously incurs a cost, a cost which must ultimately be reflected in the price passed on to consumers. By the same token, of course, it also offers a benefit in the form of a degree of respite from the consequences of volatile price movements in a series of markets whose adjustment to sudden supply shifts cannot by its nature be instantaneous. By analogy, the need for a country to amass foreign exchange or gold reserves to defend a currency position depends in part on its ability to adjust relative internal prices to restore equilibrium: "In general, a given pattern of disturbances over any time period requires fewer reserves, the more rapid and complete the adjustment mechanism".[31] This gives rise to the concept of "the social productivity of reserves," which is maximised at a point where the utility of the reserves held is equal to the opportunity costs of their best alternative use.

Examples of politicised oil trade include the Iraqi requirement that the Brazilian oil authorities recognise the PLO as a condition for continuing their supply agreement (Brazil was, until the Gulf War in the Fall of 1980, obtaining 40% of its crude from Iraq). Indian supplies were threatened by Libya after the Indian government's reluctance to supply nuclear technology. Similarly the scandal in the Italian firm ENI in late 1979, prompted the termination of the company's supply contract with Saudi Arabia. A visit to Rome by a senior Saudi minister in 1980, however, led to a restoration of good relations.

This chapter has discussed the factors which may play a part in influencing the availability of crude oil from OPEC members in the 1980s, and the extent to which the growth of control over the oil output-mix on the part of the exporting countries may affect refiners. Before finishing, one other factor has to be looked at. This is the long-discusssed notion of a long term strategy, to be adopted by all 13 OPEC members, to guide the volume and mix of oil output.

During the early 1980s considerable efforts were made at arriving at a long term strategy which could guide the members' price policy in future years. In the event the deliberations of the committee were undercut by the market itself, since the crude oil price failed to continue rising. In 1983, however, at the London meeting of OPEC's ministerial monitoring committee, the rapid changes in market conditions were pointed out. It was suggested that there would be a renewed need, in view of this volatility, to draw up some plan for the organization's long term goals. It was suggested that the long term committee resume its work, making an interim report at the Helsinki conference in July.

The long term strategy committee, chaired by Sheikh Zaki Yamani and consisting of representatives from six countries (Algeria, Iran, Iraq, Kuwait, Saudi Arabia and Venezuela) may therefore once again have a voice in OPEC. The following section considers the background of and future outlook for this committee.

The OPEC Long-Term Strategy

As foregoing chapters have argued, OPEC members will, in the 1980s, continue to face criticism from spokesmen in the rest of the world for the alleged problems that higher oil prices have wrought upon the oil-importing economies. While some of these countries' disillusion can be shown to be well-founded, and reflect real economic difficulty, to an extent their disillusion is also a consequence of the grandiose claims made by OPEC for its own goals and ambitions with respect to other countries in the Third World. It is now necessary to examine an instrument which might be used for changing the basis of these hitherto rather unsatisfactory relations. That instrument is the OPEC long term strategy.

The long term strategy is intended to deal with three areas of interest to OPEC members: long term hydrocarbon pricing policy, relations with the rest of the Third World, and relations with the developed countries. The last two areas of interest came together in OPEC official pronouncements as being parts of a renewed initiative to open "global negotiations" and to promote a form of new international economic order.

The Long Term Strategy Committee of OPEC was established in May 1978 under the chairmanship of the Saudi minister of oil, Sheikh Zaki Yamani, to investigate the possibility of a scheme whereby oil prices would be adjusted in accordance with a predetermined formula. It was anticipated that the successful implementation of such a scheme would carry with it certain advantages—both political and economic in nature—which would commend the scheme to all 13 OPEC members.

Progress towards the adoption of one of the variants defined by the committee has however not been smooth. Having brought its ideas to draft level, the committee was, broadly, well-received at the 56th (extraordinary) OPEC ministerial conference held at Taif, Saudi Arabia, on 7-8 May 1980. While Algeria, Iran and Libya voice reservations, the

152

other nine members were in principle agreeable to the scheme (Nigeria was absent, so only 12 member states were represented). The main stumbling-block at this stage was the fear, on the part of the above three members, that adherence to the formula would entail forgoing extra revenue over the longrun by not sanctioning maximum price rises. There was also a feeling that the formula should reflect conditions within OPEC states—notably their inflation rates—rather than, or to a greater extent than, conditions in the developed countries. Finally, there was no agreement on the starting-off point from which the scheme should be launched. As Sheikh Yamani announced after the meeting, "Pricing was dicussed, not prices". The 57th meeting of the Conference of OPEC, held at Algiers on June 9-11 further considered the strategy. Subsequently a Tri-Ministerial Meeting of ministers of foreign affairs, petroleum and finance, gathered in Vienna on September 15-17, to discuss the outcome of the Taif meeting.

A further meeting was to be held in London in mid-October to edge closer to agreement in preparation for the heads of state meeting due to be held in Baghdad early in November. In the event, however, the outbreak of war between two OPEC members, Iran and Iraq, resulted in both the preliminary and the substantive meetings being abandoned. As 1980 closed the regular ministerial conference, held at Bali, Indonesia on 15th December, was held without prior delegate commitment, other than at the vaguest level of principle, to any longterm strategy being implemented during 1981.

Why this delay? Examination of the scheme's outlines indicates that—as might, of course, be expected—the distribution of benefits is likely to be skewed. Due to the 13 members' having—as has been stressed several times in preceding sections of this report—frequently diverging economic and political interests, the appeal of the scheme naturally differs.

The basis of the scheme is that the oil price would be adjusted to maintain a consistent relationship with three variables: the movement of a basket of exchange-rates, the change in a chosen index of prices, and the growth-rates of a selection of developed countries. By embracing these three variables it is believed that the oil exporting countries will to a greater degree than hitherto be insulated from disadvantageous changes in the value of the currency they are paid in and from inflation in the countries where they tend to buy their imports of consumer goods and equipment. The third term, GNP growth, also attempts to capture the oil consuming nations' "ability to pay".

An example of the way the index would operate has been furnished by the OPEC secretariat. Starting from 1973, their illustration points to an oil price slightly over $21/barrel on average in 1979, rather than the $24 which prevailed by year-end. Table 6.6 shows some parts of this excercise. It is interesting that the 1975 negative growth rates recorded in OECD countries has a clear effect upon the oil price according to the formula, as does the high inflation recorded in 1974 and 1975.

Table 6.6

OIL PRICE CALCULATIONS ACCORDING TO AN
OPEC LONG-TERM STRATEGY FORMULA, 1973-1979

Year	Index of oil prices	%rise	Change in inflation rate	... of which exchange-rate basket effects	GNP changes
1973	100	—			
1974	116.8	16.8	16.8	− 0.3	0.3
1975	136.5	16.9	15.1	2.6	− 1.0
1976	143.0	4.8	4.5	− 4.8	5.2
1977	164.4	15.0	8.5	2.3	3.7
1978	195.5	18.9	3.7	10.6	3.7
1979	227.8	14.5	6.8	3.5	3.5

Source: OPEC calculations, reported in *Petroleum Intelligence Weekly,* May 12, 1980

Although, given the nature of the scheme as suggested hitherto, it is impossible to predict what price pattern will evolve from adherence to it, the strategy is expected by OPEC to yield nominal oil price rises of the order of 12-15%, giving a price of $60/barrel by 1985. This rate of increase is in fact rather above that obtained in several years between 1973 and 1979.

The critical question in practice concerns the extent to which OPEC members individually or collectively will regard the price implied by the formula as a floor or ceiling, or merely indicative of a band of prices within which their own quoted prices may legitimately fluctuate. The most likely answer to this question is that those members with large reserves of crude will prefer a slow progressive rise in oil prices; those with development programmes whose financing can be accommodated within existing oil revenues will, similarly, be less anxious to secure large price rises when the formula appears to point to a price below the revenue-maximising level.

Judged on the basis of the 1979 reserves estimates accepted by OPEC, this points clearly to Saudi Arabia, Kuwait, Iran, Iraq and UAE as the countries with the greatest interest in preserving the longevity of their reserves' economic life. These are, in other words, the producers whose long-term reserves stand the greatest chance of being undermined by oil-saving changes introduced over the coming decades. It has been clear, however, that while the bulk of this sub-group is interested in having a strategy adopted, Iran is not at all keen. The three-day meeting in Vienna in 1980 was interrupted by Iranian objections to any potentially price-suppressing scheme. In part the Iranian stance also reflected a desire to disagree on policy matters with Iraq as a matter of course. But there is also a risk, in the eye of such dissenters, that adherence to the strategy must entail forgoing the full price rises that the market could sustain at times of oil shortage. In large measure this is because any Iranian government in the 1980s will be facing substantial development needs and thus outlays.

As regards their development needs, the main characteristic of OPEC members' plans since 1973 has of course been their capacity for cost overruns unimaginable at their inception. A mere comparison of published national plans does not in itself yield a useful guide to spending and thus revenue needs since these plans are so much subject to the vagaries of new administrations, the resetting of priorities and so on. But a brief comparison of latest planning intentions gives little indication of a widespread reining-in of ambitions. The Saudi Arabian third five year plan for 1980-85, implemented from May 1980, envisages the spending of $235 billion (782.8 billion riyals) with a $50 billion contingency reseve in case of unanticipated inflation. This follows the 1975/80 plan, in which cost overruns of $33 billion, or 23%, occurred. Algeria's 1980-85 plan envisages a doubling of investment outlays. Venezuela's 1981-85 Sixth National Development Plan is premised upon a 6% annual average growth of GDP, while Nigeria's 1981-85 N82 billion ($150 billion) plan inevitably entails substantial spending upon the country's relatively newly-established 19 states' apparatus (which will take 35% of all revenues in the 1981 federal budget). Import restrictions remain, in an attempt to choke off the rapid growth of consumer durables imports that characterised the early post-1973 days. The borrowing requirement to finance the plan, at N16.9 billion, is seen by many as overly optimistic.

This quick review of spending commitment would tend to suggest that among the significant OPEC oil producers, Nigeria is likely to be pushing for higher revenues in the first half of the 1980s. Iran remains an unknown, as, to a lesser extent, will Iraq be in the immediate post-war period. It is reasonable to conclude, then, that a significant share of OPEC producers will have an interest in boosting their revenues as their domestic spending exceeds immediate funding ceilings.

This tends to suggest that the price laid down by the long-term strategy formula stands a good chance of being overridden by certain of the 13 members with revenue needs.

Quite apart from the economic aspects of the scheme as outlined above, certain OPEC members believe they stand to obtain in addition benefits of a more political nature. These benefits can be summed up as coming from the depoliticisation of trade in oils. At the same time as governments—and most notably, the governments of the major oil-exporting countries—are becoming more intimately involved in the covert regulation of crude oils trade (if not yet in oil products trade) as outlined in Chapter Six, in the strategy lies an opportunity to relieve some of the tension stemming from current public interest in oil. For if the oil price can be manoeuvered into some regularised relationship with other world economic variables—variables seen by most public opinion and politicians as being of an altogether more objective and defensible nature than the oil price—then oil prices will move off centre stage. Changes in oil prices will cease to be front-page political news, reported in terms of power-struggles between OPEC members or between various cabalistic factions within OPEC countries, and will assume instead a more strictly economic hue. Clearly, the decision to implement any strategy for oil price changes, as well as the nature of any such scheme, are intrinsically political matters. But attention will be attracted away from the means by which oil prices change, and, correspondingly, from the arguments (inevitably leaked to outside commentators) mounted over the desirable level of oil prices.

Seen in this light not the least attraction, for many OPEC members, of having such a scheme in operation is the possibility of giving the appearance of bucking the tide towards what the International Institute for Strategic Studies has called the drift to "the politicisation of resource issues".

Immediately after the first oil shock of late 1973 there was a flurry of research directed to the question of the widening politicisation of the world's raw material stocks. If the widely-reported report of the Club of Rome, *The Limits To Growth*, had aroused among many fears of the frontier of physical resource exhaustibility being closer than believed, then the subsequent fears regarding the political frontier of resource availability seems to confirm that the industrialised countries needed somehow to become active participants, at government level, in securing raw material flows. The heavy concentration of many important minerals in Southern Africa had long exercised the minds of strategic planners in the West. Indeed, in 1980 the West German government actually began planning its own stockpiles of material, although the plan was dropped soon after, for budgetary reasons as much as anything else. In the

156

meantime Dr. Henry Kissinger had spoken of the imperative of oil supply security for the USA conceivably leading to an invasion of the Mid East oilfields.

Such diplomatic advantage would probably accrue disproportionately to Saudi Arabia, which is the OPEC member whose interests, arguably, are least served by constant international focussing upon the oil price. A strategy of the sort just outlined would also help OPEC members to establish better relations with most developed country governments. It is argued elsewhere in this report that the chief characteristic of OPEC's relations with the industrialized countries is ambivalence, accompanied by confusion and hesitancy. As far as security matters are concerned, there are serious doubts whether those Western powers willing to become involved in the region do not carry with them intolerable costs. In the economic sphere, developed country representatives lay much of the blame for the state of the world economy at the door of OPEC as a matter of course. Irrespective of the verity of this claim, and notwithstanding the OPEC countries' diminishing interest in the Third World, OPEC would still prefer not to have such grievances aired so insistently and volubly. By shifting the means by which oil prices appear to be set at least a portion of this pressure will be dissipated.

Considerable uncertainties still surround the strategy, however. These arise not only because the strategy committee cannot, by its nature, make recommendations binding upon member governments, but also because each member will be keen to edge towards its own most favourable starting-point, with respect to its price and output level, before entering the scheme. Certain members tacitly acknowledge that they will be looked to to make sacrifices of various sorts to help breathe life into the scheme, so that the manoeuvring into position will be as much a problem as agreement on the scheme itself.

The overwhelming reason for life being breathed into the scheme in 1983 was the possibility that oil markets might tighten a little by year-end. With economic growth, led by the US, picking up, there was at least a chance of a resurgence in demand by 1984. This was seen as particularly likely in view of the massive inventory run downs which had characterised the industry in 1981 and the next two years.

The main difference between the time when the strategy was being thought of for the first time, and 1983, however was that many non-OPEC sources of oil have in the interim become ensconced in the market-place. Non-OPEC oil exports gained ground in OECD markets even during the slack periods of 1982. Quite how this matter should be handled, and indeed whether the organisation should concern itself with such matters at all, are no doubt questions exercising the minds of the strategy members.

There is not, at least as yet, a strategy for refined product exports from OPEC. In part, this probably reflects the difficulty which is expected to confront the membership as it grapples to deal with any strategy in the crude oil arena. But it also reflects the relative unimportance of refined products in the organisation's overall export-mix to date. At the very least, recognition has already been given to the possibility that greater refined products exports, notably from Saudi Arabia, might have an impact on other members' unprocessed crude exports. A long term strategy for gasoline is thus not in sight; nor need it be.

The lengthy discussion of the oil markets just presented is important for the products markets because, of course, the factor which above all determines the course of products prices is the market for crude oil.

The thrust of the arguments just outlined is that petroleum products prices are unlikely to be pushed up significantly by crude oil price pressures. But this is by no means to say that product prices will not rise. For there are many other factors—government intervention such as in taxes, retailers' and distributors' markups, competitive forces from other fuels, and so on—which also play a part in determining prices.

30. "Living with Crude Oil Stockpiles", *OPEC Bulletin,* March 1981, pp. 47-50. Quote is from p. 50.

31. H. G. Grubel, *The International Monetary System,* London, Penguin, 1977.

Chapter Seven

DEMAND FOR PETROLEUM PRODUCTS IN THE 1980s

HAVING considered the factors which may affect the configuration of crude oil supply for the rest of the century, this chapter looks at patterns of demand likely in the future. It begins at the aggregate level of total energy demand, then looks at demand for oil-based products, and then at the factors which may affect demand for petroleum products.

Recent Demand Patterns

Looking to a long time-frame, Table 7.1 shows petroleum consumption in the major economies from 1973 to 1982. Total consumption by International Energy Agency (IEA) members, shown in the far right column, dropped after the 1973 price rise before picking up again. In 1979 oil price increases sent demand plummeting once more, however, with the 1981 total, at 31.3 million barrels/day, the lowest in the entire period shown. 1982 will also have shown a decline, although some figures are still missing at the time of writing, since the two biggest consuming countries, the US and Japan, show large consumption declines (totalling 1.045 million barrels/day).

Table 7.1

PETROLEUM CONSUMPTION FOR MAJOR NON-COMMUNIST INDUSTRIALISED COUNTRIES

	Canada	France[1]	Italy	Japan	United Kingdom	United States	West Germany	Other IEA	Total IEA
	Thousand barrels per day								
1973	1,597	2,219	1,525	5,000	1,958	17,308	2,693	4,069	34,150
1974	1,630	2,094	1,521	4,872	1,829	16,653	2,408	4,047	32,960
1975	1,595	1,925	1,468	4,568	1,833	16,322	2,319	3,905	31,810
1976	1,647	2,075	1,503	4,786	1,601	17,461	2,507	4,215	33,770
1977	1,661	1,973	1,476	5,015	1,655	18,431	2,478	4,214	34,930
1978	1,701	2,077	1,551	5,115	1,683	18,847	2,596	4,387	35,880
1979	1,766	2,107	1,607	5,173	1,680	18,513	2,664	4,487	35,900
1980	1,730	1,965	1,602	4,680	1,420	17,056	2,360	4,152	33,000
1981	1,615	1,745	1,705	4,445	1,325	16,058	2,120	4,032	31,300
1982	1,450	1,645	1,614	4,196	1,337	15,253	NA	NA	NA

Source: US Dept. of Energy
Note: [1] Not an IEA member

162

Gasoline consumption in the OECD group as a whole has been falling consistently from 1979 (see Table 7.2). The 1982 total was 6.4% down on the 1979 figure. Some countries have not, however, witnessed consistent year-on-year declines; Australia, for example, saw consumption rise in 1981 after a fall from 1979 to 1980. Others, for instance Canada, have seen little change until very recently, while still other countries, such as Japan, have seen consumption growing. Due to its huge weight in the total (62% of total consumption in 1981), what happens in the US has a heavy bearing on total OECD performance. The fact that consumption in the US fell steeply after 1979 has therefore gone a long way to offsetting increases observed in some other countries.

Table 7.2

GASOLINE CONSUMPTION BY OECD MEMBERS, 1979-82
(Thousand Metric Tons)

	1979	1980	1981	1982 [a]
Australia	11,105	10,937	11,033	10,902
Austria	2,472	2,529	2,357	2,262
Belgium	3,284	3,125	2,807	2,872
Canada	30,984	31,214	30,272	25,394
Denmark	1,752	1,555	1,449	1,396
Finland	1,308	1,345	1,346	1,276
France	17,735	16,440	16,970	15,998
Germany	23,761	25,473	23,561	23,246
Greece	1,620	1,635	1,684	1,660
Iceland	92	88	95	72
Ireland	988	1,040	1,020	966
Italy	12,407	12,500	12,357	12,046
Japan	25,466	25,479	26,051	26,110
Luxembourg	281	282	311	284
Netherlands	3,985	5,318	4,658	4,154
New Zealand	1,679	1,666	1,663	1,710
Norway	1,515	1,476	1,474	1,396
Portugal	751	751	776	742
Spain	5,869	5,930	5,549	5,028
Sweden	3,826	3,739	3,699	3,564
Switzerland	2,602	2,751	2,858	2,768
Turkey	NA	1,953	1,958	1,792
UK	18,733	19,164	18,752	18,610
USA	304,076	285,747	281,315	281,644
OECD TOTAL	476,708	462,132	454,126	446,234

[a] First half annualised
Source: OECD, *Quarterly Energy Statistics*

The extent to which this gasoline consumption was met by domestic refining output is shown in Table 7.3. There it is apparent that the 1979-82 period has seen relatively little instability, with refinery throughputs changing less than consumption for most countries. (It should be noted that figures for refining output are necessarily somewhat imprecise. Output can be measured according to observed output—the criterion used in this table—or according to the inputs entered to the process and the output they theoretically should have yielded. The OECD calculates both measures, but in many countries the discrepancies are not large enough to merit attention.)

Table 7.3

GASOLINE OUTPUT, DOMESTIC REFINERIES, BY OECD MEMBER, 1979-82
(Thousand Metric Tons)

	1979	1980	1981	1982 [a]
Australia	10,977	10,539	10,932	11,376
Austria	1,894	1,749	1,804	1,539
Belgium	5,429	5,845	5,138	4,008
Canada	31,089	31,381	30,074	25,668
Denmark	1,447	1,102	1,159	' 928
Finland	2,162	1,941	2,051	2,060
France	19,296	17,400	16,799	17,646
Germany	21,506	21,187	20,446	20,454
Greece	1,513	1,393	1,811	1,886
Iceland	0	0	0	0
Ireland	519	487	186	0
Italy	17,132	16,048	15,169	14,688
Japan	25,434	25,228	25,874	25,098
Luxembourg	0	0	0	0
Netherlands	8,007	8,110	7,079	8,152
New Zealand	1,281	1,211	1,263	1,492
Norway	1,305	1,171	1,133	1,120
Portugal	1,042	954	1,085	852
Spain	5,469	5,556	5,279	3,976
Sweden	2,464	2,723	2,368	1,858
Switzerland	986	1,096	1,103	984
Turkey	NA	2,087	2,113	2,056
UK	16,266	16,675	17,198	17,698
USA	300,936	286,078	282,047	270,938
OECD TOTAL	476,670	459,961	452,111	434,640

[a] First half annualised
Source: OECD, *Quarterly Energy Statistics*

As Table 7.4 shows, net oil imports fell quite steeply (by 14.4%) from 947.6 million tonnes of oil equivalent (mtoe) to 810.99 mtoe. Apart from the 60 mtoe fall in US imports, a sizeable fall was also recorded by Japan. Looking to Table 7.5, which shows changes in imports by source, it is clear that, in absolute terms, the biggest falls were those recorded by Kuwait, Ecuador, Saudi Arabia, Algeria and Libya. Imports from non-OPEC sources grew appreciably. The only OPEC members whose crude oil exports to the OECD bloc rose were Gabon and Iran.

Table 7.4

OIL CONSUMPTION AND IMPORTS 1981-82
million metric tons of oil equivalent (mtoe)

| | Consumption | | Net Imports | |
	1981	1982	1981	1982
Canada	82.65	68.22	8.03	− 3.37
United States	733.35	698.94	260.90	200.69
Japan	226.66	218.03	224.27	198.65
Australia	30.64	29.96	9.66	11.42
New Zealand	3.99	4.24	3.36	3.39
Austria	11.04	10.20	9.84	8.68
Belgium	20.67	20.09	18.33	20.01
Denmark	11.45	10.47	10.05	8.51
Germany	116.62	111.28	107.98	102.10
Greece	10.99	11.18	11.11	9.56
Ireland	5.19	4.17	4.96	4.04
Italy	90.19	81.58	88.20	82.46
Luxembourg	1.04	1.03	1.03	1.02
Netherlands	27.01	23.38	22.69	21.05
Norway	8.62	8.30	− 15.70	− 15.63
Portugal	8.50	9.25	8.02	9.00
Spain	45.89	45.07	46.97	41.87
Sweden	22.16	19.36	20.71	18.53
Switzerland	11.85	11.06	11.75	10.97
Turkey	15.36	16.54	13.48	13.98
United Kingdom	74.28	76.71	− 20.63	− 31.63
IEA TOTAL	1,558.15	1,479.06	844.99	715.30
Finland	11.67	10.80	11.26	10.14
France	97.42	89.50	90.73	85.00
Iceland	0.62	0.55	0.62	0.55
OECD TOTAL	1,667.86	1,579.91	947.60	810.99

Source: OECD

Table 7.5

GROSS IEA IMPORTS OF CRUDE OIL
BY ORIGIN
Million metric tons

Origin	1981	1982	Per cent change 1981 to 1982
Canada	15.2	17.5	15.1
Norway	14.3	15.5	6.9
United Kingdom	45.7	51.2	12.0
Other IEA	5.9	13.6	130.1
Algeria	28.3	20.6	− 27.4
Ecuador	3.2	2.1	− 34.9
Gabon	3.5	3.5	0.3
Indonesia	54.0	45.4	− 15.9
Iran	21.2	50.1	136.3
Iraq	23.1	22.6	− 2.0
Kuwait	20.0	8.6	− 56.9
Libya	52.6	39.1	− 25.6
Nigeria	50.5	44.3	− 12.2
Qatar	13.7	11.0	− 19.9
Saudi Arabia	276.6	189.4	− 31.5
United Arab Emirates	49.7	44.9	− 9.6
Venezuela	32.1	25.8	− 19.5
Egypt	12.0	11.3	− 6.3
Mexico	42.6	60.1	41.0
China	9.1	9.9	8.8
USSR	15.3	23.5	53.1
Other Countries	91.2	78.9	− 13.5
TOTAL	880.0	788.9	− 10.4

Source: OECD

During 1982 the US petroleum industry was dominated by continuing demand contractions. Total consumption, at 15.3 million barrels/day, was 5% down on 1981 levels and 19% on the 1973 peak level. Imports of gasoline fell by 22% on 1981 levels, to 4.2 million b/day. This was just under half the level of imports seen in 1977. Crude oil imports to refineries

166

fell from 12.5 million b/day in 1981 to 11.8 million b/day in 1982.

Domestic production of petroleum, at 8.67 million b/day, was 1.2% above its 1981 level, directly reflecting increases in yield from Alaska's Kuparuk field. The fall in crude oil prices seen in recent years had inhibited drilling for a while, but in 1982 oil well completions rose by 7% over their 1981 figure. December 1981 was the peak month in recent years for rotary drill usage.

Net petroleum imports in 1982 stood at 4.23 million b/day, down 21.7% on their 1981 level. This was the smallest level seen since 1971. In the first eleven months of 1982, US imports from OPEC fell to 42.4% of total crude oil imports, down from 55.4% for 1981 as a whole. The costs of the nation's oil imports fell by 21% to $61.2 billion.

Motor gasoline consumption in 1982 fell for the fourth consecutive year to 6.54 million b/day. Taking into account real gasoline prices and average car fleet economy, the cost of driving one mile in the US fell by no less than 14% in 1982. Partly as a consequence, distance driven is estimated to have risen by 1.4% over the year-before level. Distillate fuel oil demand, at 2.67 million b/day, was the lowest seen for 11 years. The 1982 volume was 5.7% below its 1981 level. Conservation, the installation of more efficient fuel-burning equipment and the continuing switch by consumers to wood and natural gas, all explain the contraction of demand in the face of unusually cold weather and falling prices. Residual fuel oil consumption fell by 18.7% in 1981. Jet fuel consumption was broadly unchanged at 1 million b/day.

Table 7.7 shows gasoline demand in each state in 1980 and 1981. It can be seen that the greatest demand is in California, where nearly 11 billion gallons of gasoline were sold in both years. Second in rank is Texas, where 1981 sales volume was 8.15 billion gallons.

Petroleum Products by Sector: Motor Gasoline

As Table 7.7 shows, the annual volume of fuel consumed by each car in the USA has fallen appreciably since it peaked in 1973 at 763 gallons. By 1981 the figure had fallen by 24% to 581 gallons. The fall was particularly severe after 1978. This fall is a reflection of two forces: fewer miles being driven in each car each year, and better fuel economy per mile driven. The table shows that mileage driven fell off quite markedly after 1973, but began creeping up again before 1979, in part doubtless due to erosion of real gasoline prices. After 1979 again, however, mileage fell once more; in 1981 the average distance travelled was 94.7% of the 1967 distance.

Table 7.6

GASOLINE DEMAND IN THE US, 1980-81 BY STATE
'000 gal.

	1981 Rank	1981	1980	% Chng. '81 vs '80
Alabama............	23	1,849,225	1,961,902	−5.7%
Alaska..............	50	212,804	201,373	5.7
Arizona.............	28	1,344,438	1,328,722	1.2
Arkansas...........	33	1,137,998	1,178,800	−3.5
California...........	1	10,842,862	10,992,050	−1.4
Colorado...........	26	1,503,558	1,506,415	−0.2
Connecticut.........	29	1,296,048	1,327,582	−2.4
Delaware...........	48	290,581	293,851	−1.1
District of Columbia...	51	177,658	171,451	0.7
Florida.............	4	4,845,057	4,810,520	0.7
Georgia............	11	2,809,969	2,874,923	−2.3
Hawaii.............	47	347,848	354,529	−1.9
Idaho..............	40	471,054	488,333	−3.5
Illinois.............	6	4,598,012	4,816,780	−4.5
Indiana............	12	2,605,584	2,686,146	−3.0
Iowa...............	27	1,474,733	1,561,192	−5.5
Kansas............	31	1,268,706	1,311,181	−3.2
Kentucky...........	24	1,729,135	1,755,397	−1.5
Louisiana..........	18	2,006,034	2,081,328	−3.6
Maine.............	39	497,749	517,014	−3.7
Maryland...........	21	1,914,423	1,941,209	−1.1
Massachusetts.......	16	2,249,956	2,301,675	−2.2
Michigan...........	8	4,014,353	4,274,036	−6.1
Minnesota..........	19	1,984,982	2,045,270	−2.9
Mississippi.........	32	1,208,334	1,194,645	1.1
Missouri...........	14	2,528,616	2,602,627	−2.8
Montana...........	41	445,641	459,950	−3.1
Nebraska..........	35	798,018	847,341	−5.8
Nevada............	38	508,535	500,286	1.6
New Hampshire......	46	358,962	411,214	−3.0
New Jersey.........	9	3,150,798	3,260,992	−3.4
New Mexico.........	37	742,486	743,334	−0.1
New York..........	3	5,674,946	5,668,563	0.1
North Carolina......	10	2,874,700	2,932,274	−2.0
North Dakota.......	43	408,449	407,250	0.3
Ohio...............	5	4,766,658	4,982,574	−4.3
Oklahoma..........	20	1,922,403	1,845,259	4.2
Oregon............	30	1,271,001	1,330,612	−4.5
Pennsylvania........	7	4,471,326	4,735,849	−5.6
Rhode Island........	45	368,524	381,826	−3.5
South Carolina......	25	1,529,196	1,562,400	−2.1
South Dakota........	42	411,024	423,517	−2.9
Tennessee...........	15	2,388,466	2,417,939	−1.2
Texas..............	2	8,149,406	8,106,499	0.5
Utah...............	36	768,450	734,992	4.6
Vermont...........	49	234,114	238,842	−2.0
Virginia............	13	2,530,673	2,599,199	−2.6
Washington.........	22	1,872,364	1,882,513	−0.5
West Virginia........	34	804,632	845,242	−4.6
Wisconsin..........	17	2,067,800	2,177,363	−5.0
Wyoming...........	44	375,068	373,723	0.3
US Total...........		104,143,151	106,448,704	−2.2%

Source: National Petroleum News, May 1982

168

Table 7.7

US PASSENGER CAR EFFICIENCY

	Average Fuel Consumed per Car		Average Miles Travelled per Car		Average Miles Travelled per Gallon of Fuel Consumed	
	Gallons	Index	Miles	Index	Miles	Index
1967	684	100.0	9,531	100.0	13.93	100.0
1968	698	102.0	9,627	101.0	13.79	99.0
1969	718	105.0	9,782	102.6	13.63	97.8
1970	735	107.5	9,978	104.7	13.57	97.4
1971	746	109.1	10,121	106.2	13.57	97.4
1972	755	110.4	10,184	106.9	13.49	96.8
1973	763	111.5	9,992	104.8	13.10	94.0
1974	704	102.9	9,448	99.1	13.43	96.4
1975	712	104.1	9,634	101.1	13.53	97.1
1976	711	103.9	9,763	102.4	13.72	98.5
1977	706	103.2	9,839	103.2	13.94	100.1
1978	715	104.5	10,046	105.4	14.06	100.9
1979	664	97.1	9,485	99.5	14.29	102.6
1980	603	88.2	9,135	95.8	15.15	108.8
1981	581	84.9	9,026	94.7	15.54	111.6

Source: US Dept. of Energy

Average fuel economy has risen by 11.6% between 1967 and 1981, after reaching a low point in 1973.

As in any other area of economic activity, in the transport sector there is a normal reaction of energy volumes to price change. In the US in particular, where in 1982 some 125 million cars were on the road, the slippage in gasoline prices throughout much of 1982 and some of 1983 sent consumers—many of whom had never been fully weaned from large cars after the 1973 oil price shock—back to full-sized cars. In the first four months of the 1983 model year, for instance, gasoline-fuelled V8 engines were installed in 35% of the US-made cars sold. In the same period a year before, the proportion was only 21%. GM's sale of the fast Chevrolet Camaro were strongly affected by the shift in taste. Whereas in the first four months of the 1982 model year only 24% of the cars were ordered with V8 engines, a year later 55% were so equipped.

Industry-wide figures, as shown in Table 7.8, indicate that for much of the 1960s the 4-cylinder car was virtually unknown in the United States.

The V8 form was dominant. Even after the first scare over gasoline prices and supply, in 1973-74, demand for V8 cars picked up quite briskly, although by then it is apparent that 6-cylinder engines were also of increasing popularity. 1980 and 1981 witnessed a material decline in the V8 engine's supremacy, however, with 1980 marking the first year in decades when the V8 was not the most popular engine.

Table 7.8

US CAR PRODUCTION BY TYPE OF VEHICLE
Unit: '000 Cars

	4-cylinder	6-cylinder	V-8
1964	1.0	2,454.0	5,436.0
1968	1.0	1,150.0	7,248.0
1973	887.9	1,023.7	8,398.7
1978	939.6	2,211.5	6,102.6
1980	2,308.6	2,682.6	2,358.5
1981	2,827.6	2,358.0	1,909.2

Source: Ward's Auto World

The difference in fuel economy experienced with various engines is significant. A 4-cylinder engine in a Chevrolet Camaro yields 27 miles per gallon with highway driving. A 6-cylinder engine yields 23 mpg and a V8 engine delivers only 19 mpg. Although changes in engines' design have led to big improvements in the economy of V8 engines over the years (perhaps averaging a 20% improvement between 1978 and 1982) there is no escaping the implication that a resurgence in the V8 engine's popularity is going to have a deleterious effect on oil demand in the US transport sector.

During the first half of 1983 US manufacturers adjusted factory output to make proportionately more large-engined cars available. In late April several week-long plant shutdowns were ordered by the main producers. Diagram 7.1 shows that small cars' share in total US sales plunged from over 50% in 1981 to under 45% in 1983. At the same time large car sales built up from a low of 20% of sales in 1981.

Diesel engines had once been expected to help both producers and consumers attain greater fuel economy. A small but growing proportion of both 4 and 8-cylinder engines made in the USA were diesels. In 1981 total US production of diesel-engined cars was 441,591; this was equivalent to 6.2% of car production in that year.

170

Diagram 7.1

TRENDS IN CAR SALES BY SIZE

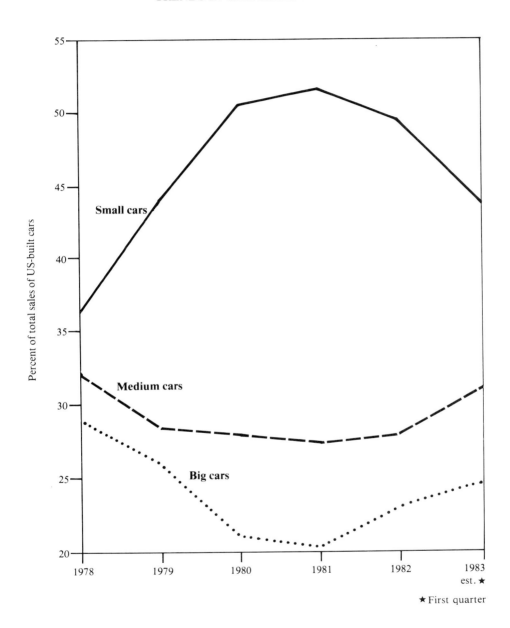

★ First quarter

Source: Business Week

This market share was still well below that obtained in European countries. In 1981 diesel engines took around 10% of total car sales; in West Germany 8%; in Belgium, 12% and in Sweden 8%. Table 7.9 shows diesel-powered car registrations in the US between 1970 and 1980, while table 7.10 shows new diesel car sales in 1981 and 1980 by manufacturer.

Table 7.9

US DIESEL NEW CAR REGISTRATIONS

Year	Diesel	% Total
1970	5,632	0.06
1971	6,285	0.06
1972	5,443	0.05
1973	6,987	0.06
1974	18,073	0.20
1975	27,018	0.30
1976	22,735	0.20
1977	37,498	0.35
1978	114,880	1.10
1979	271,052	2.60
1980	387,048	4.30

Source: Ward's Auto World

Table 7.10

US DEALER NEW DIESEL CAR SALES

	1980			1981		
	Total New Car Sales	Diesel Sales	% Diesels	Total New Car Sales	Diesel Sales	% Diesels
Oldsmobile.....	820,681	126,853	15.5%	848,739	155,859	18.4%
Volkswagen	268,064	113,943	42.5	244,190	118,674	48.6
Mercedes-Benz..	49,636	39,263	79.1	58,016	49,362	85.1
Buick..........	720,368	17,445	2.4	722,617	48,627	6.7
Cadillac	23,002	40,074	18.8	230,665	40,241	17.4
Pontiac........	614,897	19,464	3.2	552,394	34,117	6.2
Chevrolet	1,747,534	17,650	1.0	1,442,281	31,582	2.2
Peugeot........	12,807	10,226	79.8	16,725	14,351	85.8
Isuzu..........	—	—	—	17,513	10,368	59.2
Nissan.........	—	—	—	464,806	9,173	2.0
Audi	42,492	8,038	19.0	50,823	6,436	12.7
Volvo	56,339	4,147	7.4	64,103	1,998	3.1
Total sales of Makes Offering Diesels........	4,545,820	397,103	8.7	4,712,872	520,788	11.1
Total GM Lines	4,116,482	221,486	5.4	3,796,696	310,426	8.2
% Diesel of All New. Cars.......			4.4			6.1

Source: Ward's Auto World

The main problem these volatile demand shifts cause for producers—apart from great difficulties in plant scheduling, of course—is that the government's Corporate Average Fuel Economy (CAFE) rules must still be met. These rules provide for all producers' new cars averaging 26 mpg in 1983, as against 28 mpg in 1982. By 1985 all US producers must sell fleets of models whose weighted average fuel economy is at least 27.5 mpg. GM and Ford still have some way to go to meeting that requirement, whereas Chrysler and AMC are on track to achieve the limits. Unless demand for smaller cars undergoes a resurgence before 1985, then, the two biggest producers may face a difficulty.

In terms of the progress of domestic and foreign manufacturers in improving economy, the imported cars' economy is, on average, still far ahead of domestic producers'.

The US Department of Energy has carried out simulations to forecast the saving of crude oil and refined products which could result, over the course of the rest of the century, from more fuel-efficient cars being used, and from gasoline engines being replaced by diesel and engines using broadcut fuel (fuel which has neither octane nor cetane quality requirements). Table 7.11 shows a matrix of possibilities with engine types for the car population as a whole (the population is changing more slowly, of course, since the life of a car in the US on the average, in recent years, has been 7 years). Each combination and date yields a different fleet (or car population) fuel figure, the lowest being 15.3 mpg (which comes about in 1985 with no extra shift away from gasoline engines) and the highest being 24.5 mpg (achieved by having no new gasoline engines built by the year 2000). Table 7.12 then shows the savings, expressed in constant 1980 dollars, in terms of petroleum raw materials and their costs. The greatest possible raw material saving would be 370 million barrels of crude per year by the year 2000, worth $15 billion. Savings on refineries, arising from fewer being needed to produce gasoline, are estimated in the Department of Energy study to be of the order of $8 billion in 1980 dollars, if a high broadcut mix were assumed.

Obviously, a major determinant of gasoline demand for use in cars is its retail price. Average retail prices for gasoline since 1974 are shown in Table 7.13. Not until 1980 did the average price exceed $1/gallon. Thereafter it climbed to a peak close to $1.40/gallon in early 1981 before gradually sliding down again. For the US as a whole the average fell to only $1.13 in the first quarter of 1983, although in some states, such as California, by late 1982 and again in late 1983 gasoline was available at self-service stations for as little as 95 cents/gallon. The 5 cents/gallon federal tax increase imposed in April 1983 then boosted prices back to the $1.20/gallon range.

Diagram 7.2

PASSENGER CAR FUEL ECONOMY

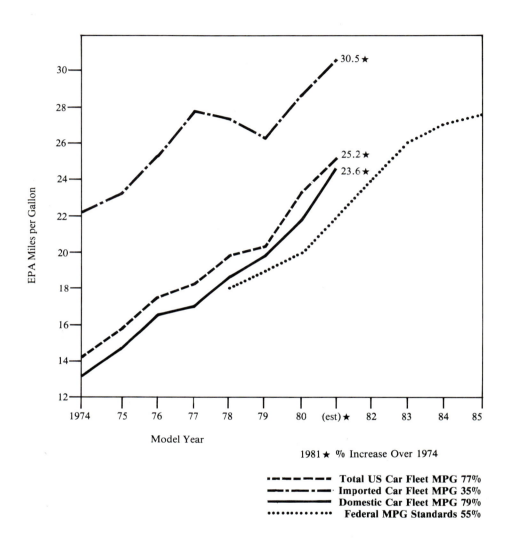

Model Year

1981 ★ % Increase Over 1974

▪ ▬ ▬ ▬ ▬ ▬	**Total US Car Fleet MPG 77%**
▪ ▬ ▪ ▬ ▪ ▬	**Imported Car Fleet MPG 35%**
▬▬▬▬▬▬	**Domestic Car Fleet MPG 79%**
••••••••••••••	**Federal MPG Standards 55%**

Source: Motor Vehicle Manufacturers Association

174

Table 7.11

VEHICLE PRODUCTION AND POPULATIONS BY FUEL TYPE

Case, Year	Manufacture of New Passenger Cars (% By Fuel Type)			Populations of Passenger Car Fleet % By Fuel Type			Average Fleet MPG*
	Gasoline	Diesel	Broad-Cut	Gasoline	Diesel	Broad-Cut	
Baseline							
2000	90.6	9.4	0.0	90.9	9.1	0.0	19.9
1995	90.6	9.4	0.0	91.4	8.6	0.0	19.2
1985	90.6	9.4	0.0	96.3	3.7	0.0	15.3
Least Cost							
2000	0.0	38.4	61.6	23.0	35.0	42.0	24.5
1995	4.2	73.7	22.0	39.0	52.7	8.3	22.7
High Diesel							
2000	30.8	69.2	0.0	37.5	62.5	0.0	23.3
1995	15.0	85.0	0.0	39.8	60.2	0.0	22.6
1985	64.1	35.9	0.0	91.9	8.1	0.0	15.5
Standardized							
2000	18.1	45.6	36.2	33.8	41.5	24.7	23.6
1995	24.0	45.6	30.4	55.1	33.5	11.4	21.5
1985	80.0	20.0	0.0	94.5	5.5	0.0	15.4
Limited Broadcut							
2000	6.3	62.9	30.8	22.2	56.8	21.0	24.5
Limited Diesel							
2000	0.0	23.9	76.1	26.0	22.1	51.9	24.2
Less Efficient Diesel							
2000	0.0	9.4	90.6	29.1	9.1	61.8	24.0
Less Efficient Broadcut							
2000	0.0	63.8	36.2	17.7	57.6	24.7	21.7
High Broadcut							
2000	0.0	9.4	90.6	29.1	9.1	61.8	24.0
1995	14.6	9.4	76.0	62.8	8.6	28.6	21.1
Restricted High Diesel							
2000	43.5	56.5	0.0	48.9	51.1	0.0	22.5

* For passenger car and light truck fleet combined.
Source: US Dept. of Energy

Table 7.12

POTENTIAL PETROLEUM SAVINGS DUE TO ALTERNATIVE AUTOMOTIVE FUEL DEMAND MIXES FOR THE YEAR 2000

		Estimated Savings	
	Miles/Gallon	Barrels	Dollars
Least Cost	24.5	370	15,000
Standardized	23.6	300	13,000
High Diesel	23.3	240	10,000
Limited Diesel	24.2	360	15,000
High Broad-Cut	24.0	340	15,000
Limited Broad-Cut	24.5	320	13,000
Less Efficient Broad-Cut	21.7	240	10,000
Restricted High Diesel	22.5	180	8,000

Source: US Dept. of Energy

Table 7.13

US CITY AVERAGE RETAIL PRICES FOR MOTOR GASOLINE

	Leaded Regular	Unleaded Regular	Leaded Premium	Average for All Types
		Cents per gallon, incl. tax		
1974	53.2	NA	56.9	NA
1975	56.7	NA	60.9	NA
1976	59.0	61.4	63.6	NA
1977	62.2	65.6	67.4	NA
1978	62.6	67.0	69.4	65.2
1979	85.7	90.3	92.2	88.2
1980	119.1	124.5	128.1	122.1
1981	131.1	137.8	143.9	135.3
1982	122.2	129.6	141.7	128.1
1983 March	106.4	115.1	127.4	113.5

Source: US Dept. of Energy

The spread between leaded regular, unleaded regular, and leaded premium grades has widened in absolute terms, from a few cents per gallon in the mid-1970s to nearly 20 cents for 1982 on average.

It is interesting to compare the cost to end-users of motor gasoline over the 1977-82 period with that of other fuels. Table 7.14 shows how gasoline prices have changed vis à vis prices for residential heating oils, residential natural gas and electricity for households. It can be seen that there has been relatively little change in any of these fuels, although the price of residential electricity jumped upwards somewhat during 1982. Motor gasoline prices, measured on a dollars per million BTU basis, rose from 3.21 to 4.84 between 1978 and 1980, as a consequence of the big crude oil price rises of early 1979, but 1982 saw some decline.

The gasoline market has evolved over the 1970s in more ways than just in volume and price—very significant though those changes have been. There has also been an important shift in the composition of demand, towards regular unleaded and away from premium leaded and regular leaded. Diagram 7.3 shows these changes in the US market over the 1975-82 period. The biggest underlying change in the composition of demand has been in the collapse of premium brand sales, from some 40% of the total in 1971 to only 2% in 1981 and perhaps 1% by 1983. This fall has mirrored the fall in the number of cars in the vehicle fleet for which premium fuel is recommended, from 30% in 1971 to around 3% in 1983.

177

Table 7.14

COST OF FUELS TO END USERS IN CONSTANT (1972) DOLLARS

	Leaded Regular Motor Gasoline cent/gal	Residential Heating Oil cent/gal	Residential Natural Gas cent/mn cu ft	Residential Electricity cent/kWh
1973	NA	NA	121.2	2.30
1974	45.1	29.4	121.4	2.63
1975	44.1	29.3	132.8	2.73
1976	43.4	29.8	145.4	2.74
1977	42.9	31.8	162.2	2.80
1978	40.1	31.7	164.4	2.76
1979	49.4	37.8	171.5	2.67
1980	60.5	49.7	186.9	2.72
1981	60.4	55.7	209.7	2.85
1982	53.0	51.4	239.7	2.97

Source: US Dept. of Energy

Compression ratios, which are a determinant of this have edged down for the US car fleet on average, from 9.4 in 1965 to 8.5 in 1983.

Data on the relative profitability of premium brands is difficult to obtain, but it may well be the case that there is a lower degree of price-elasticity (which, from a marketing standpoint, means consumer resistance to price increases) for premium grades, given that they are already, by definition, more expensive than the other grades. It may be that the basis of competition between gasoline retailers is largely price and location for regular gasoline grades but non-price considerations for premium grades. Such considerations could include perceived quality of the gasoline vis à vis other brands, or some other factors. To the extent that there does exist such a dichotomy of competitive bases, retailers will presumably be acting upon it and attempting to use what little remains of the premium market to extract proportionately higher per unit profits there. As the volume of the market shrinks, on the other hand, premium grades may eventually become a liability for certain types of retailers, particularly the lower-volume ones. This is because the cost of trading a particular type of gasoline (in terms of reordering, handling and inventory cost, plus the cost of forgone sales of regular grades arising from forecourt space being taken up by premium pumps) may become unreasonably high relative to the revenue generated. At the same time, as earlier comments have pointed out, selling gasoline is not the sole source of revenue for the typical retailer in the business today, and—depending

Diagram 7.3

HOW GASOLINE MARKET HAS CHANGED, 1975-1982

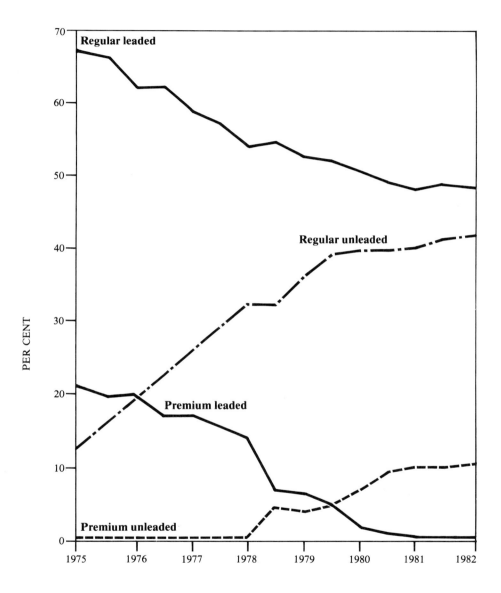

Source: Lundberg Letter, December 1982

179

on circumstances—there may be a case for stocking small-volume brands if they add (or are thought to add) to non-gasoline, associated sales such as through attached convenience stores.

Other Motor Fuel Products

Gasahol sales have been rising strongly in the US market in recent years, although even after a massive 165% rise in sales between 1981 and 1982—to 2.2 billion gallons—it only accounted for 2.5% of the gasoline market. What factors have accounted for interest in this product?

The overwhelming attraction of gasahol (which is a mixture of 90% gasoline and 10% ethyl alcohol or ethanol) is its price. Exemption from the usual state tax on gasoline allows the product to be sold for rather less than gasoline, although in states where exemption is not offered (such as Oregon) the product still sells. Local taxes are thus a major factor shaping the regional pattern of demand for the fuel. In California a 4 cents/gallon exemption has spurred sales to 4.2% of the state's 7.6 billion gallon annual gasoline market. In Nebraska, during the first nine months of 1982 gasahol took no less than 9.3% of the gasoline market. In Oklahoma it took 6.1% of sales; Michigan, 3.8% and Iowa, 3.5%.

As Table 7.15 shows, gasahol consumption by state in 1982 varied very widely, with Iowa selling most (close to 500 million gallons) closely followed by California. Other states, such as Utah, recorded no sales whatever. This does not, however, mean that there were no sales; merely that in some states gasahol and gasoline are taxed in the same manner and their sales are not recorded separately. This tax exemption will continue until 1992, so that in some localities at least there is a chance of the fuel becoming a significant if still minor element in the demand mix. By the end of 1983 capacity was in place to produce 350 million gallons of gasahol annually.

A factor which might undermine this growing interest is the fall in gasoline prices seen in 1982 and 1983, for ethanol becomes less economic as gasoline prices fall.

A different alternative fuel which is now being promoted by a major car manufacturer—Ford Motor Co.—is *propane*. In 1982-83 Ford built about 5,000 Granada and Mercury Cougar models fitted to take propane fuel. It is estimated that up to 350,000 vehicles were converted privately—at a cost which averages about $1,500—during 1981. This is in addition to the substantial fleet of farm vehicles and long-distance trucks already using propane. The fact that it retails for 55 cents/gallon makes it slightly under

180

Table 7.15

GASAHOL CONSUMPTION BY STATES—1982
Unit: '000 Gallons
Total

Alabama........	—
Alaska.........	—
Arizona........	5,096
Arkansas........	8,462
California.......	464,004
Colorado	27,231
Connecticut	4,461
Delaware........	—
Dis. of Col......	9
Florida	103,053
Georgia.........	148
Hawaii	368
Idaho...........	1,989
Illinois..........	—
Indiana	—
Iowa	499,502
Kansas..........	6,094
Kentucky	18,872
Louisiana	—
Maine	—
Maryland	107
Massachusetts ...	290
Michigan........	206,794
Minnesota.......	4,653
Mississippi	—
Missouri.	—
Montana........	10,170
Nebraska	89,698
Nevada	964
New Hampshire..	—
New Jersey......	—
New Mexico.....	186
New York.......	—
North Carolina..	7,456
North Dakota...	6,499
Ohio	64,927
Oklahoma.......	155,053
Oregon	—

...continued

Table **7.15** contd.

Total

	Total
Pennsylvania	—
Rhode Island	22
South Carolina . . .	59,688
South Dakota	13,991
Tennessee	—
Texas	38,142
Utah	—
Vermont	—
Virginia	30,834
Washington	7,230
West Virginia	—
Wisconsin	—
Wyoming	248
T o t a l	1,836,241

Source: Federal Highway Administration, Monthly Gasoline Reports, 1982.

half the price of gasoline—an advantage which will attract considerable attention in future years. In Europe there is already a considerable fleet of propane-fuelled cars: 750,000 are estimated to exist in Italy and 350,000 in the Netherlands.

The fact that storage in service stations has to be handled rather more carefully than for gasoline, since propane is kept pressurized, may slow down the fuel's acceptance by retailers and thus motorists' willingness to entrust themselves to the distribution network, but in time this difficulty is expected to be overcome.

Industrial Gas

Gas used in industry in the US fell at an annual average rate of 1.1% between 1978 and 1981, but gas nonetheless increased its market share because other fuels' demand fell faster. There was some fall due to lower industrial production, of course—US industrial output in 1982 was below its average 1981 level—but some ground was also lost to residual fuel oil. It is estimated by the American Gas Association that conservation measures led to a 3.6% fall in demand for gas in American industry, other things being equal. The effect of this can be seen in diagram 7.4, where actual gas demand from 1978 to 1982 is shown. A range of three forecasts suggests that gas demand may not regain its 1978 level until 1987.

Despite a number of cancellations and deterrents, such as the Alaskan Highway Gas Pipeline Project, 1983 and the following few years are expected to see substantial growth in the world's network of pipelines for carrying crude oil, refined products, natural gas, natural gas liquids (NGL), slurry and so on. Plans at mid-1983 called for 92,400 miles of pipe to be laid around the world, of which 21,600 miles are to be in the Communist bloc and China. For 1983, world-wide expansion to pipeline networks is forecast at 18,760 miles, costing $8.95 billion. Of this total, 8,710 miles in the US will cost $2.33 billion and 10,050 in the non-Communist bloc altogether will cost $6.62 billion. Among the more noteworthy projects going ahead around the world are the North Sea's 527 mile gas system called Statpipe; the Argentina-Uruguay pipeline, 171 miles long, from Gas del Estado; four major Sonatrach pipelines in Algeria, and others in Denmark, Scotland, Egypt, Czechoslovakia, Australia and elsewhere.

Gas distribution in the US is facilitated by a million mile distribution network, falling gas prices (in 1983 at least) and technical changes, such as the use of plastic pipe, which make extensions to the distribution systems easier than before. Nonetheless substantial expenditures must be made to maintain the system: in 1983 nearly $0.9 billion will be spent to keep the 856,000 miles of main gasoline and 550,000 miles of gas service lines in good condition. This maintenance cost is equivalent to around 34% of all capital expenditure by the industry. Total expenditure in 1983 is forecast at $2.53 billion, up by 7.4% over 1982 levels.

Residual Fuel Oil

Demand for resid in the USA peaked in 1977 at 3.07 million b/day and since then it has dropped by nearly 50% to an estimated 1.55 million b/day in 1983. The chief factors at work depressing demand are those of substitution, away from resid and towards coal, nuclear and hydro-power sources in the electric utility industry. Utilities' total demand for resid in the US has collapsed from 1.066 million b/day in 1980 to about 550,000 b/day in 1983—a 48% fall. In industry too demand has trailed off sharply, with 610,000 b/day consumed in 1980 but only 420,000 b/day in 1983. Few analysts expect to see any change in this process, and expect to see investment made to have more light products yielded in refineries, further contracting the long-term supply of resid.

After such a long period of contracting demand, the fuel oil industry is making an effort to stem the drift away to other fuels—primarily this

Diagram 7.4

ACTUAL AND PROJECTED GAS DEMAND, USA, 1978-87

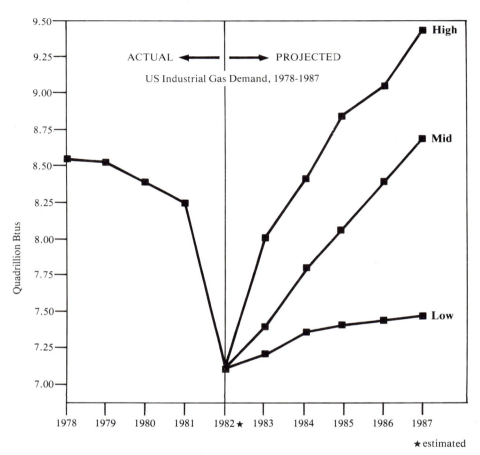

★ estimated

High projection assumes gas use per unit of output will return to 1981 levels in 1983 and beyond. Mid case assumes gas use per unit of output remains unchanged after 1982. Low case assumes that 1982 gas use per unit out output declines at an annual rate of 3% during 1983-87 due to fuel switching and conservation.

Source: May 1983, Pipeline & Gas Journal

184

means gas and electricity—and to start convincing people that there are good reasons for selecting fuel oil. A lot of the renewed push, particularly in the USA, is going to come through a vigorous marketing campaign. The essence of the campaign is targeting and service. For one thing, recent investigations have revealed that in the US oil heating market the main customer segment is now the 25 to 44 age group rather than the over-44 group as had once been thought. Also, a target market is being made of those customers who do not have flame retention burners, which only about one in five homes currently possesses. The servicing component of the effort will focus on the following areas:

—offer furnace or burner servicing.

—develop ties with local builders, to try to have oil heating installed in houses as they are built. In 1983 only 3% of new houses had oil fired heating installed, as against 50% for electricity and 40% for gas.

—strengthen the delivery of kerosene for those with burners, who have been drifting away to gas appliances.

—be ready to capitalise upon any technological change, such as the anticipated oil-fired furnace, expected to have an efficiency rate of 95-97%.

—be more active in promoting the sale of oil furnaces to take market share back from gas. In 1983 600,000 gas furnaces were shipped by manufacturers, as against a mere 34,000 oil furnaces.

—the greatest fall in demand for fuel oil was suffered in the electrical utilities segment. There is little that can be done about that, but the residential segment (which accounted for about 18.5% of total fuel oil demand in 1982) promises some opportunity.

Naphtha Jet Fuel

Demand for this largely comes from the military. In the first half of 1983 demand averaged 220,000 b/day, or 6.8% more than a year earlier. For 1983 overall, total demand is expected to be of the order of 210-220,000 b/day, a 3.4% increase over 1982.

Kerosene Jet Fuel

This is used primarily by commercial airlines. Demand averaged 800,000 b/day in the first half of 1983. For the year as a whole it is forecast to average around 815,000 b/day, which would be a slight rise on the 1982 level of 801,000 b/day. The prospects for long-term demand here are affected by:

—growing fuel efficiency of modern aircraft.
—smaller aircraft being substituted for larger ones on little-used routes.
—improved scheduling by airlines, to increase capacity utilization and cut fuel use per thousand passenger kilometres travelled.

The Outlook for Demand

This section contains some forecasts for the pattern of demand in the next year or so. The factors felt to have been important in shaping demand over the course of 1982 and 1983 are discussed to present some idea of the likely course of the market for refined products during 1984 and 1985. As far as US forecasts are concerned, they draw on the mid-1983 forecast exercise carried out by the Department of Energy. Certain amendments to these figures have to be made, in the light of developments in the US economy which could not be fully anticipated before the summer.

The Department of Energy foresaw retail motor gasoline prices falling by 6% between 1982 and 1983. Retail heating oil prices were expected to fall by 2%, while wholesale heating oil prices were forecast to register a 5% decline. Retail residual fuel oil prices would probably fall by about 8%, in part due to refiners' acquisition costs falling. Among the other major forces expected to be at work during 1983 was a continuation in the big draw-down of stocks held throughout the distribution system. High interest rates, weak profits, weak product prices and (in early 1983, at any rate) the possibility of further falls in posted crude oil prices all made holding inventories of oil products very unattractive. The unusually low inventory build up in the second half of 1982 (stocks are normally accumulated gradually before winter demand sets in) carried over in 1983, so that the industry began and continues the year with little stock held.

This level of demand, coupled with a likely level of domestic oil output of 9 million b/day, suggests that net oil imports might be in the region of 4.7 million b/day. Forecasts for fuels are as follows: motor gasoline demand is expected to fall by 2.8% to 6.36 million b/day; distillate fuel oil use is expected to rise by 3% to 2.75 million b/day (due to an anticipated 6% price fall); residual fuel oil demand is expected to grow by 1.8% and jet fuel demand is expected to remain unchanged at its 1982 level.

These figures are, however, likely to understate somewhat the actual change in demand during 1983, because of the unexpectedly brisk recovery of the US economy from the recession. The US Department of Energy was forecasting a year-on-year GDP change of only 1.7%, and a first half of 1983 to first half of 1984 change of 4.3%, when it put together its 1983 fuel use forecasts. By late 1983, however, it became apparent that the economy would expand faster than this. Indeed, in the second quarter GNP grew at an annualized rate of 9.4%, considerably in excess of most earlier forecasts.

Forecasts of gasoline demand are to a substantial extent dependent upon forecasts of demand for the car population and car usage. Forecasting the level of car sales from year to year is an extremely difficult task—and one with which the main car manufacturers have had only mixed success. However, the number of cars already in existence fluctuates less from year to year than does the level of new sales, and forecasts of the number of vehicles in operation (the "parc", as it is known) can thus be made more confidently. The number of privately owned cars in the US has grown at a fairly steady rate since 1950. In 1981 nearly 125 million cars were in use in the USA, and, if trends established over the last three decades continue, by 1990 some 150 million cars would be in use. Furthermore, around 40 million trucks and 600,000 buses would be in use if their trend rates of growth were sustained to 1990.

A number of forces are likely to be at work in the market for *resid*. First, the fact that overall demand for crude oil has fallen has prompted exporting countries to sell the highest-priced crude varieties they had available. For the most part this has meant that light crudes have become relatively more abundant on world markets than have heavier crudes. Second, investments in facilities to convert resid into lighter product means that demand for resid is higher than would otherwise have been the case. Finally, however, demand for heavy fuel is ebbing away to coal and natural gas—and, in some areas, to nuclear fuel, so that over the long run demand for it is unlikely to grow strongly. A study from Petroleum Industry Research Associates published in 1983 suggests that, taking into account all three factors, demand for resid is most likely to grow after 1985, and before then will only experience a partial bounce-back to its pre-1979 level. Diagram 7.5 summarizes these trends.

The *home heating oil* market in the US has been altered by two forces. First, the volume of oil consumed in an average home has fallen steeply. Fuel Oil News estimates that in 1973 oil consumption per household was 1,750 gallons annually. By 1981 that figure had dropped to 941 gallons annually. This fall reflected the usual factors at work in the 1980s; the shift to other fuels, and the greater use of insulation, thermostats set lower and simple economy of use.

Diagram 7.5

HOW DEMAND FOR RESID WILL CHANGE

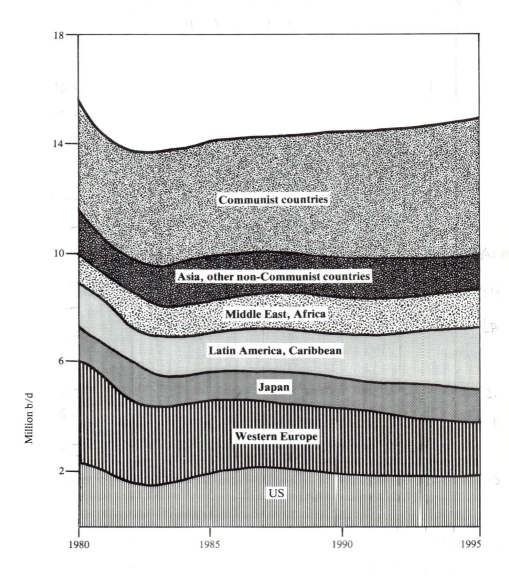

Source: Petroleum Industry Research Associates Inc. and Chem Systems Inc.

For distributors, this means that twice as many customers need to be supplied to maintain the same volume as in the early 1970s. Moreover, gross profit margins in the business are said to have fallen, in the case of dealers in the North-east of the USA, from 40 cents per gallon in 1973 to 23 cents in 1983. Exacerbating this problem has been the growing tendency for dometic oil consumers to form or join buying cooperatives, which give greater bargaining strength to purchasers. In New York in mid-1983 the landed or harbour price of heating oil was 86 cents/gallon. Cooperatives pay, on average, 15 cents/gallon more than this. Normal retail sales would have the customers paying $1.10-$1.15/gallon.

Airlines' use of *aviation fuel* (20 billion gallons in 1982, at an average cost of close to $1/gallon) is expected to rise once again as the recession ends. 1983 passenger route miles may rise by 6 or 8% over their 1982 levels, in the opinion of industry specialists. But it is not clear whether the 6.2% annual average increase seen between 1974 and 1982 can be sustained throughout the 1980s; demand, both by passengers and by firms moving freight, is a function of many variables of which fuel prices, and their impact upon our travel costs is merely one. Forecasts by Airbus Industries point to the number of billion passenger kilometres travelled on scheduled air services rising from 928 in 1980 to 1,561 in 1990 and 2,802 in 2000. The greater part of the growth would be from developing country travel and the growth of travel in Asia. Continuing efforts at cutting fuel use by aircraft will lead to only a minor pickup in jet-fuel use, however. In the US market, for instance (which was 48% of the world market in 1980) over the 1973-86 period passenger miles flown may double, but fuel use should grow by only 6%, to 870,000 barrels/day. The new series of Boeing 747 jets give 10% better fuel economy than the late 1970s models, which were in turn an improvement on earlier versions. Diagram 7.6 shows revenue passenger miles per gallon of fuel use as forecast to 1986 by Data Resources Inc.

Fuel Use in Electrical Utilities

As Table 7.16 shows, oil inputs to the electrical utility sector in the USA have declined from a peak of 636 million barrels in 1978 to only 249.7 million barrels in 1982. This extremely large fall took place while the output of the utilities sector as a whole grew from 1,922.6 billion kWh in 1978 to 1,932.5 billion kWh in 1982. Peak output was in 1981, when it stood at 2,034.1 billion kWh. This increased energy output was largely made possible by higher coal inputs and, until 1982 at least, by higher volumes of gas input too. Coal demand from utilities rose from 351 mn tonnes in 1972 to 482 mn tonnes in 1978 then on to 594 mn tonnes in 1982—a rise since 1978 of 24% to be compared with the fall in oil demand

Diagram 7.6

PASSENGER REVENUE MILES PER GALLON 1973-1986

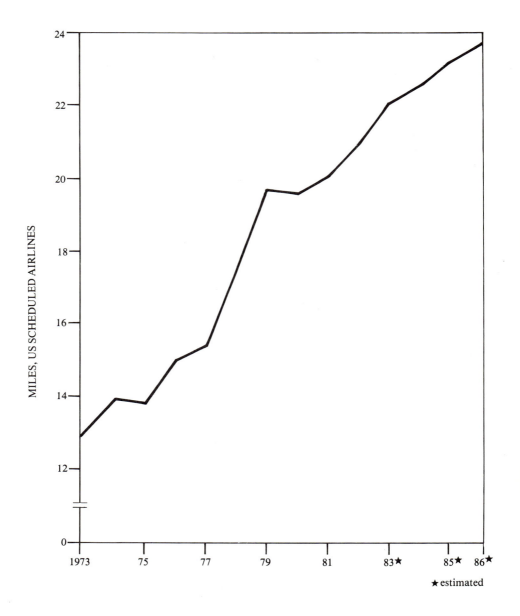

★estimated

Source: Business Week

over the same period of 82%. Gas used fell from the 1972 figure of 3,977 billion cu ft to 3,188 bn cu ft in 1978 before rising to an average of 3,661 bn cu ft over 1980-81. 1982 registered a major fall, however, to 2,227 bn cu ft. In aggregate, coal led oil as a utility input by a ratio of 8.1 to 1, by 3.9 to 1 over gas and 4.2 to 1 over nuclear fuel. As Table 7.17 shows, the total amount of electric power generated by oil-fired facilities fell from a peak of 364 bn kWh in 1978 to only 146.4 bn kWh in 1982. At the same time nuclear fuel's share of output increased from 54 bn kWh in 1972 to 283 bn kWh in 1982. Coal generation rose from 771 bn kWh in 1972 to 1,193 bn kWh in 1982. Gas generation fell from 376 bn kWh in 1972 to 305 bn kWh in 1982. Thus the share of oil in total power generation, which in 1972 had stood at 19%, had by 1982 been reduced to 8%. Meanwhile the share of coal had grown from 53% to 62% over the same period.

Table 7.16

CONSUMPTION BY TYPE OF FUEL: ELECTRICAL UTILITIES

| | Coal, oil and gas consumed | | |
	Coal millions of tons	Oil millions of barrels	Gas billions of cu ft
1972	351.2	496.6	3,976.9
1973	388.7	562.5	3,660.2
1974	391.2	539.1	3,443.4
1975	405.9	506.6	3,157.7
1976	448.3	556.5	3,080.9
1977	477.0	624.2	3,191.2
1978	481.6	635.8	3,188.4
1979	527.1	523.3	3,490.5
1980	569.5	420.2	3,681.6
1981	596.8	351.1	3,640.2
1982	594.1	249.7	3,226.6

Source: Electrical World, Sept. 1983.

Forecasts of utility output in the USA point to growth averaging 3% per year in the 1980s with total output reaching 3,041 bn kWh in 1995 and 3,426 bn kWh in the year 2000. At this level it would be 78% higher than it was in 1982.

Table 7.17

FUEL SOURCES FOR US ELECTRICAL UTILITY SECTOR, 1972-82
Energy generated, billions of kWh

	Nuclear	Coal	Oil	Gas	Total
1972....	54.1	771.1	274.3	375.7	1,477.0
1973....	83.5	847.7	314.3	340.9	1,588.6
1974....	114.0	828.4	300.9	320.1	1,566.1
1975....	172.5	852.8	289.1	299.8	1,617.6
1976....	191.1	944.4	320.0	294.6	1,754.0
1977....	250.9	985.2	358.2	305.5	1,903.8
1978....	276.4	976.6	364.2	305.4	1,922.6
1979....	255.2	1,075.0	303.0	329.5	1,967.0
1980....	251.1	1,162.0	245.6	346.2	2,010.4
1981....	272.7	1,203.2	206.1	345.8	2,034.1
1982....	282.8	1,192.5	146.4	305.3	1,932.5

Source: Electrical World

The Long-Term Outlook for Demand

Having looked at the factors which are likely to affect short-term variations in demand for various petroleum products, this final section looks at the long term. It attempts to identify the factors which are likely to be at work in shaping the pattern and the volume of demand over the present decade and beyond.

A major focus of interest and research into energy issues in the last ten years or so has involved the concept of the energy/GDP ratio, which measures the amount of energy inputs needed to yield an x% change in GDP. The ratio can be redefined as an oil/GDP ratio, still at the aggregate level, or indeed can be redefined at any number of sectoral levels, to illustrate, for instance, the input of crude oil needed to yield x% more ethylene, polystyrene, furniture, glass or alumina.

This type of measurement, the pitfalls of which will be discussed below, came to be of interest not only because of the boost to oil prices during 1973, but also through the attention generated by such studies as *The Limits to Growth*. By stressing the finite supply of the presently indispensable raw materials used in industrial economies, these studies directed attention towards the possibility of GDP stagnation over sustained periods due to lack of certain production inputs. The general rise in primary product prices along with crude oil prices also served at

192

about that time to generate interest in the relationship between raw material inputs and the gross output of an economy.

Investigation revealed quite quickly that there were many problems in both collating and interpreting the results of such work. Six major difficulties can be distinguished.

First, it is important to distinguish aggregate and individual fuel-demand ratios. Thus, the total energy/GDP coefficient may reveal little about an economy's immediate dependence upon oil if agents within the economy are able to bring about rapid inter-fuel substitution, and if the importance of oil among all fuel sources is already minor. Second, the length of time during which transitions between fuel use may take place is vital in making sense of a ratio reflecting energy use at a given moment in time. The malleability of the ratio must be considered when interpreting it. Third, the impact of tax and other regulatory factors upon fuel use must be taken into account: to the extent that these buttress the direction and size of the original price change, they naturally amplify its effect. Fourth, the movement of energy use and GNP is similar to that between any factor of production and output, in that proportionate input changes cannot always be accommodated due to indivisibilities. During a recession, for instance, work of a reduced intensity still requires space heating and certain irreducible energy inputs to be applied to the production process. Maximum energy efficiency will typically be attained at close to full capacity utilization, therefore, and due account must be given to this fact in interpreting energy coefficients. Fifth, the more aggregate forms of the energy/output elasticity are likely to be misleading due to the use of non-market fuels in some sectors and in certain countries. As is the case with all national accounting systems, the accounts can only capture those parts of the economy which are characterized by monetary exchange: barter and non-commercial transactions are excluded. To the extent that firewood or other indigenous fuels may be being used but not recorded the coefficient will accordingly understate the true use of energy in the sector or economy. Finally, the discretionary component of fuel use—particularly in sophisticated economies—renders the energy coefficient still less useful as a tool of prediction. Given that a significant proportion of the gasoline used in driving or the energy consumed in the typical home is not, strictly speaking, needed to achieve the desired objectives, and can be saved by such actions as careful driving and thoughtful use of time-switches or insulation, the use of energies could be expected to vary quite markedly and quickly.

Nothwithstanding the various analysical difficulties which attend the construction of energy coefficients, it is still possible to gauge roughly the scale of the efforts at improved efficiency being made in the major oil-using sectors of the world economy.

Across a wide range of manufacturing and processing industries efforts are underway to reduce significantly, quickly and permanently the input of oil (and, more generally, energy of all sorts) per unit of output. The years between 1973 and 1979 did not see a great deal of effort in this direction, partly because the 1973 oil price rises were increasingly seen as an irrepeatable aberration, and, as if in corroboration of this view, real prices were falling for much of that period. After 1979, by contrast, most energy-intensive sectors of industrialized economies have been characterized by a massive and relentless search for energy efficiency. The "energy audit"—a careful accounting for all energy passing around in a plant—has become common-place, with energy managers increasingly hired to institute and maintain a high degree of awareness of such matters. As a major US trade journal aptly summed up the post-1979 position: "With each jump in oil costs comes a flurry of energy-saving ideas".

Among the industries which are currently seeking most aggressively an improvement in their use of energy are glassmaking, aluminium smelting, copper refining, paper and board and timber processing. The impact of energy cost changes has precipitated responses which have, naturally, tended to differ according to the industry. Among aluminium firms, for instance, the policies adopted have ranged from government-assisted closure (as with the Japanese government, which rationalized away half a million tons of aluminium smelting capacity), to relocation (as in US firms, which are presently thinking of moving away from the smelter area of the Pacific North West, due to large impending electricity price rises, towards the natural gas supplied area of the Texas Coast). The long-standing effort to improve the energy efficiency of the Hall-Heroult method for smelting goes on. While in 1920 it took 12 kWh to produce a pound of aluminium, by 1979 it required only 6 kWh. Furthermore, attempts to improve the process of gaining alumina from clay—a procedure which should ultimately render considerable energy savings—are proceeding. At present this method tends to be more energy-intensive than beginning from bauxite itself. In the iron and steel industry the path to energy efficiency is different, and is unlikely to entail any significant breakthroughs in new technology, in the opinion of the trade press. In the blast furnace method perhaps the limit of energy efficiency is already within sight: and the move towards direct reduction, using gas or oil inputs rather than coke, will spread. Electric furnaces, allowing lower initial capital costs than conventional blast furnaces as well as making better use of the energy value contained in scrap; plasma furnaces; continuous melting of scrap, sponge iron and the marge materials in a shaft furnace connected to an electric furnace for refining (the Klockner-Youngstown method); iron ore pelletising; using turbines to collect power

194

from blast furnaces; and continuous casting are further areas in which energy-cutting research will proceed.

In all these areas and more, the talents of the developed countries' energy audit managers and engineers will continue to lower unit energy inputs.

There is a second aspect of this thrust towards greater energy efficiency among firms in industries such as those just mentioned. This is the fact that there should be a considerable amount of relocation of plants towards places where energy costs are lower, if, of course, energy costs assume sufficient weight in total costs and other input prices, as well as transport costs for inputs and the finished product, are favourable. An illustration again is the aluminium industry, whose producers "will have to switch production to new cheap power sources round the world as best they can"—one of a series of "radical changes" forced upon the industry by the importance of energy in its total costs.

This tendency to enhanced conservation within industrialized countries will, however, to an increasing extent be counteracted by rising total energy consumption within certain developing countries. As more and more such countries approach that stage of development (typically between $500 and $1,200 per capita) when industry accounts for an accelerating share of GDP, energy needs must grow. This will be underwritten by the tendency for 'heavy' industry to account for an increasing share of the overall industrial structure at this stage of development. Furthermore, a substantial portion of the technologies installed in developing countries are likely to be of less than the latest vintage, so that they will not embrace all recent energy-saving characteristics. Clearly, the relationship between income per capita in a country and the extent of industrialization is neither immutable nor exact. The pattern of industrialization is one obvious factor which has independent influence upon the energy used in the sector—a point amply demonstrated by the exceptional rise in domestic energy needs forecast in many OPEC states in the 1980s, as explained above.

Summing up

Having looked at short-term and long-term demand projections, it is now time to draw together the thoughts presented in the first seven chapters of this report to highlight the most significant influences on the refined petroleum products industry. After this, Chapter Eight will draw together some of the more important institutional points, such as the future of the oil companies, of OPEC as a major oils supplier, and thoughts on the general political economy of the industry.

From the preceding chapters, the following points stand out.

i   Supplies of crude oil from OPEC countries may well have peaked, on the grounds that:
—non-OPEC sources have been gaining ground, and as exploration, in the US (especially offshore from Southern California), the North Sea, and elsewhere goes ahead, further substitution may take place;
—some OPEC members are attempting to substitute refined for crude oil exports to an increasing extent.

ii Demand for refined products in the developed countries is likely to grow again as the 1982-83 recession ends, but that demand will accelerate more slowly than in previous recoveries, the reasons being:
—continuing efforts at conservation, both in domestic and industrial settings, as a continuing response to price rises over the last ten years:
—tax increases, which will tend to push the real costs of energy to final consumers up, or at least maintain its level in real terms;
—shifts away from oil use entirely—for instance, in the electricity generating sector, towards other fuel inputs, notably coal;
—changes in the composition of the automobile fleet such that overall fuel economy delivered by the mid-1980s will be about twice what it was fifteen years before.

iii Demand for refined products in advanced communist countries will continue to grow faster than in advanced non-communist countries. Even this rapid growth, which has been observed since the early 1950s, will however decelerate once price shifts within these economies are accomplished to reflect more fully real resource scarcities. Factors pushing up oil demand in these countries will be:
—sluggish reaction by officials to changing world oil prices, and thus delays in confronting domestic industrial users with the need to economize;
—albeit from a very low base, growth of the private automobile fleet, which will eventually come to trigger a substantial though still latent source of gasoline demand;
—the fact that the communist states' industrial structure is still more heavily weighted by heavy industries (e.g. steel, chemicals) than light ones. This tends to increase the energy intensity of the whole economy above what it would otherwise be.

iv Demand for refined products in LDCs other than the oil exporters will probably roughly double over the 1980-1990 period, this very brisk increase being fuelled by:

196

—as in some communist states, a reluctance by officials to let fuel prices rise in line with world prices, or by an amount necessary to encourage conservation. Some LDCs still subsidise oil consumption;

—rapid industrial output growth, as industry's share in GDP moves up toward the 25-35% share associated with more developed economies. This will tend to drag in more fuel inputs, especially as older, less fuel-efficient, technologies will tend to be those predominantly used;

—explosive growth in private automobile demand, encouraged via the tax system by some LDCs' governments so as to court popularity with the urban middle-class and also in the hope that the auto industry can act as some type of growth-pole for a broader industrial development.

v Demand for refined products in OPEC member and other oil-exporting LDCs is increasingly difficult to forecast. Until 1982-83, when the impact of contracting oil export volumes and real oil prices on OPEC members' balances of payments began to become apparent, growth had been explosive, as a result of:

—low fuel prices, often with subsidies;

—massive urbanization, accompanied by auto purchases by the growing middle class.

Since the balance of payments problems have surfaced, however, the prospects are for:

—stricter import controls on autos;

—higher domestic fuel prices;

—lower rates of economic growth overall, depressing growth of demand.

These factors in turn imply that:

—demand may indeed continue to grow very quickly, at about the rate expected for the rest of the Third World, but it will be driven more by industrial demand, to the extent that the more advanced OPEC members are able to broaden their industrial base.

vi As an alternative to refined petroleum products, demand for other fuels will probably grow much faster. Coal is likely to see its recent resurgence in demand continuing, as will natural gas.

vii Within petroleum products, the overwhelming shift, which is likely to remain in force for the foreseeable future, is the lightening of the product barrel, so that light distillate supply increases faster than supply of heavy distillates. Of course, this shift has implications for the need to reconfigure many of the world's oil refineries— a course of action that is now well underway, although recently sceptics

197

have queried the economic basis of these (very costly) investments, as the price differential favouring heavy crudes over light has collapsed. The factor pushing towards a lighter barrel is above all a higher income elasticity of demand for light products such as gasoline, which means that as GDP rises, demand for such items rises faster than does demand for items such as resid.

Table 7.18 shows energy demand and supply, arranged by energy type, and by country groupings, for 1970 and 1980 and forecasts for 1990 drawn from the World Bank. The following points should be noted:

(i)   In developed market countries, the importance of petroleum as an energy source is expected to decline markedly, from very nearly 50% of energy demand in 1970 to around 43% in 1990. Primary electricity is expected to become much more significant.
(ii) In Communist industrialized countries, total energy demand is forecast to grow much more sharply than in developed market economies, with total energy consumption more than doubling between 1970 and 1990. Petroleum use is forecast to rise more than twofold over that period, compared with growth of less than 33% in the developed market countries.
(iii) Proportionately, the biggest growth in energy use is foreseen in the oil-exporting countries (shown as 'capital surplus'), whose consumption is expected to grow to 1.7 million barrels equivalent.
(iv) Total world energy demand, in these forecasts, is expected to grow by 37% between 1980 and 1990 after growing by the same percentage over the preceding decade. Some forecasters would certainly look to energy demand growing more slowly in the second time-period, although there are grounds for the World Bank estimates.
(v)   By 1990, the structure of world petroleum demand would be as follows:

	%
industrial market economies	51.6%
non-market industrial economies	23.9
oil exporting states: capital surplus	0.2
other	7.6
oil importing LDCs	16.7
TOTAL WORLD	100.0

Table 7.18
COMMERCIAL PRIMARY ENERGY PRODUCTION AND CONSUMPTION,
BY COUNTRY GROUP, 1970-1990
(millions of barrels a day oil equivalent)

Country group	1970 Pro-duction	1970 Con-sumption	1980 Pro-duction	1980 Con-sumption	1990 Pro-duction	1990 Con-sumption
Industrial market economies	**43.2**	**60.6**	· **50.6**	**72.4**	**64.3**	**87.0**
Petroleum	12.7	29.9	14.5	35.0	16.4	37.4
Natural gas	13.0	12.8	13.8	15.0	13.2	16.2
Solid fuels	13.0	13.3	13.9	14.0	20.4	19.1
Primary electricity	4.5	4.6	8.4	8.4	14.3	14.3
Nonmarket industrial economies	**28.8**	**27.6**	**45.2**	**43.0**	**63.4**	**62.1**
Petroleum	8.0	7.2	13.7	13.1	17.9	17.3
Natural gas	3.8	3.8	7.7	7.0	12.6	12.3
Solid fuels	16.1	15.7	21.8	20.9	29.8	29.4
Primary electricity	0.9	0.9	2.0	2.0	3.1	3.1
Capital-surplus oil exporters	**12.8**	**0.3**	**18.6**	**0.9**	**21.7**	**1.7**
Petroleum	12.7	0.2	18.3	0.7	20.4	1.1
Natural gas	0.1	0.1	0.3	0.2	1.3	0.6
Solid fuels	—	—	—	—	—	—
Primary electricity	—	—	—	—	—	—
Developing countries oil exporters	**13.7**	**2.8**	**16.7**	**5.5**	**25.2**	**10.0**
Petroleum	12.7	1.8	14.2	3.6	18.3	5.5
Natural gas	0.7	0.7	2.0	1.4	5.9	3.5
Solid fuels	0.1	0.1	0.1	0.1	0.3	0.3
Primary electricity	0.2	0.2	0.4	0.4	0.7	0.7
Oil importers	**4.7**	**7.8**	**7.5**	**13.7**	**15.1**	**24.3**
Petroleum	1.2	4.2	1.5	7.3	2.8	11.2
Natural gas	0.3	0.3	0.5	− 0.7	1.6	1.6
Solid fuels	2.3	2.4	3.5	3.7	5.6	6.4
Primary electricity	0.9	0.9	2.0	2.0	5.1	5.1
Total world	**103.2**	**99.1**	**138.6**	**135.5**	**189.7**	**185.1**
Petroleum	47.3	43.3	62.2	59.7	75.8	72.5
Natural gas	17.9	17.7	24.3	24.3	34.6	34.2
Solid fuels	31.5	31.5	39.3	38.7	56.1	55.2
Primary electricity	6.5	6.6	12.8	12.8	23.2	23.2
Bunkers and others		2.9		3.1		4.6

	Average annual growth rate of world supplies (percentages)	
	1970-80	1980-90
Total world	3.0	3.2
Petroleum	2.8	2.0
Natural gas	3.1	3.6
Solid fuels	2.2	3.6
Primary electricity	7.0	6.1

Note: Total world consumption refers to apparent domestic consumption only. Total world requirements of primary energy equal total world consumption plus bunkers and others. Synthetics from coal are not included in solid fuels.
Source: World Bank

199

Chapter Eight

SUMMARY AND CONCLUSIONS

New Thinking on Oil Matters

IN the past, it has tended to be the case that interest in oil and oil markets has been inversely related to the rate of change of price. During periods of rapid build-up of prices, such as was seen in the first half of 1979, interest in oil approached panic. Reports, articles and conferences flooded out. Experts roamed the world. But in periods of relative quiescence in the market, such as characterized some of 1982, and in periods of price falls, such as early 1983, journalists and consultants tend to look elsewhere for material. All the more curious, therefore, that now that oil demand has been falling for three consecutive years and OPEC is sailing through fairly calm waters, a slew of new reports are stressing the possibility of renewed crisis later in the decade.

One of the more doom-laden reports is that of the Department of Energy of the American government. Entitled *The 1982 Annual Energy Outlook, With Projections To 1990,* the report looks at the possibility of the recent weakness in real oil prices triggering a return to higher volumes of oil use. The Department of Energy believes that by 1985 oil demand in the West could be back at its peak level of 1978. In the USA, for instance, demand for oil-derived products could rise to 18.4 million barrels/day (b/day), as against 15.6 million b/day in 1982. The record level was 18.8 million b/day, achieved in 1978. The result would be a huge rise in imports, since domestic American oil supply would quickly be overwhelmed by this surge in demand. Crude and products imports could rise to 8.45 million b/day, up from the 3.46 million b/day averaged in 1982. Total world demand, according to the Department, would grow to 50.8 million b/day, as against 45.4 million b/day last year. If non-OPEC output does not rise significantly during the mid-1980s (a point which another study, discussed below, makes much of) this means a direct stimulus to demand for OPEC oil. OPEC oil exports would then comfortably exceed their recent 17.5 million b/day level and rise towards 25 million b/day. The consequence of this resurgence in demand would soon be a pickup in real oil prices. From a level of $25/barrel in 1985 (measured in 1982 prices) oil prices would reach $37/barrel by 1990—almost a 50% increase.

The fact that exports from outside OPEC have been far less badly affected than have exports from OPEC during the 1980-83 shrink in oil demand suggests strongly that the bulk of any demand resurgence will be felt directly by the 13 members of OPEC. A report called *Middle East Crude Oil Potentials From Known Deposits,* also from the US Department of Energy, makes the point that OPEC's influence over world oil markets is bound to grow, simply by virtue of its members' huge oil reserves. Of the non-Communist world's proven reserve of 585 billion barrels, 75% are located in OPEC countries. Middle Eastern members alone possess 60% of the total.

From these figures the US agency infers that, over the long run, the importance of OPEC must grow. One can, however, raise objections to this line of reasoning even without questioning the accuracy of the department's geological work. First, the question of what constitutes a "proven" reserve in the oil business has, of course, long been a contentious matter. Some experts, such as Professor Peter Odell of Erasmus University, have long argued that reserves are usually counted far too conservatively by the major institutions in the business. Moreover, the issue of the size of reserves is also a function of the techniques available to exploit oil. Depending on the extent to which tertiary recovery is used, for instance, whereby more oil than usual is ultimately recovered from a well, the amount of oil that can economically be obtained from an area varies quite substantially. A second point is that what determines market share in the short term in the oil business is the price and the volume offered, as much as ultimate reserve holdings. Over the next five years at least, consumers will make their purchases on the basis of who offers how much at what price. Only in the long run will exporting countries' prices differ materially by the size of their reserves.

Taking a different tack in the analysis of the energy outlook, Esso's publication *Energy In Europe: Looking Forward To The Year 2000* looks at the extent to which economic growth necessarily requires growth in energy inputs. The ratio of energy use to national income, usually known as the energy coefficient, is of course a highly aggregated and therefore not entirely satisfactory indicator. But as a rough rule of thumb it has its uses. Esso's estimate is that in 1960-1970 each $1,000 of GDP required 0.68 tonnes of oil equivalent per capita. By 1970-1980 the figure had fallen to 0.62. For 1990 Esso forecasts a figure of 0.55 and for the year 2000 a figure of 0.49. This change, between 1960 and 2000, is fairly substantial, and points to a significant decline in the amount of extra energy needed for marginal increases in national income. Following from these forecasts, Esso's view is that there might be some upward pressue on real energy prices during the late 1980s and into the 1990s, as a consequence of GDP starting to return to its longtime trend rate of growth.

Another oil company, BP, in its annual *Statistical Review of World Energy,* 1982 issue (published in London in June 1983) focusses on the ending of the long worldwide oil stock rundowns, and the consequent probability of there being a slight pickup in world oil demand in 1984. Following a 3% fall in oil consumption in 1982 from 1981, to a level of 2,819 million tonnes, 1983 and 1984 might witness sufficient growth in demand for OPEC's output to exceed 20 million b/day once again. Over the rest of the century, however, energy demand will more and more be shifting towards energy forms other than oil and natural gas. Among the many figures presented are some showing how oil use moved in each geographical region last year. Although, as mentioned above, world oil demand fell by 3%, in Western Europe the fall was 5%. In North America demand fell by 6.2% and in Japan, by 7.6%.

Taken together, these reports suggest that those who are professionally involved in the oil business continue to view the long-term outlook with concern. They recognize that demand for oil, particularly for oil from OPEC members, has fallen severely in the last three years. But their argument, which refers primarily to the mid-1980s and beyond, is that eventually the world's oil needs will begin to pick up. While the pace at which it does so will almost certainly be slower than in previous economic recoveries—due to the change in the energy coefficient—there is no avoiding a rise. This will probably take many market observers by surprise. Even with OPEC output hovering at a mere 17-18 million barrels per day, and with plenty of spare capacity in hand, there is a possibility of panic setting in. To the extent that reports such as those discussed above help avoid any surprise when oil demand picks up again, they are to be welcomed. But insofar as many still pose as definitive accounts of an extraordinarily volatile industry, their forecasts are to be taken with a pinch of salt, and their warnings with wariness. The one thing that is known about the oil business is that nobody knows how it may look ten years from now.

In the face of this uncertainty about the long-term availability of crude oil, the major oil companies are obliged to keep up their cost-cutting exercises in their home markets, in a bid to retain or possibly enhance their market share.

Cost Cutting Continues

On the forecourts of the service stations of the developed world, the continuous search for new products with which to lure customers goes on. With the 20 or so biggest oil companies in the US more and more keen to withdraw within newly-drawn boundaries of regional autonomy, with fewer competitors within each, and with shorter spans of management control and communications, there is still scope for the independent. Not

202

that that necessarily has to mean gasoline retailing—one of the faster growing architecture and design practices in the US advises on how to turn disused gas stations into plant shops, restaurants, and art galleries. In the gasoline business, although refiner and distributor margins have become very thin recently, the trade press is always full of stories of independents who have seen an opportunity and acted upon it.

In the Third World, motorization is bound to proceed very quickly, despite all the problems of urban crowding, high taxation and real car purchase costs well in excess of those pertaining in most developed countries. This will also proceed, by and large, against the efforts of governments to use tax and trade policy to suppress the demand for private transport, at least at this comparatively early stage of economic development. So long as owning a car continues to be unmistakeably the definitive signal of arrival in the middle class in developing countries, however, so it will be desired. This must eventually lead to a huge demand for retail gasoline distribution networks, particularly in the physically large countries like Brazil, Nigeria and Mexico. The mere existence of burgeoning demand will not in itself, of course, guarantee an effectively-functioning supply apparatus. In Nigeria, for instance, during the mid-1970s, when oil exploration and exporting were booming, even in the capital city of Lagos there were persistent troubles in finding properly-stocked service stations. In Brazil huge logistical problems have confronted the state energy agency Petrobras in its effort to throw a supply network across the country, both for gasoline and gasohol. In addition to the usual problems of refinery balance and demand forecasting, then, oil agencies in developing countries will typically be facing problems in building up their own first infrastructures. On the other hand, these agencies will tend to have a relatively free hand insofar as most of them will be monopolies, and many of them public monopolies at that. This will eliminate the exigencies of competitive positioning which bedevil firms in, say, Europe or the US, although this very lack of competition is unlikely to work to the great advantage of these firms' consumers. Recent experience with state energy agencies in Indonesia (Pertamina), Mexico (Pemex) and in Venezuela do, however, also suggest that the free hand given to them and their (often politically-appointed) managers may not continue to be a central characteristic. Too frequently, very substantial amounts of public funds have been absorbed by these firms with relatively little, other than debts, to show for it. An era of more sober financial management in developing countries' oil agencies is therefore to be expected.

In oil companies based in developed countries, the same conclusion will very probably hold, although for different reasons. As discussed in Chapter Four, opening up the divisions of an intergrated firm, so as to allow easier identification of performers and non-performers, is the path most likely to be followed, given that independents, enjoying low overheads and the benefits of flexibility of input-mix (via buying from different sources as it suits them) are now doing very well in certain parts of the industry. The continuing pressure from developing country governments to enforce various supply conditions is making the comfortable assumptions of assured supply more open to question, while the appeal of exploration and drilling as a focus of activity is certainly not diminishing.

The Politics of Oil: Still Uncertain

In one sense it was politics which precipitated the wave of interest in energy that made it among the most written-about topics of the 1970s and early 1980s. Depletion of the world's proven oil reserves was beginning—but only beginning—to become an issue around the world. But many authorities were not convinced that there was any real cause for alarm. For one thing, estimates of crude oil availability have always tended to be on the conservative side; for another, in a free market economy there must always be "enough" of a resource so long as prices are allowed to adjust to reflect changing conditions of supply and demand. What led to the first major oil price rises, many analysts would now argue, was a mood of bitterness over the future of Middle Eastern borders—in the arc of crisis, as one political scientist has put it—and a concomitant mood of hopelessness on the part of oil consuming countries' governments and their advisers. To the extent that this interpretation holds true, one of the most significant aspects of the post-1982 energy situation has been the erosion of the self-confidence of the oil exporting countries governments in the political arena. Perhaps here lies the greater issue?

Before looking at this question in detail, it is worth reviewing the diplomatic aspects of OPEC. Shortly after the 1973 oil price rises, a major effort was made by several OPEC governments to present their action as defensible and legitimate, and indeed as part of any coherent developing country strategy for growth. There was talk of OPEC being a "pioneering group of developing countries" and of the oil price rises as being "a practical illustration of what the developing world can do." The implicit premise was that other raw material exporting groups could and should be formed, and moreover that these could and should emulate what the oil exporters had achieved.

As the 1970s progressed, this implicit premise was refined. Financial support for other raw materials cartels was only briefly provided, and only then by Venezuela. While the OPEC multilateral aid effort did swing into action quite impressively, it has faltered since 1980.

The twenty-first anniversary conference of OPEC, held in Vienna in November 1981, heard a number of speeches about the renewed effort of the organization to channel its aid more and more to the poorest oil-importing LDCs, and the possibility of OPEC using its diplomatic power as leverage on such significant multilateral issues as the common fund or debt renegotiation. In retrospect, however, it seems more likely that that meeting was as much the epitaph as the rallying-call of the once-vaunted OPEC push for Third World leadership.

Since the possibility of extra aid from oil-exporting countries was always part of the implicit contract made between the former and their new poor allies of the 1970s—the relationship which historian Ali Mazrui called "the political shrinking of the Sahara"—perceived difficulties among aid donors will naturally alert recipients to the need to look elsewhere. And as if to confirm this drifting away of the poorer LDCs' governments, especially those in black Africa, from their new-found Arab allies of the mid to late 1970s, in late 1982 the Israeli government opened up that diplomatic front a second time. With the first official visit by an Israeli official to a black African country since 1973, foreign minister Yitzhak Shamir's trip to Zaire has signalled to all those who are interested that Israel is back in business. After suffering the loss of diplomatic relations with 27 black African countries in 1973 (only Swaziland, Lesotho and Malawi are currently aligned) Israel has a lot of ground to make up. But it seems unlikely that a few years hence Israel will not have regained all its diplomatic ties in Africa.

Nigeria, Indonesia and Venezuela are among the members with the greatest short-term economic problems. In Nigeria's case the manifest failure of successive governments to come to grips with the disbursement of revenues across their vast and fragmented country is exacerbated by the fears caused by the possibility of the August 1983 general elections triggering unrest. In that unhappy country the very real possibiity of an aborted push to capital-intensive industrialization now hangs over politicians' heads. The situation in Iraq too is a tightening of budgets, both public and private. The other combatant, Iran, is also witnessing a heavy state-imposed austerity policy, although there, curiously enough, prime minster Hussein Musari in late 1982 claimed that Iran would shortly be offering oil on concessional terms to other developing countries.

But money is not the only factor at work. Among the other factors that are eroding this relationship are the increasing preoccupation of the Middle Eastern OPEC members with the Iran-Iraq war, and what it may mean for them when it ends. Although the Saudi Arabian foreign ministry is hard at work on enterprising initiatives (the king made the first official Saudi visit to China, and the Moscow link still simmers gently) the North African governments' foreign preoccupations lie with the war. There, their horror that Khomeini's forces might actually try to push towards Jerusalem is matched by their ambivalence over the occasionally violent and despotic régime they are keeping in power in Baghdad.

The inference from all this is fairly clear. Through economic necessity it is likely that most OPEC members will henceforth adopt more parochial goals. To the extent that foreign relations reflect in part the leverage that aid yields the donor, OPEC members will probably look first to their own back yards. Venezuela's oil-concession scheme (run in conjuction with non-OPEC member Mexico) for Central American oil-importing countries is a possible model. Nigeria may similarly retreat from the grand stage of international relations and devote more of its diminished effort into Ecowas and other West African regional organizations. Indonesia may look more to its Asian neighbours which are, after all, its main trade partners. If such matters as apartheid in South Africa or the interests of Costa Rica ever did carry much weight in the corridors of Jakarta, they are likely to carry rather less now.

All this is not to say that the OPEC members' foray into international relations after 1974 has been a fiasco. Rather, it has been undergoing a process of finessing. Shortage of cash has simply determined the timing of what would almost certainly have happened in any case—a less ambitious and more circumspect definition of what one group of developing countries can reasonably expect to do for some others.

Future Outlook

The foregoing chapters have suggested that the single most important factor facing the petroleum products business will be uncertainty. There will continue to be uncertainty about the amount of crude oil that exists in the world, and about the amount that can be realistically recovered under different sets of economic circumstances. The World Petroleum Congress in 1983 heard the results of a five-nation study group which had been looking into the question of devising commonly-agreed terms such as recoverable, likely and established reserves. As yet, however, no universally accepted nomenclature has emerged. There will continue to be uncertainty about the extent and nature of national oil corporations' intentions regarding oil discovery, exploitation, transportation, and other factors. After a period of ebullience in the 1970s, some countries which became major oil exporters were beginning, by the early 1980s, to feel that they might have made conditions for foreign cooperation too unattractive. The later part of the 1980s might therefore see some resurgence of foreign firms' involvement in overseas drilling and exploration. Next, refiners and distributors will continue to face uncertainty over demand for their products. Prices in many developed country markets moved fairly erratically in the first three years of the 1980s; this tended to obscure fundamental demand shifts and made long-range forecasting unusually hazardous. The mix of demand between fuels, and between energy sources more widely defined, is far from settled. As the world economy begins to recover from the 1980-83 recession, different petroleum products are likely to face rising demand at different speeds. Automotive demand may push up gasoline use faster than jet fuel use is rekindled by the air transport industry, for instance. To the extent that this is the case—and it would be surprising if it were not—further production scheduling difficulties will arise. Over the long-term, moreover, refiners still face uncertainty over the size of the energy conservation efforts that have still to show up. It is unlikely that all of the impact of post-1973 fuel conservation measures have even now shown themselves; the problem is, of course, that nobody can accurately estimate the magnitude of the effects to come. A further difficulty which those in the gasoline industry will face is competition over market shares. Some fairly vigorous corporate strategy changes were in progress in 1983, and most of them had the aim of boosting firms' market share, although often redefined in a smaller geographical area. How far diversification into other areas, such as retailing, confectionery and tyres can help offset volatile business volume and tight margins remains unclear in the US. As for European distributors, competition from servicing and repair chains threatens to take away some of their higher-margin business. What can be put in the place of this type of business—other than closure and consolidation—also remains unknown. For all these reasons, petroleum products will continue to be a risky business to be in. But it also seems likely to continue being a lucrative one—for those that get it right.

FACT FILE

The World Market for Petroleum and Petroleum Products

Table 1

PROVEN CRUDE OIL RESERVES 1970-1983
Unit: million tonnes of oil equivalent

Region and country	1970	1976	1980	1981	1982	1983
North America	6,478	5,471	4,543	4,475	5,059	5,021
USA	5,280	4,503	3,615	3,602	4,063	4,063
Canada	1,198	968	928	873	996	958
South America	3,981	4,826	7,704	11,891	12,019	11,140
Venezuela	2,012	2,415	2,438	2,660	2,769	2,933
Mexico	819	1,297	4,263	6,650	8,030	8,030
Ecuador	103	336	151	116	192	191
Argentina			337	363	355	
Brazil			178	182	240	
Middle East	46,181	40,793	49,379	49,396	49,501	50,380
Abu Dhabi	2,182	4,024	3,820	3,956	4,175	4,162
Saudi Arabia	19,987	20,710	22,712	22,924	22,900	22,554
Iraq	3,752	4,680	4,229	4,093	4,052	5,594
Iran	7,503	8,799	7,913	7,845	7,776	7,545
Kuwait	10,164	9,713	9,349	9,267	9,240	9,160
Oman	682	805	327	319	351	372
Qatar	750	798	513	489	468	467
United Arab Emirates	1,751	4,274	4,166	4,408	4,432	4,414
Africa	6,778	8,348	7,786	7,546	7,672	7,889
Algeria	1,091	1,006	1,151	1,119	1,102	1,288
Egypt	160		290	396	400	454
Gabon	68	300	68	83	75	80
Libya	4,775	3,561	3,206	3,138	3,083	2,933
Nigeria	682	2,756	2,374	2,278	2,251	2,285
Western Europe	243	3,477	3,203	3,157	3,362	3,127
United Kingdom	1	2,183	1,200	1,125	1,050	900
Norway		955	784	905	782	1,028
Centrally Planned Economies	8,185	14,051	12,278	11,774	11,711	11,612
USSR	6,139	10,969	9,140	8,595	8,595	8,595
China	2,045	2,728	2,729	2,797	2,714	2,658
Asia/Oceania	1,792	2,897	2,641	2,678	2,613	2,695
Indonesia	1,228	1,910	1,310	1,296	1,337	1,303
India	98	125	355	352	365	466
Australia	341	232	291	322	233	221
OPEC	59,498	60,136	59,473	59,494	63,186	60,748
OPEC percentage share	81	67	67	65	67	66
WORLD TOTAL	73,638	89,863	87,534	90,917	91,937	91,864

Source: Statistiques de l'Industrie Petroliere

Table 2

CRUDE OIL PRODUCTION 1970-1983
Unit: million metric tonnes

Region and country	1970	1975	1980	1981	1982	1983
North America	603.7	554.6	554.2	546.2	541.2	554.1
USA	533.7	469.3	475.6	475.6	478.6	479.2
Canada	70.0	85.3	78.6	70.6	62.6	74.9
South America	267.8	225.3	296.1	315.8	329.8	327.3
Venezuela	193.2	122.1	114.8	111.7	100.0	89.71
Mexico	24.1	40.0	106.3	126.8	149.6	147.0
Ecuador	0.2	8.0	10.4	10.7	10.6	11.7
Argentina	20.0	20.8	25.2	25.2	24.8	24.0
Brazil	8.0	8.4	9.4	11.0	13.5	
Middle East	713.8	980.6	915.3	773.1	611.1	560.0
Abu Dhabi	33.3	67.3	64.5	54.5	40.8	39.0
Saudi Arabia	190.2	352.0	494.8	484.0	321.7	251.2
Iraq	76.6	110.1	129.9	43.9	45.2	46.8
Iran	191.7	266.7	74.1	66.3	98.9	122.8
Kuwait	150.8	104.8	83.5	56.7	41.7	48.4
Oman	16.6	17.1	14.0	15.9	15.9	18.8
Qatar	18.1	21.9	22.8	20.1	15.9	14.1
Ecuador	0.2	8.0	10.4	10.7	10.6	11.7
Africa	174.6	232.3	301.3	232.7	231.5	222.6
Algeria	47.3	45.8	51.5	46.4	45.4	44.1
Egypt	16.4	14.8	30.1	31.8	35.0	37.0
Gabon	5.4	11.2	8.9	7.7	7.7	7.8
Libya	159.2	72.4	88.5	53.5	55.4	54.3
Nigeria	53.4	88.0	101.9	70.9	63.8	62.1
Western Europe	16.2	24.2	118.4	126.1	143.8	162.3
United Kingdom	0.1	1.4	80.0	89.4	103.4	115.9
Norway	0	9.3	24.3	23.5	24.5	31.0
Centrally Planned Economies	393.1	590.9	731.0	732.8	737.7	745.8
USSR	352.7	490.8	603.0	609.0	613.0	620.9
China	23.9	77.0	106.0	101.2	101.7	106.0
Romania	13.4	14.6	11.5	11.6	11.7	
Asia/Oceania	67.0	109.0	134.6	136.0	129.3	131.9
Indonesia	42.1	65.5	77.4	78.7	65.7	65.0
India	6.8	8.3	9.4	14.9	19.0	19.0
Australia	8.5	20.5	18.8	18.2	17.5	20.1
Malaysia	0.9	4.7	13.2	12.6	14.5	
OPEC	1,160.8	1,349.3	1,341.0	1,123.4	930.9	869.1
EEC		11.2	90.6	100.4	117.2	127.9
OECD		577.1	701.1	700.5	722.0	739.9
OPEC percentage share	49.7	49.7	44.0	39.2	34.2	32.0
WORLD TOTAL	2,336.2	2,716.9	3,050.9	2,862.7	2,724.4	2,718.4

Sources: Statistiques de l'Industrie Petroliere
OECD/IEA Oil Statistics Quarterly
Petroleum Economist
OPEC Statistical Bulletin
BP Statistical Review of World Energy
UN Yearbook of World Energy Statistics
Own Estimates & Calculations

212

Table 3

TOTAL REFINERY THROUGHPUT 1970-1983
Unit: million metric tonnes

Region and country	1970	1975	1980	1981	1982	1983(est.)
North America	626.4	726.1	766.5	728.3	688.5	655.5
USA	565.5	641.5	675.0	645.6	616.6	588.8
Canada	60.9	84.6	91.5	82.7	71.9	66.7
South America	132.9	134.0	158.0	151.5	140.1(est)	
Venezuela	67.7	45.2	49.0	45.4	45.7	
Mexico	23.6	33.0	56.4	61.5	59.1	
Ecuador	1.2	2.2	4.7	4.5	4.6	
Brazil	25.1	43.4	54.4	52.5		
Argentina	21.6	22.6	26.7	26.7		
Middle East	87.3	103.8	122.3	110.3	112.3	
United Arab Emirates	—	—	0.6	4.7	8.0	
Saudi Arabia	29.2	27.0	37.4	37.9	33.1	
Iraq	3.9	6.8	8.4	8.3	7.6	
Iran	25.5	32.2	29.1	27.0	28.7	
Kuwait	20.2	16.8	19.0	16.6	15.5	
Qatar	—	0.2	0.6	0.6	0.6	
Israel	5.9	7.0	6.6	6.7		
Syria	2.0	2.5	6.3	6.2		
Africa	29.2	50.6	72.9	74.4	84.2	
Algeria	2.5	5.6	11.7	14.6	21.0	
Egypt	6.4	8.6	12.6	13.6		
Gabon	0.8	0.9	1.2	1.2	1.1	
Libya	0.4	1.9	6.1	7.0	7.5	
Nigeria	1.0	2.3	4.5	9.8	9.9	
Morocco	1.5	2.5	2.9	3.8	4.1	
Tunisia	1.1	1.1	1.5	1.5	1.5	
South Africa	8.3	14.2	13.0	15.0		
Western Europe	582.2	575.3	597.6	540.3	508.1	491.7
West Germany	98.4	90.8	100.1	92.6	89.4	85.0
France	95.6	102.9	108.4	91.4	79.4	60.1
Italy	112.2	95.6	89.1	86.7	81.1	75.7
United Kingdom	94.7	86.6	79.2	72.0	70.7	70.6
Belgium	28.2	27.7	31.5	27.4	23.4	21.0
Netherlands	58.5	54.0	47.7	37.7	36.3	39.4
Spain	32.0	42.7	49.1	47.4	44.5	42.9
Greece	4.3	11.2	13.7	15.3	14.5	13.5
Sweden	11.4	9.1	17.1	13.4	12.5	13.1
Turkey	7.0	12.1	11.8	12.5	15.4	14.9
EEC	504.1	478.9	477.9	429.7	401.0	386.2

...contd. overleaf

Table 3 contd.

TOTAL REFINERY THROUGHPUT 1970-1983
Unit: million metric tonnes

Region and country	1970	1975	1980	1981	1982	1983(est.)
Centrally Planned Economies		551.3	666.0	671.0	678.8	
USSR	289.7	404.2	488.0	496.0	500.0	
China	23.3	60.7	82.3	76.2		
Romania	15.0	19.2	24.0	23.5		
East Germany	10.2	17.8	21.3	19.1		
Czechoslovakia	9.2	15.5	18.2	17.2		
Yugoslavia	6.4	9.9	13.7	14.0		
Asia/Oceania		336.0	386.5	378.0	359.5	
Indonesia	11.4	15.2	24.5	27.1	25.0	
India	18.5	21.8	25.1	29.5		
Australia	24.2	26.5	27.4	28.1	28.1	26.9
Japan	159.2	201.9	199.5	180.3	165.0	149.2
Singapore	10.9	18.9	32.6	34.7		
Malaysia	5.3	4.0	5.5	5.6		
OPEC	160.9	·154.4	195.0	201.6	202.8	
OECD		1,496.4	1,693.6	1,569.1	1,472.5	1,403.4
WORLD TOTAL	2,300.0	2,664.4	3,017.0	2,869.6	2,778.1	

Sources: Statistiques de l'Industrie Petroliere
OECD/IEA Oil Statistics Quarterly
BP Statistical Review of World Energy
UN Yearbook of World Energy Statistics
OPEC Annual Statistical Bulletin
Petroleum Economist
Own Estimates and Calculations

Table 4

TOTAL REFINERY CAPACITY 1970-1982
Unit: million tonnes per annum

Region/country	1970	1975	1980	1981	1982	% of total	Millions 1982 capacity in barrels per day
North America	710.5	864.0	1,014.8	1,042.4	1,006.7	24.57	20.2
USA	643.0	760.0	895.8	927.2	890.8	21.74	17.9
Canada	67.5	104.0	119.0	115.2	115.9	2.83	2.3
South America	282.3	376.0	431.0	437.7	433.1	10.56	8.6
Venezuela	68.2	77.8	76.0	76.0	76.0	1.85	1.4
Mexico	29.6	39.3	56.8	63.5	63.5	1.55	1.3
Ecuador	1.7	2.2	4.8	4.8	4.8	0.12	0.1
Brazil	27.3	51.7	69.8	69.8	70.1	1.71	1.4
Argentina	24.0	31.3	33.7	33.8	33.8	0.82	0.7
Middle East	122.8	137.5	209.2	230.3	207.1	5.05	4.2
United Arab Emirates	—	—	0.8	0.8	6.2	0.16	0.1
Saudi Arabia	33.7	35.0	38.0	58.6	58.6	1.43	1.2
Iraq	5.8	9.2	15.0	15.0	15.0	0.37	0.3
Iran	30.4	40.3	63.2	63.7	38.2	0.69	0.6
Kuwait	21.9	30.5	29.7	29.7	29.7	0.72	0.6
Qatar	0.03	0.4	0.5	0.5	0.5	0.01	0.01
Israel	6.0	10.5	9.7	9.5	9.5	0.23	0.2
Syria	3.0	3.7	11.1	11.1	11.6	0.28	0.2
Africa	38.1	64.6	92.6	111.2	111.7	2.73	2.3
Algeria	2.9	5.8	6.3	20.3	20.3	0.50	0.4
Egypt	5.2	8.8	12.5	14.5	14.5	0.35	0.3
Gabon	0.9	1.2	2.3	2.3	2.3	0.06	0.04
Libya	0.5	3.0	6.5	6.5	6.5	0.16	0.1
Nigeria	2.8	3.0	7.5	12.2	12.2	0.30	0.2
Morocco	1.7	3.6	3.6	3.6	3.6	0.09	0.1
Tunisia	1.2	1.2	1.7	1.7	1.7	0.04	0.03
South Africa	10.4	19.0	23.8	23.3	21.1	0.51	0.4
Western Europe	756.2	1,032.3	1,027.4	1,002.3	976.5	23.83	19.6
West Germany	120.3	153.9	153.9	150.4	143.4	3.50	2.9
France	116.5	169.5	166.8	166.1	158.3	3.86	3.2
Italy	168.0	207.5	210.0	195.6	196.1	4.78	3.9
United Kingdom	114.8	146.4	133.3	130.5	125.4	3.06	2.5
Belgium	35.9	48.8	54.3	52.3	52.3	1.28	1.1
Netherlands	68.5	103.1	91.0	91.0	85.0	2.08	1.7

...contd..

Table 4 contd.

TOTAL REFINERY CAPACITY 1970-82
Unit: million tonnes per annum

Region/country	1970	1975	1980	1981	1982	% of total	Millions 1982 capacity in barrels per day
Spain	34.9	57.7	75.1	78.1	76.1	1.86	1.5
Greece	5.1	12.9	21.5	21.3	21.3	0.52	0.4
Sweden	13.1	21.2	22.8	22.5	23.5	0.57	0.5
Turkey	7.9	14.0	16.8	16.8	18.3	0.45	0.4
EEC	642.7	855.9	844.3	820.7	795.3	19.41	16.0
Centrally Planned Economies	420.6	612.5	756.0	790.6	800.6	19.54	16.1
USSR	304.0	420.0	545.0	567.7	577.7	14.10	11.6
China	27.6	60.5	79.7	90.1	90.1	2.20	1.8
Romania	16.0	20.0	30.4	30.9			
East Germany	10.7	17.8	24.0	24.0			
Czechoslovakia	11.0	16.0	22.0	22.8			
Yugoslavia	12.0	12.0	14.8	14.8	14.8	0.36	0.3
Asia/Oceania	272.9	472.8	537.1	551.4	562.1	13.72	11.3
Indonesia	14.7	20.8	23.5	23.5	23.5	0.57	0.5
India	23.3	27.8	27.7	27.7	27.7	0.68	0.6
Australia	30.0	35.0	36.1	36.1	36.1	0.88	0.7
Japan		282.8	297.0	297.0	297.0	7.25	6.0
Singapore	18.4	46.0	45.9	53.2	54.6	1.34	1.1
Malaysia	6.8	6.8	8.6	8.8	8.7	0.21	0.2
OPEC	183.5	229.2	274.1	313.9	283.8	6.94	5.6
OECD	1,507.8	2,231.6	2,395.8	2,398.3	2,338.3	57.1	47.0
WORLD TOTAL	2,626.6	3,550.1	4,068.0	4,166.0	4,097.7	100.00	82.2

Sources: Statistiques de l'Industrie Petroliere
BP Statistical Review
UN Yearbook of World Energy Statistics

Table 5

CRUDE OIL EXPORTS 1970-1983
Unit: million metric tonnes

Region/country	1970	1975	1980	1981	1982	1983 (est.)
North America	33.3	34.4	19.7	16.2	20.3	22.4
USA	0.7	1.1	4.8	3.5	3.8	3.6
Canada	32.6	33.3	14.9	12.7	16.5	18.8
South America	128.5	95.2	119.1	131.5	141.5	
Venezuela	127.6	76.7	67.5	65.9	55.1	
Mexico		4.8	41.3	54.8	74.1	
Ecuador	0.04	7.3	5.5	6.3	5.4	
Middle East	607.1	887.4	816.8	678.5	511.4	
United Arab Emirates	37.7	82.1	82.1	70.6	57.3	
Saudi Arabia	159.5	328.2	461.9	449.8	281.3	
Iraq	73.3	101.9	120.9	36.6	39.8	
Iran	165.1	233.7	40.2	36.0	81.8	
Kuwait	129.4	90.9	63.8	42.9	19.4	
Qatar	17.4	20.7	22.5	19.0	15.7	
Africa	283.7	224.0	244.0	180.2	143.6	
Algeria	46.4	40.6	33.3	34.6	15.2	
Gabon	4.5	10.4	7.7	6.0	5.7	
Libya	159.5	69.2	81.8	48.5	45.6	
Nigeria	51.2	84.9	96.8	64.4	51.5	
Egypt	14.7	5.2	8.0	10.0		
Western Europe	2.1	12.9	68.8	77.0	94.1	104.4
United Kingdom	1.1	0.8	38.5	51.0	60.2	64.4
Norway	0.6	7.8	20.4	18.9	19.2	25.0
Eastern Europe	67.8	94.5	123.5	120.8	129.8	
USSR	66.8	93.1	122.0	116.0	119.0	
Asia/Oceania	43.1	72.2	89.4	88.4	79.8	
China		8.3	13.1	14.5		
Indonesia	29.9	52.5	50.9	50.1	41.6	
OPEC	1,001.6	1,199.2	1,134.8	930.7	715.4	
OPEC percentage share	86.0	84.5	76.7	71.9	63.8	
WORLD TOTAL	1,164.7	1,418.6	1,480.3	1,293.4	1,120.6	

Sources: OPEC Statistical Bulletin
OECD/IEA Quarterly
UN Yearbook of World Energy Statistics
Own Estimates

Table 6

CRUDE OIL IMPORTS 1970-1983
Unit: million metric tonnes

Region/country	1970	1975	1980	1981	1982	1983 (est.)
North America	95.6	244.5	289.5	244.1	197.8	182.6
USA	66.2	203.1	261.4	218.3	181.1	170.6
Canada	29.4	41.4	28.1	25.8	16.7	12.0
South America	126.4	139.0	142.0	133.5	109.0	
Brazil	17.4	35.7	43.6	42.3		
Middle East/		23.7	27.1	27.6	30.5	
Africa	21.7	33.4	32.9	34.7	33.2	
South Africa	8.8	15.0	15.0	17.0		
Western Europe	596.4	630.5	579.8	502.1	456.5	421.6
Belgium	29.8	28.4	32.0	27.5	23.7	21.1
France	100.2	106.1	109.5	90.3	75.5	65.4
Germany	98.8	90.0	97.9	79.6	72.5	65.5
Greece	4.6	12.9	17.2	18.5	14.6	11.5
Italy	113.8	93.8	88.6	85.5	79.6	70.6
Netherlands	60.3	55.2	50.0	38.2	37.2	38.9
Spain	32.1	41.5	47.4	46.4	43.4	41.8
Sweden	11.8	12.1	18.1	15.2	13.1	14.7
Turkey	3.8	9.6	10.5	11.5	14.0	13.6
United Kingdom	100.8	87.2	44.8	36.9	33.8	31.0
EEC	521.1	484.1	448.8	383.0	355.2	322.3
Eastern Europe		84.1	101.2	92.7	85.7	
USSR	3.5	6.5	7.0	7.0	6.0	
Asia/Oceania		335.1	358.7	336.6	309.0	
India	11.7	13.6	16.0	15.6		
Japan	27.1	224.9	216.8	196.9	184.1	167.2
Singapore	11.0	21.4	30.9	36.0		
Australia	16.2	7.7	8.0	9.1	11.8	9.6
OECD	738.1	1,110.8	1,123.8	985.2	877.5	806.9
OECD percentage share	63.3	77.8	73.3	71.9	71.8	
WORLD TOTAL	1,166.5	1,427.2	1,532.5	1,369.4	1,221.7	

Sources: UN Yearbook of World Energy Statistics
  OPEC Statistical Bulletin
  OECD/IEA Quarterly Statistics
  ENI—Energy and Hydrocarbons
  Own Estimates

Table 7

TOTAL PRODUCTION OF REFINED PETROL PRODUCTS 1970-1983
Unit: million metric tonnes

Region/country	1970	1979	1980	1981	1982	1983 (est.)
North/Central America	717.4	814.6	869.5	834.4		
USA	531.8	606.2	635.4	613.4	643.6	615.8
Canada	60.2	82.0	90.5	83.3	76.4	70.7
Mexico	21.1	30.2	52.4	57.2	55.0	
Netherlands Antilles	45.6	24.0	25.0	23.7		
US Virgin Islands	13.1	24.6	24.1	20.2		
South America	122.3	124.1	144.0	136.5		
Argentina	19.6	19.6	23.7	23.3		
Brazil	23.5	40.6	47.3	47.1		
Ecuador	1.1	1.9	4.5	4.3	4.5	
Venezuela	61.6	42.3	46.7	40.6	39.4	
North Africa/Middle East	119.4	132.8	175.1	172.5		
Algeria	2.3	5.1	8.9	8.9	12.8	
Libya	0.4	2.1	5.0	5.4	6.2	
Egypt	3.2	8.4	13.1	13.2		
Bahrain	11.2	9.7	10.5	11.2		
Iran	24.6	31.4	28.7	22.9	24.1	
Iraq	3.3	4.5	8.1	6.5	6.5	
Kuwait	18.1	13.3	17.0	13.4	16.2	
Qatar		0.3	0.5	0.5	0.5	
Saudi Arabia	26.9	22.6	33.0	35.4	32.0	
United Arab Emirates			1.3	3.2	5.1	
Africa	18.8	28.5	31.9	32.2		
Gabon	0.9	0.9	1.2	1.2	1.2	
Nigeria	1.0	2.2	6.2	6.5	6.6	
South Africa	7.7	13.3	12.7	13.1		
Western Europe	537.4	534.9	551.3	502.6	545.2	527.6
Belgium	26.0	25.7	28.9	24.9	24.8	22.3
France	88.8	96.7	97.3	85.5	84.5	64.0
West Germany	94.3	86.0	91.8	85.5	94.7	90.0
Italy	103.6	90.0	85.4	82.6	86.9	81.1
Netherlands	51.6	51.2	49.8	42.3	45.7	49.5
United Kingdom	90.6	84.2	77.2	69.4	76.2	76.0
Spain	29.3	39.4	44.2	42.7	45.4	43.7
EEC	471.5	454.0	451.6	411.5	434.3	418.2
Eastern Europe	303.4	430.5	502.1	503.2		
USSR	245.9	343.7	395.6	404.6		

...contd..

Table 7 contd.

TOTAL PRODUCTION OF REFINED PETROL PRODUCTS 1970-1983
Unit: million metric tonnes

Region/country	1970	1979	1980	1981	1982	1983 (est.)
East Germany	10.8	16.4	18.8	18.2		
Romania	14.0	17.2	24.1	21.4		
Far East	234.3	329.3	382.4	367.5		
China	21.8	54.8	76.0	71.5		
India	14.3	16.9	19.4	22.4		
Indonesia	10.4	14.9	23.7	23.4	23.5	
Japan	146.1	188.0	184.8	170.0	176.7	159.8
Republic of Korea	8.8	14.3	21.9	21.3		
Singapore	10.7	15.8	26.7	28.2		
Oceania	25.6	29.6	32.8	31.3		
Australia	22.2	25.7	28.6	27.2	28.3	27.1
New Zealand	2.8	2.8	2.7	2.5	2.3	2.4
OPEC	150.6	141.5	184.8	172.2	178.6	
OECD	1,300.5	1,439.6	1,493.3	1,399.0	1,472.5	1,403.4
OPEC percentage share	7.2	5.8	6.9	6.7		
OECD percentage share	62.6	59.4	55.5	54.2		
WORLD TOTAL	2,078.6	2,424.2	2,689.1	2,580.2		

Sources: UN Yearbook of World Energy Statistics
OECD/IEA Oil Statistics Quarterly
OPEC Statistical Bulletin
Statistiques de l'Industrie Petroliere
Own Estimates

Table 8

TOTAL REFINED PETROL PRODUCTS EXPORTS 1970-1983
Unit: million metric tonnes

Region/country	1970	1975	1980	1981	1982	1983 (est.)
North/Central America	83.8	79.2	75.0	71.9	87.3	
USA	4.4	1.8	13.0	16.2	29.6	29.7
Canada	2.5	7.2	6.1	5.9	4.3	6.1
Netherlands Antilles	39.6	24.8	21.8	21.0		
Puerto Rico	3.4	3.6	3.1	2.4		
Trinidad and Tobago	18.2	9.6	9.3	6.3		
US Virgin Islands	9.1	21.5	18.7	14.9		
South America	52.9	31.7	35.5	31.6	28.5	
Ecuador	0	0	1.1	1.1	1.0	
Venezuela	50.5	28.9	29.2	22.8	23.8	
North Africa/Middle East	50.1	40.8	56.6	54.0	59.5	
Iran	10.8	8.3	6.3	4.6	4.7	
Iraq	0.1	0.7	3.2	2.3	2.1	
Kuwait	13.2	8.3	12.5	8.4	12.4	
Qatar	0	0.1	0.1	0.2	—	
Saudi Arabia	12.6	10.4	15.0	17.6	19.7	
United Arab Emirates	0	0	0.8	1.3	—	
Bahrain	9.9	8.4	9.5	10.1		
Algeria	0.5	2.1	4.9	4.8	10.0	
Libya	0.1	1.0	1.8	1.7	2.2	
Egypt	0	0.5	0.5	0.5		
Africa	3.3	3.9	4.1	4.1	5.9	
Nigeria	0	0.3	0.5	0.5	0.1	
Gabon	0.5	0.5	0.6	0.7	0.7	
Western Europe	102.0	87.5	126.0	127.5	131.8	137.7
Belgium	7.9	10.5	17.6	16.9	15.3	15.8
France	8.1	10.0	13.6	15.2	11.8	10.1
West Germany	5.8	4.2	7.1	7.3	8.0	7.7
Italy	23.9	11.8	11.8	14.2	13.9	13.2
Netherlands	28.8	27.8	40.9	39.9	45.8	49.8
Spain	4.6	1.8	2.5	3.0	4.2	5.2
United Kingdom	16.2	12.4	14.1	12.3	12.6	13.2
EEC	93.3	81.5	113.8	114.8	116.4	118.9
Eastern Europe	34.5	47.7	56.1	56.3	63.0	
USSR	25.6	37.0	41.0	43.0	44.0	
Romania	4.9	5.8	8.8	8.0		
East Germany	1.2	2.5	2.8	2.7		

...contd..

Table 8 contd.

TOTAL REFINED PETROL PRODUCTS EXPORTS 1970-1983
Unit: million metric tonnes

Region/country	1970	1975	1980	1981	1982	1983 (est.)
Far East	20.0	19.1	31.6	30.8	30.5	
Indonesia	0.9	2.7	5.6	5.3	4.0	
Japan	0.3	1.3	0.4	0.5	1.7	2.8
Singapore	12.0	11.6	20.6	19.1		
Malaysia	1.2	0.1	0.1	0.2		
Oceania	1.5	2.8	3.4	2.8	4.6	
Australia	1.4	2.4	1.5	2.4	4.3	4.4
New Zealand	0.1	0	0.2	—	—	—
OPEC	89.2	63.3	81.6	71.3	80.7	
OPEC percentage share	25.6	20.6	21.0	18.8	19.6	
OECD	110.7	100.2	147.1	152.4	171.6	180.5
OECD percentage share	31.8	32.6	37.9	40.2	41.7	
WORLD TOTAL	348.1	307.7	388.3	379.0	411.1	

Sources: OPEC Statistical Bulletin
      OECD/IEA Statistical Quarterly
      Statistiques de l'Industrie Petroliere
      UN Yearbook of World Energy Statistics
      Own Estimates

Table 9

TOTAL REFINED PETROL PRODUCTS IMPORTS 1970-1983
Unit: million metric tonnes

Region/country	1970	1975	1980	1981	1982	1983 (est.)
North/Central America	127.2	116.4	93.7	84.8	84.8	
USA	103.1	93.1	76.1	68.5	49.9	54.8
Canada	9.3	1.4	1.4	1.5	2.1	1.8
Mexico	1.8	2.8	0.3	0.2	0.2	
Netherlands Antilles	1.8	6.5	2.1	2.2		
Panama	4.7	3.9	2.4	2.2		
Puerto Rico	0.7	0.7	1.5	1.5		
South America	4.2	3.4	6.3	5.1	7.7	
Argentina	1.1	1.0	0.6	0.6		
North Africa/Middle East	5.0	7.0	14.2	12.4	10.7	
Lebanon	0.2	0.2	0.1	0.2		
Oman	0.4	1.1	0.8	0.7		
Syria	0	1.2	0.8	1.0		
Egypt	2.1	0.1	0.2	0.1		
Tunisia	0.1	0.3	0.8	1.2		
Africa	11.1	8.7	9.5	8.5	9.1	
Western Europe	116.8	115.5	171.3	167.1	189.2	192.5
Belgium	4.9	6.9	10.0	9.3	12.4	12.6
Denmark	10.5	10.3	7.9	6.2	6.4	5.2
France	4.9	5.9	12.7	13.8	18.9	20.5
West Germany	25.8	29.6	37.5	36.4	38.7	43.5
Italy	2.2	9.4	18.8	17.6	17.1	15.8
Netherlands	7.7	5.0	28.5	32.8	36.4	37.8
United Kingdom	15.9	9.8	9.2	9.1	12.5	9.5
Norway	3.6	2.4	3.3	3.0	3.2	2.6
Spain	0.9	1.6	4.3	4.8	7.5	9.9
Sweden	19.8	15.1	12.9	10.4	11.1	9.0
Switzerland	7.3	7.6	8.6	7.9	7.2	8.5
EEC	72.7	79.7	131.7	132.0	151.7	152.9
Eastern Europe	7.5	6.0	8.6	8.9	6.9	
Bulgaria	2.5	1.9	2.0	2.0		
Hungary	1.0	0.8	1.6	1.5		
Poland	1.9	1.6	3.1	3.4		
Far East	54.1	38.8	59.5	57.4	62.2	
Hong Kong	3.8	4.6	6.5	6.8		
India	0.7	2.2	6.4	5.3		
Japan	26.7	13.9	24.6	27.0	29.9	32.9

...contd..

223

Table 9 contd.

TOTAL REFINED PETROL PRODUCTS IMPORTS 1970-1983
Unit: million metric tonnes

Region/country	1970	1975	1980	1981	1982	1983 (est.)
Singapore	9.4	4.9	5.4	4.1		
Oceania	4.7	7.5	7.5	5.9	6.9	
Australia	2.2	3.3	3.0	2.8	2.2	2.6
New Zealand	0.7	1.1	1.3	1.0	1.2	1.2
OECD	258.8	228.3	254.8	249.2	274.6	285.9
OPEC	1.9	5.0	9.6	8.6		
WORLD TOTAL	331.8	304.1	370.6	350.1	377.5	

Sources: OPEC Statistical Bulletin
OECD/IEA Statistical Quarterly
UN Yearbook of World Energy Statistics
Statistiques de l'Industrie Petroliere
Own Estimates

Table 10

TOTAL PRODUCTION OF REFINERY FEEDSTOCKS 1975-1983
Unit: thousand metric tonnes

Region/country	1975	1980	1981	1982	1983 (est.)
North/Central America					
USA	—	—	—	10,542	4,420
Canada	—	—	958	1,450	1,496
South America					
North Africa/Middle East					
Africa					
Western Europe	3,606	20,415	24,848	31,886	35,135
West Germany	—	12,462	15,420	16,951	20,356
France	—	—	—	2,348	2,212
Italy	2,844	2,150	2,139	1,730	1,681
United Kingdom	142	2,005	2,486	3,163	2,487
Belgium	—	—	—	206	—
Netherlands	—	2,452	3,718	6,396	7,261
Norway	—	274	185	202	188
Sweden	115	433	152	557	584
Switzerland	—	—	—	—	—
Spain	—	485	243	324	596
EEC	3,482	18,966	23,720	30,794	33,767
East Europe					
Far East					
Japan	16,269	21,383	15,695	4,920	4,707
Oceania					
Australia	—	—	—	—	—
New Zealand	—	—	—	—	—
OECD WORLD TOTAL	21,355	41,798	41,501	48,798	45,757

Source: OECD/IEA Oil Statistics Quarterly
Own Estimates

225

Table 11

REFINERY FEEDSTOCKS (FUEL GAS OR REFINERY GAS) EXPORTS 1975-1983
Unit: thousand metric tonnes

Region/country	1975	1980	1981	1982	1983 (est.)
North America	1,077	—	—	—	—
USA	1,077	—	—	—	—
Canada	—	—	—	—	—
South America					
Middle East/					
Africa					
Western Europe	—	1,749	1,184	1,644	1,672
West Germany	—	142	—	—	—
France	—	—	—	—	—
Italy	—	87	550	524	432
United Kingdom	186	797	272	504	828
Belgium	—	1	86	318	
Denmark	46	394	30	38	5
Norway	—	230	153	144	185
Sweden	333	98	93	116	176
EEC	405	1,421	938	1,384	1,311
Eastern Europe					
Asia/Oceania					
Japan	—	—	—	—	—
Australia	—	24	52	36	11
OECD WORLD TOTAL	1,567	1,773	1,236	1,680	1,683

Source: OECD/IEA Oil Statistics Quarterly
　　　　Own Estimates

Table 12

REFINERY FEEDSTOCKS (FUEL GAS OR REFINERY GAS) IMPORTS 1975-1983
Unit: thousand metric tonnes

Region/country	1975	1980	1981	1982	1983 (est.)
North America	1,016	1,663	2,875	5,232	8,084
USA	1,016	1,663	2,875	5,232	8,084
Canada	—	—	—	—	—
South America					
Middle East/					
Africa					
Western Europe	14,203	18,569	17,589	19,967	24,017
West Germany	1,823	2,280	—	—	—
France	—	4,062	4,841	4,516	3,555
Italy	3,292	4,609	5,430	5,355	6,380
United Kingdom	2,379	3,455	3,783	6,063	8,352
Austria	102	268	135	118	297
Belgium	921	1,474	1,464	1,608	3,047
Denmark	45	1,028	637	819	680
Norway	137	306	175	483	420
Portugal	9	57	115	354	81
Sweden	1,087	—	214	272	889
Switzerland	338	772	745	330	169
EEC	12,203	17,120	16,205	18,410	22,160
Eastern Europe					
Asia/Oceania					
Japan	—	—	—	—	—
Australia	—	196	1,701	2,399	3,208
New Zealand	676	650	445	295	239
OECD WORLD TOTAL	17,781	21,078	22,610	27,893	35,548

Source: OECD Oil Statistics Quarterly
      Own Estimates

227

Table 13

TOTAL PRODUCTION-LIQUID PETROLEUM GAS 1970-1983

Unit: thousand metric tonnes

Region/country	1970	1975	1980	1981	1982	1983 (est.)
North/Central America	50,102	55,675	61,059	63,211		
USA	45,123	47,859	48,229	50,090	8,698	10,261
Canada	3,569	5,818	8,549	8,319	2,443	2,639
Mexico	1,134	1,620	3,763	4,297		
South America	3,402	5,417	6,080	5,502		
Ecuador	5	5	72	60		
Venezuela	1,295	1,842	1,647	1,251		
Brazil	972	1,940	2,553	2,300		
North Africa/Middle East	3,765	7,880	15,881	17,682		
Algeria	107	546	896	880	2,322	
Libya	—	304	383	380		
Iran	550	949	850	800		
Iraq	22	310	260	210		
Kuwait	1,101	1,114	2,400	1,409		
Qatar	0	141	114	161		
Saudi Arabia	1,461	3,415	8,349	10,647		
United Arab Emirates	0	0	781	1,274		
Africa	106	193	200			
Gabon	6	5	5	5		
Nigeria	1	9	20	20		
Western Europe	10,739	11,897	13,063	12,199	11,962	12,455
France	2,566	2,988	3,133	2,831	2,523	
West Germany	2,048	2,030	2,408	2,278	2,264	2,112
Italy	2,064	2,283	2,069	1,908	1,855	1,832
Netherlands	704	929	932	754	1,039	1,208
Spain	1,062	749	1,094	968	934	1,059
United Kingdom	1,181	1,450	1,385	1,402	1,417	1,640
EEC	9,248	10,335	10,776	10,038	9,909	10,119
Eastern Europe	5,722	8,141	10,194	10,437		
USSR	4,925	7,107	8,521	8,995		
Far East	7,328	9,225	9,299	9,271		
Indonesia	16	30	27	29		
Japan	6,880	8,147	8,022	7,811	5,555	3,800
Oceania	313	1,497	2,029	1,974		
Australia	312	1,492	2,616	1,929	363	327
OPEC	4,564	8,670	15,804	17,126		
OECD	66,623	75,215	79,888	80,389	29,021	29,481
OPEC percentage share	5.6	8.7	13.4	14.2		
OECD percentage share	81.8	75.3	67.8	66.7		
WORLD TOTAL	81,477	99,925	117,800	120,476		

Sources: OECD/IEA Oil Statistics Quarterly
UN Yearbook of World Energy Statistics
Own Estimates

228

Table 14

LIQUID PETROLEUM GAS—EXPORTS 1970-1983
Unit: thousand metric tonnes

Region/country	1970	1975	1980	1981	1982	1983 (est.)
North America	2,665	3,762	5,627	1,636	218	209
USA	855	815	673	1,319		
Canada	1,810	2,947	4,956	317	218	209
South America	719	1,112	1,376	825		
Venezuela	655	900	751	558		
Middle East/Asia	2,880	5,504	12,007	14,074		
Iran	258	500	550	300		
Iraq	—	221	160	160		
Kuwait	1,101	1,099	2,220	1,209		
Saudi Arabia	1,420	3,284	7,905	10,450		
United Arab Emirates	0	0	769	1,260		
Africa	15	411	726	735		
Algeria			421	615	877	
Western Europe	2,068	2,069	2,568	3,111	3,279	3,599
Belgium	131	87	244	307	297	279
France	528	657	658	681	692	644
Germany	459	267	430	542	589	747
Italy	234	330	232	159	203	285
Netherlands	418	445	546	847	838	679
United Kingdom	105	176	364	427	452	581
EEC	1,932	2,003	2,501	3,043	3,157	3,327
Eastern Europe	182	346	399			
USSR	132	258	275			
Oceania	147	1,080	1,501	787		
Australia	147	1,080	1,501	787	1,377	1,432
OECD	4,923	6,911	19,696	15,534	4,877	5,244
WORLD TOTAL	8,680	14,341	24,615	26,200		

Sources: OECD/IEA Oil Statistics Quarterly
UN Yearbook of World Energy Statistics
Own Estimates

Table 15

LIQUID PETROLEUM GAS—IMPORTS 1970-1983
Unit: thousand metric tonnes

Region/country	1970	1975	1980	1981	1982	1983 (est.)
North America	1,646	3,497	6,788	7,642		
USA	1,624	3,496	6,788	7,642		
Canada	22	1				
South America	1,760	1,711	1,046	944		
Mexico	762	942	265	166		
Argentina	378	416	309	232		
Brazil	376	28	131	136		
Middle East/Asia	1,895	6,091	10,640	11,386		
Japan	2,683	5,683	9,725	10,073	11,861	10,627
Africa	226	346	673	538		
Algeria	50	52	160	160		
Egypt	93	120	245	100		
Western Europe	1,932	2,455	5,672	7,745	7,687	6,688
Belgium	215	266	267	353	391	329
Denmark	115	91	143	143	155	119
France	216	244	549	959	1,140	875
Germany	259	247	623	671	817	675
Italy	69	175	337	638	537	607
Netherlands	88	54	994	2,060	1,634	1,015
Portugal	212	237	268	267	296	264
Spain	444	826	1,518	1,590	1,322	1,588
Turkey	73	149	348	424	485	440
United Kingdom	175	54	141	200	363	264
EEC	1,218	1,269	3,246	5,215	5,240	4,049
Eastern Europe	77	101	1,788	387		
Oceania	11	16	22	19		
OECD	6,264	11,636	22,188	25,460	19,548	17,315
WORLD TOTAL	8,547	14,217	26,629	28,661		

Sources: OECD/IEA Oil Statistics Quarterly
UN Yearbook of World Energy Statistics
Own Estimates

230

Table 16

TOTAL PRODUCTION MOTOR GASOLINE 1970-1983
Unit: million metric tonnes

Region/country	1970	1975	1980	1981	1982	1983 (est.)
North/Central America	282.4	325.7	335.0	329.1		
USA	245.4	280.0	280.2	275.0	278.4	274.8
Canada	19.2	26.0	28.5	27.3	26.5	25.3
Mexico	5.9	7.8	13.7	15.1	14.5	
South America	19.2	25.1	26.2	28.6		
Ecuador	0.4	0.7	1.0	0.9		
Venezuela	3.1	4.7	7.0	7.0		
Brazil	7.1	10.5	8.2	10.2		
North Africa/Middle East	9.5	15.0	20.3	20.3		
Algeria	0.5	0.8	1.1	1.1		
Libya	0.1	0.2	0.5	0.6		
Egypt	0.5	1.3	2.0	2.1		
Iran	2.1	3.1	3.3	2.5		
Iraq	0.4	0.5	1.3	1.3		
Kuwait	0.4	0.5	0.9	1.0		
Qatar	0	0.1	0.1	0.1		
Saudi Arabia	0.7	1.0	3.5	3.5		
United Arab Emirates	0	0	0.2	0.4		
Africa	4.3	6.3	8.2	8.4		
Gabon	0.1	0.1	0.1	0.1		
Nigeria	0.2	0.5	2.4	2.5		
South Africa	2.3	3.8	3.7	3.9		
Western Europe	70.4	86.4	105.5	102.0	104.5	107.9
France	13.1	16.5	17.4	16.8	17.9	17.2
West Germany	13.9	16.6	21.2	20.4	20.4	20.2
Italy	12.9	14.0	16.0	15.2	15.6	15.4
Netherlands	4.6	6.4	8.1	7.2	8.4	9.0
United Kingdom	11.3	13.9	16.7	17.2	19.2	20.2
EEC	61.8	74.8	88.2	85.1	88.6	90.5
Eastern Europe	57.4	80.1	91.5	91.9		
USSR	46.9	65.5	73.5	75.0		
Far East	26.4	36.4	47.6	47.5		
China	3.0	7.9	10.5	9.5		
Indonesia	1.4	1.9	2.8	3.0		
Japan	15.4	19.9	25.3	25.8	26.2	25.6
Oceania	8.5	10.4	12.2	12.1		
Australia	7.3	9.2	10.5	10.9	11.3	10.9

...contd..

231

Table 16 contd.

TOTAL PRODUCTION MOTOR GASOLINE 1970-1983
Unit: million metric tonnes

Region/country	1970	1975	1980	1981	1982	1983 (est.)
New Zealand	1.2	1.2	1.3	1.3	1.2	1.3
OPEC	9.4	14.1	24.2	24.0		
OECD	358.9	422.7	451.3	442.3	448.1	445.8
OPEC percentage share	2.0	2.4	3.7	3.8		
OECD percentage share	75.1	72.2	69.8	69.1		
WORLD TOTAL	478.1	585.4	646.5	639.9		

Sources: UN Yearbook of World Energy Statistics
OECD/IEA Oil Statistics Quarterly
OPEC Statistical Bulletin
Statistiques de l'Industrie Petroliere
Own Estimates

Table 17

MOTOR GASOLINE EXPORTS 1970-1983
Unit: thousand metric tonnes

Region/country	1970	1975	1980	1981	1982	1983 (est.)
North America	156	530	750	706	1,989	1,568
USA	54	88	56	82	1,410	509
Canada	102	442	694	624	579	1,059
South America	8,888	8,295	8,997	8,687		
Netherlands Antilles	3,556	1,237	2,600	2,600		
Puerto Rico	2,490	1,951	1,651	1,412		
Trinidad & Tobago	2,109	1,489	1,016	800		
US Virgin Islands	178	2,931	3,038	2,500		
Brazil	0	112	250	1,050		
Middle East/Asia	4,740	3,567	5,189	5,210		
Iran	1,200	960	1,150	1,000		
Singapore	1,321	798	1,866	1,630		
Bahrain	937	1,058	898	1,030		
Africa	740	854	713	611		
Western Europe	11,377	14,152	21,234	21,227	21,592	24,715
Belgium	1,795	2,647	3,932	3,625	2,381	2,673
France	1,315	1,616	1,736	1,327	999	849
West Germany	879	904	1,321	1,578	1,801	1,375
Italy	2,663	2,803	3,497	3,688	3,511	3,824
Netherlands	1,997	3,348	7,062	6,903	8,342	9,808
United Kingdom	942	1,155	906	1,169	2,047	2,700
EEC	9,996	13,081	19,615	19,453	20,275	22,583
Eastern Europe	5,570	7,873	12,025	11,500		
USSR	3,900	5,675	8,000	8,000		
Romania	647	1,464	2,693	2,500		
Oceania	172	198	245	306		
OECD	11,755	14,866	22,187	22,229	23,906	26,699
WORLD TOTAL	31,643	35,469	49,153	48,247		

Sources: UN Yearbook of World Energy Statistics
OECD/IEA Oil Statistics Quarterly
Own Estimates

Table 18

MOTOR GASOLINE IMPORTS 1970-1983
Unit: thousand metric tonnes

Region/country	1970	1975	1980	1981	1982	1983 (est.)
North America	3,495	7,915	2,192	4,164	6,753	9,806
USA	2,861	7,912	2,056	4,069	6,667	9,519
Canada	634	3	136	95	86	287
South America	1,326	1,851	3,006	2,675		
Middle East/Asia	3,060	3,543	3,221	3,708		
Africa	1,947	2,155	2,540	2,224		
Nigeria	220	498	585	450		
Western Europe	13,512	15,363	18,653	21,228	21,799	23,325
Belgium	392	728	1,425	968	1,203	1,352
France	392	562	664	1,374	2,202	2,527
Germany	2,505	4,482	3,791	5,777	6,516	7,404
Netherlands	498	297	2,171	2,958	2,857	3,941
Sweden	1,834	1,877	1,744	1, 585	1,679	1,312
Switzerland	1,303	1,616	1,684	1,832	1,800	1,916
United Kingdom	3,919	2,660	2,702	2,174	729	440
EEC	8,661	10,304	13,588	16,513	16,802	18,651
Eastern Europe	2,372	1,400	669	2,408		
USSR	600	470	400	500		
Oceania	881	1,119	1,191	1,130		
Australia	445	300	318	371	206	439
OECD	17,664	24,037	21,631	26,130	29,198	34,073
WORLD TOTAL	26,593	32,346	35,440	35,757		

Sources: OECD/IEA Oil Statistics Quarterly
UN Yearbook of World Energy Statistics
Own Estimates

Table 19

TOTAL PRODUCTION OF AVIATION FUEL (GASOLINE) 1975-1981
Unit: thousand metric tonnes

Region/country	1970	1975	1980	1981
North/Central America	3,021	2,040	1,998	1,820
USA	2,288	1,592	1,489	1,339
Canada	137	169	171	172
Netherlands Antilles	484	190	190	180
South America	155	107	109	100
Venezuela	22	27	33	40
North Africa/Middle East	711	514	317	193
Iran	667	470	250	130
Saudi Arabia	4	0	0	0
Africa	—	14	7	7
Western Europe	830	656	737	999
France	50	36	29	30
Germany	343	410	359	422
Italy	153	64	193	147
Netherlands	221	130	87	91
United Kingdom	49	16	66	57
EEC	817	656	737	999
Eastern Europe	22	16	11	11
Far East	127	48	38	32
Indonesia	20	21	18	14
Japan	58	26	19	17
Oceania	29	37	67	65
Australia	29	37	67	65
OPEC	713	518	301	184
OECD	3,342	2,480	2,483	2,592
OPEC percentage share	15	15	9	6
OECD percentage share	68	72	76	80
WORLD TOTAL	4,895	3,432	3,284	3,227

Source: UN Yearbook of World Energy Statistics

Table 20

AVIATION FUEL (GASOLINE) EXPORTS 1970-1981
Unit: thousand metric tonnes

Region/country	1970	1975	1980	1981
North America	108	12	0	0
USA	107	12	0	0
Canada	1	0	0	0
South America	545	228	248	244
Netherlands Antilles	482	193	189	179
Middle East/Asia	895	526	237	131
Iran	655	450	180	110
Singapore	233	75	57	21
Africa	0	2	3	7
Western Europe	394	339	376	318
West Germany	21	149	257	233
Netherlands	259	141	118	85
EEC	392	324	376	318
East Europe	0	0	0	0
Oceania	12	18	9	8
Australia	12	17	7	7
OECD	517	369	383	325
WORLD TOTAL	1,954	1,125	873	708

Source: UN Yearbook of World Energy Statistics

Table 21

AVIATION FUEL (GASOLINE) IMPORTS 1970-1981
Unit: thousand metric tonnes

Region/country	1970	1975	1980	1981
North America	14	2	0	0
USA	0	0	0	0
Canada	14	2	0	0
South America	419	358	427	398
Puerto Rico	105	131	220	200
Middle East/Asia	1,574	480	271	198
Singapore	249	229	51	32
Africa	163	120	145	137
Western Europe	504	297	167	275
France	43	38	21	20
Germany	63	75	35	176
United Kingdom	86	62	28	18
EEC	354	199	120	232
East Europe	19	4	4	2
Oceania	148	152	116	87
Australia	38	63	17	15
New Zealand	23	20	37	7
OECD	579	468	221	297
WORLD TOTAL	2,841	1,413	1,130	1,097

Source: UN Yearbook of World Energy Statistics

237

Table 22

TOTAL PRODUCTION OF AVIATION KEROSENE (JET FUEL) 1970-1983
Unit: thousand metric tonnes

Region/country	1970	1975	1980	1981	1982	1983 (est.)
North & Central America	47,187	49,736	56,711	54,342		
USA	38,883	40,950	47,081	45,524	47,117	48,853
Canada	1,831	3,215	3,934	3,832	2,251	2,409
Mexico	397	718	1,299	1,319		
Netherlands Antilles	3,601	1,866	1,600	1,450		
South America	5,005	3,718	5,726	5,221		
Ecuador	79	60	139	148		
Venezuela	3,494	1,062	1,580	1,180		
Brazil	663	1,413	2,098	1,950		
North Africa/Middle East	7,324	6,095	6,636	7,076		
Algeria	100	185	370	380		
Libya	—	346	599	600		
Iran	1,641	1,612	400	300		
Iraq	76	140	507	360		
Kuwait	105	136	538	750		
Qatar	0	29	63	70		
Saudi Arabia	1,775	462	44	50		
Bahrain	1,789	1,095	1,783	1,790		
Africa	861	1,450	1,433	1,511		
Gabon	86	75	75	80		
Nigeria	—	—	40	45		
Western Europe	14,655	18,813	22,733	21,373	19,695	20,264
Belgium	1,064	1,040	1,707	1,756	1,366	1,121
France	1,856	3,464	4,621	4,366	4,051	4,440
West Germany	1,149	1,046	941	1,038	1,021	1,019
Greece	325	644	1,369	1,398	1,345	1,115
Italy	1,485	2,260	1,840	1,630	1,039	943
Netherlands	2,055	2,742	3,301	2,917	2,752	2,668
United Kingdom	3,414	4,193	5,198	4,559	4,457	4,888
EEC	12,482	15,499	19,006	17,677	16,053	16,224
Spain	1,346	2,150	2,289	1,965	1,977	2,143
East Europe	151	320	340	325		
Far East	4,923	7,391	9,763	9,794		
Indonesia	118	118	241	223		
Japan	1,941	2,698	3,719	3,642	3,377	3,332
Singapore	690	1,540	2,473	2,375		
Oceania	1,267	2,377	2,786	2,562		
Australia	1,127	1,722	1,996	1,892	1,938	1,835
OPEC	7,474	4,225	4,596	4,511		
OECD	58,437	67,398	79,463	74,585	74,378	76,693
OPEC percentage share	9	5	4	4		
OECD percentage share	72	75	75	73		
WORLD TOTAL	81,373	89,900	106,128	102,204		

Sources: OECD/IEA Oil Statistics Quarterly
UN Yearbook of World Energy Statistics
Own Estimates

Table 23
AVIATION KEROSENE (JET FUEL) EXPORTS 1970-1983
Unit: thousand metric tonnes

Region/country	1970	1975	1980	1981	1982	1983 (est.)
North America	305	163	177	198	499	426
USA	279	79	61	113	468	315
Canada	26	84	116	85	31	111
South America	8,634	4,598	4,982	4,124		
Netherlands Antilles	3,572	1,836	1,750	1,650		
Trinidad and Tobago	1,493	376	328	100		
US Virgin Islands	206	1,078	867	500		
Venezuela	3,145	397	647	620		
Middle East/Asia	7,705	3,446	3,768	3,662		
Bahrain	1,661	931	1,237	1,284		
Singapore	1,934	1,460	1,951	1,718		
Japan	0	—	—	—	1,186	1,992
Africa	205	379	415	514		
Western Europe	3,795	5,831	6,963	6,532	7,311	6,852
Belgium	780	707	1,034	1,114	1,104	961
France	621	753	1,185	1,280	1,057	1,043
Greece	74	249	1,509	1,180	1,195	788
Netherlands	1,433	2,139	2,336	1,987	2,287	2,049
Italy	86	756	54	14	584	471
United Kingdom	659	805	622	439	375	397
EEC	3,764	5,483	6,802	6,119	6,733	5,771
East Europe	30	37	794	818		
Oceania	195	664	671	654		
Australia	194	283	174	224	630	549
OECD	4,294	6,277	7,314	6,954	9,626	9,819
WORLD TOTAL	20,869	15,118	17,770	16,502		

Sources: OECD/IEA Oil Statistics Quarterly
UN Yearbook of World Energy Statistics
Own Estimates

Table 24

AVIATION KEROSENE (JET FUEL) IMPORTS 1970-1983
Unit: thousand metric tonnes

Region/country	1970	1975	1980	1981	1982	1983 (est.)
North America	7,108	6,301	2,963	1,382	799	992
USA	6,703	6,249	2,877	1,366	794	992
Canada	405	52	86	16	5	—
South America	754	597	714	647		
Middle East/Asia	2,421	1,679	3,354	3,603		
Saudi Arabia	0	130	1,118	1,130		
Hong Kong	348	568	602	778		
Japan	—	301	390	709	475	700
Africa	889	815	861	860		
Western Europe	3,517	4,885	6,367	5,566	5,884	4,908
Denmark	505	724	597	514	503	547
West Germany	578	1,004	1,805	1,713	1,725	1,692
Greece	262	333	1,291	567	681	549
Sweden	399	433	256	210	250	159
Switzerland	417	490	562	547	480	519
United Kingdom	628	950	350	521	694	496
France	4	11	52	32	117	28
Italy	0	0	25	23	0	0
EEC	2,433	3,503	4,770	4,054	4,747	3,959
East Europe	0	0	0	0		
Oceania	767	935	1,063	916		
New Zealand	214	235	294	274	256	233
Australia	76	76	85	7	44	81
OECD	10,915	11,798	9,257	6,797	7,458	6,915
WORLD TOTAL	15,456	15,212	16,151	13,392		

Sources: OECD/IEA Oil Statistics Quarterly
UN Yearbook of World Energy Statistics
Own Estimates

240

Table 25

TOTAL PRODUCTION OF KEROSENE 1970-1983
Unit: thousand metric tonnes

Region/country	1970	1975	1980	1981	1982	1983 (est.)
North/Central America	18,751	13,855	13,965	12,155		
USA	12,321	7,170	6,447	5,609	53,049	53,913
Canada	2,773	3,252	3,116	2,394	4,373	4,061
Mexico	1,461	1,666	2,411	2,422		
South America	3,745	3,897	3,701	3,324		
Ecuador	65	184	292	284		
Venezuela	554	421	594	560		
North Africa/Middle East	6,569	9,166	11,895	10,885		
Algeria	102	266	45	50		
Libya	40	34	132	140		
Egypt	491	1,142	1,565	1,560		
Iran	2,301	3,477	4,700	4,400		
Iraq	578	560	700	550		
Kuwait	613	662	1,426	945		
Qatar	4	5	4	5		
Saudi Arabia	885	1,130	1,506	1,520		
United Arab Emirates	0	0	35	40		
Africa	955	1,328	1,890	1,946		
Gabon	27	27	20	25		
Nigeria	151	294	800	820		
Western Europe	9,460	6,491	4,911	5,481	26,432	26,219
France	49	44	136	172	4,172	4,656
Germany	92	33	55	46	1,104	1,100
Italy	4,348	2,988	2,419	2,471	4,059	3,315
Netherlands	1,200	595	472	396	3,159	3,376
United Kingdom	2,682	2,299	2,034	1,904	6,310	6,432
EEC	8,678	6,090	5,237	5,056	21,612	21,225
East Europe	25,523	32,292	38,933	38,721		
USSR	24,000	30,600	36,000	37,000		
Romania	969	1,018	868	770		
Far East	26,143	36,979	47,851	47,393		
China	4,300	11,340	16,500	15,500		
India	2,912	2,323	2,343	2,844		
Indonesia	1,914	3,200	4,391	4,611		
Japan	14,172	16,705	19,309	18,836	21,852	21,465
Singapore	400	1,263	2,375	2,680		
Oceania	650	857	616	385		
Australia	650	857	616	385	2,290	2,211
OPEC	7,234	10,260	14,645	13,950		
OECD	39,376	34,475	31,599	32,705	107,996	107,869
OPEC percentage share	7.9	9.8	11.8	11.6		
OECD percentage share	42.9	32.9	25.5	27.2		
WORLD TOTAL	91,796	104,865	123,762	120,290		

Sources: OECD/IEA Oil Statistics Quarterly
   UN Yearbook of World Energy Statistics
   Own Estimates

Table 26

KEROSENE EXPORTS 1970-1983
Unit: thousand metric tonnes

Region/country	1970	1975	1980	1981	1982	1983 (est.)
North/Central America	1,289	838	1,105	868		
USA	16	7	64	114	508	325
Canada	16	7	145	85	34	187
Netherlands Antilles	363	123	40	30		
Trinidad and Tobago	700	492	330	200		
US Virgin Islands	126	125	423	350		
South America	131	36	34	98		
North Africa/Middle East	2,295	2,490	3,210	2,549		
Algeria	37	145	0	0		
Bahrain	128	366	62	120		
Iran	700	188	300	170		
Iraq	0	146	250	150		
Kuwait	513	551	1,086	595		
Saudi Arabia	760	960	1,386	1,360		
Africa	162	161	182	155		
West Europe	2,295	1,558	9,397	9,063	9,564	
Italy	1,183	734	1,963	2,104	2,143	1,552
Belgium	110	42	1,052	1,121	1,131	997
France	2	114	1,193	1,316	1,078	1,052
Netherlands	197	304	2,637	2,223	2,612	2,640
United Kingdom	450	307	736	531	493	563
EEC	2,088	1,533	9,156	8,580	8,785	7,681
East Europe	1,782	2,019	3,053	3,550		
USSR	1,600	1,989	3,000	3,500		
Far East	988	974	2,263	1,816		
China	5	95	350	400		
Japan	153	8	28	10	1,196	1,992
Singapore	528	612	1,837	1,358		
Oceania	10	23	30	52		
OECD	2,489	1,600	9,902	9,500	11,870	13,708
WORLD TOTAL	8,952	8,099	19,274	18,151		

Sources: OECD/IEA Oil Statistics Quarterly
UN Yearbook of World Energy Statistics
Own Estimates

Table 27

KEROSENE IMPORTS 1970-1983
Unit: thousand metric tonnes

Region/country	1970	1975	1980	1981	1982	1983 (est.)
North & Central America	661	312	2,289	1,814		
USA	187	138	2,877	1,436	1,008	1,209
Canada	167	36	87	32	5	15
South America	66	104	180	152		
North Africa/Middle East	487	266	139	165		
Africa	784	497	833	786		
Nigeria	66	82	415	380		
Western Europe	1,791	1,208	6,529	5,509	6,532	5,681
West Germany	23	30	1,850	1,771	1,873	1,787
Denmark	119	62	652	548	540	588
Netherlands	437	153	574	464	849	439
Switzerland	10	5	567	552	485	524
United Kingdom	515	267	387	543	754	533
EEC	1,258	860	5,189	4,348	5,321	4,389
East Europe	416	539	545	548		
USSR	80	181	150	150		
Far East	3,253	3,266	4,244	4,856		
Japan		12	110	809	1,091	700
China	10	15	25	25		
India	327	679	1,934	2,017		
Indonesia	299	466	895	598		
Pakistan	165	336	474	420		
Vietnam	1,548	770	120	130		
Oceania	153	185	108	87		
New Zealand	29	44	29	12	267	247
OECD	2,234	1,515	9,992	8,472	9,095	7,933
WORLD TOTAL	7,611	6,376	15,967	13,917		

Sources: OECD/IEA Oil Statistics Quarterly
UN Yearbook of World Energy Statistics

Table 28

TOTAL PRODUCTION OF GAS/DIESEL OIL 1970-1983
Unit: million metric tonnes

Region/country	1970	1975	1980	1981	1982	1983 (est.)
North/Central America	156.8	177.4	189.9	183.6		
USA	124.1	134.0	134.7	132.1	133.0	120.8
Canada	17.7	21.4	25.3	22.6	19.4	18.2
Mexico	4.2	7.6	12.1	13.3	11.6	
Netherlands Antilles	3.6	3.4	3.5	3.4		
US Virgin Islands	1.0	2.9	6.0	4.6		
South America	21.4	26.6	39.0	35.9		
Ecuador	0.3	0. 4	0.8	0.7		
Venezuela	7.6	6.9	9.3	7.4		
Argentina	4.8	5.5	7.7	7.6		
Brazil	5.8	10.3	16.3	15.5		
North Africa/Middle East	22.6	29.2	39.1	40.7		
Algeria	1.0	1.7	3.6	3.7		
Libya	0.1	0.5	1.2	1.3		
Egypt	0.6	1.6	2.5	2.6		
Iran	4.7	6.5	5.3	4.8		
Iraq	0.8	1.1	2.4	2.0		
Kuwait	3.7	3.3	3.7	3.3		
Qatar	0	0.1	0.2	0.2		
Saudi Arabia	3.0	3.5	6.2	6.2		
United Arab Emirates	0	0	0.2	0.8		
Bahrain	2.4	2.8	3.3	3.6		
Africa	4.7	8.5	11.8	10.9		
Gabon	0.2	0.3	0.3	0.4		
Nigeria	0.2	0.5	1.7	1.8		
South Africa	2.0	4.2	4.6	4.8		
Western Europe	144.0	147.7	168.8	151.6	168.5	163.8
Belgium	9.4	9.2	10.9	8.8	7.9	7.7
France	9.5	10.1	13.6	12.1	28.4	25.7
West Germany	38.3	36.2	41.6	36.3	35.5	33.4
Italy	23.1	23.5	25.2	25.0	24.3	23.4
Netherlands	17.1	16.1	17.5	14.0	13.5	13.6
United Kingdom	22.5	23.3	22.2	20.4	20.6	20.8
Spain	7.0	8.7	10.8	10.5	10.1	9.8
EEC	125.1	125.0	137.9	123.4	136.9	131.8
East Europe	88.0	126.1	148.6	148.9		
USSR	68.3	98.2	115.0	118.0		

...contd..

Table 28 contd.

TOTAL PRODUCTION OF GAS/DIESEL OIL 1970-1983
Unit: million metric tonnes

Region/country	1970	1975	1980	1981	1982	1983 (est.)
Far East	40.1	67.6	89.0	89.3		
China	6.4	14.5	18.5	17.5		
India	4.8	7.1	8.1	9.9		
Indonesia	1.2	2.9	4.7	4.5		
Japan	18.8	29.7	36.9	35.7	33.9	33.1
Singapore	1.7	3.9	6.2	6.7		
Oceania	5.1	6.4	8.2	8.0		
Australia	4.3	5.8	7.3	7.0	7.2	6.6
New Zealand	0.7	0.6	0.7	0.7	0.6	0.6
OPEC	25.2	27.7	39.6	37.1		
OECD	309.6	339.2	373.7	349.7	362.5	343.3
OPEC percentage share	5.2	4.7	5.7	5.5		
OECD percentage share	64.2	57.6	53.8	52.3		
WORLD TOTAL	482.6	589.4	694.2	669.0		

Sources: UN Yearbook of World Energy Statistics
OECD/IEA Oil Statistics Quarterly
Statistiques de l'Industrie Petroliere
Own Estimates

Table 29

GAS/DIESEL OIL EXPORTS
Unit: thousand metric tonnes

Region/country	1970	1975	1980	1981	1982	1983 (est.)
North/Central America	8,832	10,183	14,000	12,122		
USA	124	37	162	204	4,038	3,701
Canada	456	815	1,798	1,465	709	1,528
Bahamas	350	870	1,300	1,400		
Netherlands Antilles	3,764	3,161	3,150	3,000		
Trinidad & Tobago	1,871	1,242	1,517	1,150		
US Virgin Islands	750	2,619	4,800	3,500		
South America	5,654	3,878	4,792	4,434		
Venezuela	5,402	3,477	3,897	3,082		
North Africa/Middle East	12,421	10,565	15,174	15,018		
Algeria	169	371	1,893	1,880		
Bahrain	2,228	2,478	3,124	3,370		
Kuwait	3,520	2,940	3,153	2,738		
Saudi Arabia	2,340	2,606	4,416	4,450		
Africa	703	958	851	824		
Western Europe	39,660	33,945	40,191	38,934	38,562	40,924
Belgium	2,895	3,573	5,777	5,333	5,104	5,457
France	3,045	3,594	3,055	2,899	2,848	2,327
West Germany	1,513	1,204	1,405	1,203	1,032	908
Greece	41	808	2,504	2,537	2,724	1,969
Italy	10,248	4,394	2,803	4,121	3,998	4,095
Netherlands	12,531	11,356	16,440	14,010	13,563	15,560
United Kingdom	6,210	5,751	4,737	4,615	4,435	4,599
EEC	37,105	31,758	37,191	35,247	34,067	35,547
East Europe	15,323	19,530	26,100	26,664		
USSR	11,373	15,890	21,000	22,000		
Romania	2,548	2,024	2,641	2,500		
Far East	3,719	3,842	5,743	6,610		
Singapore	2,342	2,819	4,693	5,043		
Oceania	315	311	618	692		
OECD	40,559	35,095	42,679	41,367	44,246	47,183
WORLD TOTAL	86,627	83,212	102,299	99,348		

Sources: OECD/IEA Oil Statistics Quarterly
UN Yearbook of World Energy Statistics
Petroleum Economist
Own Estimates

Table 30

GAS/DIESEL OIL IMPORTS 1970-1983
Unit: thousand metric tonnes

Region/country	1970	1975	1980	1981	1982	1983 (est.)
North & Central America	13,624	12,376	10,556	11,714		
USA	7,456	7,840	7,180	8,436	1,667	5,557
Canada	2,535	187	84	192	34	339
South America	861	1,072	1,525	1,599		
North Africa/Middle East	1,463	2,772	4,354	3,655		
Africa	3,901	3,046	3,024	2,654		
Western Europe	51,238	48,960	57,384	51,282	58,519	63,757
Belgium	1,835	3,276	4,046	4,034	4,897	4,825
Denmark	4,235	3,785	3,270	2,567	2,874	2,164
France	3,147	1,940	3,578	4,024	6,066	6,719
West Germany	18,802	19,472	15,937	14,586	14,606	16,865
Greece	323	164	2,450	1,490	2,289	1,556
Italy	291	1,005	2,692	1,758	4,526	3,112
Netherlands	4,087	2,704	8,042	8,192	7,641	12,505
Sweden	7,216	5,293	4,267	3,206	4,112	2,177
Switzerland	4,580	4,837	5,295	4,523	4,052	5,292
United Kingdom	1,888	1,572	1,418	1,211	1,517	1,003
EEC	35,281	34,973	42,781	39,493	45,994	50,148
East Europe	2,223	1,979	3,630	3,115		
USSR	179	123	100	100		
Far East	8,596	9,599	15,059	13,777		
Japan	2,173	3,196	1,626	1,942	1,823	1,576
India		543	3,267	2,389		
Oceania	1,188	1,545	1,775	1,484		
OECD	64,131	60,968	60,867	58,132	63,043	71,984
WORLD TOTAL	83,503	81,509	97,307	89,280		

Sources: UN Yearbook of World Energy Statistics
OECD/IEA Statistics Quarterly
Own Estimates

Table 31

TOTAL PRODUCTION OF RESIDUAL FUEL OIL 1970-1983
Unit: million metric tonnes

Region/country	1970	1975	1980	1981	1982	1983 (est.)
North & Central America	129.9	158.6	174.4	152.6		
USA	38.9	68.1	87.4	72.5	69.8	54.2
Canada	12.8	19.0	17.1	15.0	11.4	9.3
Mexico	7.2	9.9	17.5	19.1	19.3	
Netherlands Antilles	33.9	17.2	17.0	16.0		
US Virgin Islands	11.5	17.3	13.5	12.0		
South America	68.0	57.0	60.3	55.2		
Brazil	8.4	14.8	16.8	15.8		
Ecuador	0.4	0.5	2.3	2.2		
Venezuela	44.9	27.0	25.5	22.2		
North Africa/Middle East	68.9	67.8	84.2	77.6		
Algeria	0.5	1.5	2.8	2.8		
Libya	0.2	0.8	2.1	2.4		
Egypt	1.6	4.2	6.7	6.7		
Iran	12.6	15.3	13.9	10.0		
Iraq	1.5	1.8	2.9	2.1		
Kuwait	12.1	7.6	8.0	6.1		
Qatar	0	0	0	0		
Saudi Arabia	19.1	13.1	13.4	13.5		
United Arab Emirates	0	0	0.1	0.4		
Africa	7.8	10.8	10.1	9.7		
Gabon	0.4	0.4	0.7	0.7		
Nigeria	0.4	0.9	1.2	1.3		
West Europe	170.6	246.2	220.9	196.5	161.5	143.0
Belgium	11.3	10.4	10.1	9.0	8.2	5.5
France	56.1	60.7	55.3	46.0	20.8	17.7
West Germany	33.6	25.7	20.8	21.5	19.2	15.8
Italy	56.7	42.4	36.0	34.4	32.0	28.2
Netherlands	25.4	21.5	17.0	14.6	13.0	14.9
United Kingdom	46.4	36.1	27.2	22.1	18.8	16.1
Spain	15.7	22.0	22.9	22.3	18.6	16.7
EEC	237.5	206.3	175.9	156.9	120.4	105.9
East Europe	117.2	170.1	197.8	196.9		
USSR	93.9	131.0	149.0	152.0		
Romania	4.2	6.0	10.2	9.1		
Far East	127.6	164.0	167.5	155.4		
China	8.1	21.0	30.5	29.0		

...contd..

Table 31 contd.

TOTAL PRODUCTION OF RESIDUAL FUEL OIL 1970-1983
Unit: million metric tonnes

Region/country	1970	1975	1980	1981	1982	1983 (est.)
Indonesia	5.5	5.0	6.1	6.6		
Japan	87.4	105.1	85.7	72.3	59.2	53.0
Republic of Korea	5.7	9.6	13.7	13.1		
Singapore	7.0	8.1	13.5	14.2		
Oceania	8.4	6.1	5.1	4.3		
Australia	7.1	4.7	3.8	3.3	3.2	2.9
New Zealand	0.9	1.0	0.8	0.6	0.4	0.3
OPEC	97.6	73.9	79.0	70.3		
OECD	417.7	464.1	405.7	360.2	305.5	262.7
OPEC percentage share	12.2	8.4	8.6	8.3		
OECD percentage share	52.3	52.7	44.1	42.5		
WORLD TOTAL	798.3	880.4	920.3	848.3		

Sources: UN Yearbook of World Energy Statistics
OECD/IEA Statistics Quarterly
Own Estimates

Table 32

RESIDUAL FUEL OIL EXPORTS 1970-1983
Unit: thousand metric tonnes

Region/country	1970	1975	1980	1981	1982	1983 (est.)
North & Central America	55,641	51,329	40,359	40,633		
USA	2,993	807	1,848	6,531	11,189	9,892
Canada	76	2,948	2,533	2,292	1,616	1,809
Netherlands Antilles	27,852	18,181	14,000	13,500		
Trinidad & Tobago	12,027	6,015	6,068	4,000		
US Virgin Islands	7,920	14,700	9,600	8,000		
South America	43,114	25,949	28,188	24,323		
Venezuela	41,105	23,988	23,800	18,702		
North Africa/Middle East	26,977	18,105	21,599	16,813		
Bahrain	4,967	3,520	4,099	4,065		
Iran	4,000	4,700	4,100	3,000		
Kuwait	7,910	3,603	5,800	3,520		
Saudi Arabia	6,450	3,526	1,300	1,350		
Africa	1,581	2,012	2,345	3,148		
West Europe	42,342	29,585	31,979	33,641	36,283	36,375
Belgium	2,174	3,423	4,431	4,802	4,906	4,241
Italy	9,373	2,787	1,871	2,548	2,301	2,251
France	2,603	3,205	5,033	6,822	4,065	3,436
Germany	2,817	1,644	1,805	2,076	2,578	2,793
Netherlands	11,888	10,028	7,493	7,563	10,647	11,265
United Kingdom	7,815	4,171	5,121	3,654	3,356	3,091
EEC	38,033	27,232	27,795	29,811	30,424	30,337
East Europe	11,678	12,870	17,733	17,581		
USSR	8,637	8,162	12,000	12,500		
Romania	1,532	2,258	3,419	3,000		
Far East	7,382	9,974	17,862	17,052		
Indonesia	451	2,435	5,616	5,266		
Singapore	5,634	5,633	9,971	9,131		
Oceania	691	512	343	277		
OECD	46,075	35,114	35,657	41,133	50,081	49,279
WORLD TOTAL	189,406	150,336	160,408	153,468		

Sources: UN Yearbook of World Energy Statistics
OECD/IEA Oil Statistics Quarterly
Own Estimates

250

Table 33

RESIDUAL FUEL OIL IMPORTS 1970-1983
Unit: thousand metric tonnes

Region/country	1970	1975	1980	1981	1982	1983 (est.)
North & Central America	97,544	82,678	62,901	54,131		
USA	84,256	67,444	51,890	43,889	36,004	32,928
Canada	5,560	1,109	1,056	1,138	1,261	579
South America	2,085	1,041	2,292	1,240		
North Africa/Middle East	1,717	1,825	5,039	4,267		
Africa	3,761	3,673	5,911	5,718		
West Europe	44,328	42,537	46,897	48,892	57,551	52,981
Belgium	2,306	2,499	2,702	2,721	4,051	4,533
France	1,081	3,084	3,397	2,491	4,013	3,197
West Germany	3,566	4,248	4,600	6,608	7,440	9,037
Italy	1,790	7,866	8,774	10,233	9,352	9,913
Netherlands	2,360	1,640	7,509	9,528	12,730	9,741
United Kingdom	8,710	4,163	2,777	2,927	6,567	4,843
EEC	27,543	30,606	35,219	38,293	47,571	43,836
East Europe	3,089	2,498	2,763	3,393		
Far East	33,111	16,270	26,140	23,017		
Japan	21,577	4,603	9,166	6,664	5,819	6,775
Oceania	1,590	3,504	3,176	2,155		
Australia	1,093	2,397	1,767	1,559	922	1,352
OECD	156,814	118,118	96,220	93,326	101,557	94,615
WORLD TOTAL	187,225	153,067	155,119	142,813		

Sources: UN Yearbook of World Energy Statistics
OECD/IEA Oil Statistics Quarterly
Own Estimates

251

Table 34

TOTAL CRUDE OIL CONSUMPTION 1970-1983
Unit: million metric tonnes

Region	1970	1975	1980	1981	1982	1983 (est.)
North & Central America	728.2	826.1	899.4	838.1		
South America	132.7	133.6	156.4	151.0		
North Africa/Middle East	112.7	133.6	166.3	162.1		
Africa	19.7	29.7	32.4	39.7		
West Europe	608.2	581.1	611.3	534.6	500.9	485.8
East Europe	289.7	404.2	488.2	496.0	500.9	
Far East	240.9	390.2	456.2	432.2	412.8	
Oceania	27.2	30.1	32.3	31.9	32.6	
OPEC	164.1	149.4	185.8	178.2		
OECD	1,410.3	1,543.3	1,628.1	1,470.6	1,377.7	1,316.2
OPEC percentage share	7.2	5.6	6.2	6.3		
OECD percentage share	62.1	58.3	54.5	52.1		
WORLD TOTAL	2,271.1	2,647.3	2,986.5	2,821.1		

Source: Own Calculations

Table 35

OIL DEMAND PER CAPITA 1970-1982
Unit: barrels of oil equivalent per capita

Region	1970	1975	1980	1981	1982
OPEC	1.45	1.93	2.52	2.56	2.52
CEP's	2.12	2.92	3.36	3.29	3.26
LDC's	1.15	1.33	1.57	1.55	1.55
OECD	15.95	16.59	16.69	15.19	14.22
USA	25.07	26.00	25.19	23.01	21.48
Europe	11.33	11.23	11.80	10.58	10.02
Japan	13.08	15.16	15.03	13.79	12.67
Rest	19.82	21.28	21.21	19.84	18.19
World	4.42	4.76	4.92	4.61	4.40

Source: OPEC

Table 36

CONSUMPTION OF CRUDE OIL 1970-1983
Unit: million metric tonnes
NORTH and CENTRAL AMERICA

	1970	1975	1980	1981	1982	1983 (est.)	1981 kgs per capita
Canada	58.7	80.9	88.7	80.9	72.3	64.7	3.3
USA	539.9	615.5	664.3	614.9	579.4	555.1	2.7
Bahamas	2.5	10.8	9.0	8.4			33.9
Barbados	0.1	0.2	0.2	0.2			0.7
Costa Rica	0.3	0.3	0.5	0.5			0.2
Cuba	4.3	6.0	6.3	6.4			0.7
Dominican Republic		1.3	1.4	1.5			0.3
El Salvador	0.2	0.6	0.7	0.7			0.1
Mexico	21.8	30.3	50.6	55.3	56.6		0.7
Guatemala	0.8	1.0	0.9	0.9			0.1
Netherlands Antilles	47.7	25.6	26.7	25.2			96.6
Honduras	0.7	0.6	0.5	0.5			0.1
Jamaica	1.5	1.5	1.0	1.0			0.4
Nicaragua	0.5	0.6	0.6	0.6			0.2
Panama	3.7	3.9	1.9	2.0			1.0
Puerto Rico	9.1	9.9	8.2	7.0			2.2
Trinidad & Tobago	21.6	11.8	12.1	10.4			8.8
US Virgin Islands	14.4	24.7	25.4	21.5			188.6
TOTAL	728.2	826.1	899.4	838.1			2.2

Cont'd.

Table 36 continued

SOUTH AMERICA

	1970	1975	1980	1981	1981 kgs per capita
Argentina	21.7	22.6	27.2	26.9	1.0
Bolivia	0.6	0.9	1.2	1.0	0.2
Brazil	25.2	43.0	54.3	52.4	0.4
Chile	3.6	3.9	5.0	4.3	0.4
Colombia	7.1	8.1	7.5	8.0	0.3
Ecuador	1.2	2.1	4.7	4.5	0.5
Paraguay	0.2	0.2	0.2	0.3	0.1
Peru	4.1	5.7	7.3	6.9	0.4
Uruguay	1.8	1.9	1.9	1.7	0.6
Venezuela	67.3	45.1	47.1	45.0	3.1
TOTAL	132.7	133.6	156.4	151.0	0.6

Cont'd.

Table 36 continued

NORTH AFRICA and MIDDLE EAST

	1970	1975	1980	1981	1982	1983	1981 kgs per capita
Bahrain	12.5	10.6	11.8	12.8			39.8
Cyprus	0	0.3	0.6	0.5			0.8
Yemen	6.8	1.6	2.0	2.1			1.1
Iran	26.2	33.9	33.9	29.2			0.7
Iraq	3.8	4.8	9.2	8.1			0.6
Israel	6.0	7.2	6.6	6.7			1.7
Jordan	0.5	0.8	1.8	2.2			0.6
Kuwait	20.6	14.6	17.4	14.6			10.0
Lebanon	2.0	2.1	1.9	1.8			0.7
Oman							
Qatar		0.2	0.5	0.4			1.8
Saudi Arabia	29.2	23.2	29.6	30.1			3.2
Syria	2.0	2.5	6.3	6.2			0.7
Turkey	7.2	13.0	12.8	13.4	16.4	15.9	0.3
United Arab Emirates			0.6	2.4			3.2
Algeria	1.5	4.3	4.4	4.4			0.2
Egypt	3.4	8.8	14.2	14.5			0.3
Libya	0.4	1.9	7.0	6.1			2.0
Morocco	1.6	2.6	4.1	4.8			0.2
Tunisia	1.2	1.2	1.6	1.6			0.2
TOTAL	112.7	133.6	166.3	162.1			0.5

Cont'd.

256

Table 36 continued

AFRICA

	1970	1975	1980	1981	1981 kgs per capita
Angola	0.7	0.8	1.2	1.3	0.2
Congo	0	0	0.1	0.4	0.3
Ethiopia	0.6	0.5	0.6	0.6	0.02
Gabon	0.9	0.9	1.2	1.5	2.7
Ghana	0.8	1.2	1.0	1.0	0.1
Ivory Coast	0.7	1.5	1.5	1.1	0.1
Kenya	2.2	2.9	2.7	2.3	0.1
Liberia	0.4	0.5	0.6	0.6	0.3
Madagascar	0.6	0.8	0.4	0.4	0.04
Mozambique	0.8	0.4	0.6	0.5	0.05
Nigeria	1.3	2.5	4.7	6.8	0.1
Reunion			0.01	0.01	0.02
Senegal	0.6	0.7	0.8	0.8	0.1
Sierra Leone	0.3	0.4	0.3	0.3	0.1
Somalia			0.3	0.3	0.1
South Africa	8.3	14.2	13.0	17.0	0.5
Sudan	0.8	1.1	1.1	1.1	0.1
Togo			0.4	0.5	0.2
Cameroon			1.1	2.6	0.3
Tanzania	0.7	0.7	0.8	0.8	0.04
Zaire	0.7	0.6	0.4	0.4	0.1
Zambia		0.8	0.8	0.7	0.1
TOTAL	19.7	29.7	32.4	39.7	0.2

Cont'd.

Table 36 continued

WESTERN EUROPE

	1970	1975	1980	1981	1982	1983 (est.)	1981 kgs per capita
Austria	6.2	8.2	10.3	8.8	7.5	6.5	1.2
Belgium	29.9	29.4	33.6	29.3	25.2	22.6	3.0
Denmark	10.2	8.1	6.8	6.3	6.1	7.0	1.2
Finland	8.2	8.6	12.8	11.1	9.6	10.6	2.3
France	102.5	109.3	114.4	97.3	85.7	77.3	1.8
Federal Republic of Germany	107.2	94.2	105.4	85.5	81.6	78.4	1.4
Greece	5.1	11.6	14.3	15.9	15.1	14.0	1.6
Ireland	2.8	2.6	2.0	0.6	0.5	1.2	0.2
Italy	114.9	98.4	92.1	85.6	78.9	74.0	1.5
Netherlands	62.2	56.9	50.1	40.4	41.8	45.3	2.8
Norway	5.9	6.9	7.4	7.2	6.9	7.1	1.8
Portugal	3.6	5.7	7.6	8.0	7.9	8.0	0.8
Spain	30.8	43.0	49.5	48.8	45.4	44.3	1.3
Sweden	11.8	11.4	17.7	14.2	12.2	13.4	1.7
Switzerland	5.5	4.7	4.6	4.0	4.1	4.1	0.6
United Kingdom	101.4	92.1	82.7	71.6	72.4	72.0	1.3
EEC	536.2	492.6	501.4	432.5	407.3	391.8	1.3
TOTAL	608.2	581.1	611.3	534.6	500.9	485.8	

Cont'd.

Table 36 continued

EASTERN EUROPE

	1970	1975	1980	1981	1982	1981 kgs per capita
Albania	1.4	2.6	2.4	2.2		0.8
Bulgaria	5.9	10.6	13.3	12.8		1.4
Czechoslovakia	10.1	16.0	19.3	18.9		1.2
German Democratic Republic	10.4	17.1	21.9	20.6		1.2
Hungary	6.0	9.5	9.5	9.2		0.9
Poland	7.5	13.5	16.7	13.8		0.4
Romania	15.7	19.7	27.5	24.7		1.1
Yugoslavia	7.3	11.1	15.2	13.7	13.4	0.6
USSR	289.7	404.2	488.2	496.0	500.9	1.9
TOTAL	354.0	504.3	614.0	611.9	615.4	

Cont'd.

Table 36 continued

FAR EAST

	1970	1975	1980	1981	1982	1983 (est.)	1981 kgs per capita
Bangladesh		0.8	1.2	1.3			0.02
Brunei	0.04	0.2	0.3	0.3			1.1
Burma	1.1	1.0	1.3	1.3			0.04
China	24.3	66.7	93.3	87.2	84.7		0.1
Kampuchea	0.4	0	0	0			0
India	18.5	21.8	25.1	29.5			0.04
Indonesia	11.9	15.8	25.5	25.1			0.2
Japan	169.7	223.8	220.6	196.4	178.1	164.1	1.7
Korea, Democratic People's			1.8	2.0			0.1
Korea, Republic	9.8	16.3	25.2	24.8			0.6
Malaysia	5.3	4.0	5.5	5.6			0.4
Mongolia	0.002	0.01	0	0			0
Pakistan	4.4	3.4	4.3	4.2		0.1	
Philippines	9.1	9.3	9.8	10.0			0.2
Singapore	10.9	18.9	32.6	34.7			14.2
Sri Lanka	1.8	1.5	1.8	1.6			0.1
Thailand	3.7	7.1	8.5	8.8			0.2
TOTAL	240.9	390.2	456.2	432.2	412.8		

Cont'd.

Table 36 continued

	1970	1975	1980	1981	1982	1983	1981 kgs per capita
			OCEANIA				
Australia	23.8	25.9	28.1	28.1	29.1	28.5	1.9
Guam	0.6	1.2	1.6	1.5			14.4
New Zealand	2.8	3.1	2.6	2.3	2.0	2.1	0.7
TOTAL	27.2	30.1	32.3	31.9	32.6		1.4

Sources: UN Yearbook of World Energy Statistics
OECD/IEA Oil Statistics Quarterly
Statistiques de l'Industrie Petroliere
Own Estimates and Calculations

TABLE 37

CONSUMPTION OF TOTAL REFINED PETROL PRODUCTS 1970-1983
Unit: million metric tonnes

Region	1970	1975	1980	1981	1982	1983 (est.)
North/Central America	725.1	812.7	838.1	796.4	748.2	
South America	67.6	90.5	111.7	106.3	106.7	
North Africa/ Middle East	39.9	68.9	94.8	94.2	99.3	
Africa	19.0	27.2	31.0	30.5	35.0	
West Europe	499.9	524.1	550.2	511.1	489.6	463.7
East Europe	274.5	390.8	447.7	449.5	454.2	
Far East	240.6	322.4	381.1	372.9	378.0	
Oceania	24.9	29.7	31.6	30.6	29.7	
OPEC	35.0	60.8	93.1	91.8	95.7	
OECD	1,360.1	1,494.7	1,533.1	1,440.5	1,363.5	1,307.8
OPEC percentage share	1.8	2.7	3.7	3.8	4.1	
OECD percentage share	71.7	65.7	61.3	59.8	58.3	
WORLD TOTAL	1,896.2	2,274.6	2,502.2	2,409.0	2,340.7	

Source: Own Calculations

Table 38

CONSUMPTION OF REFINED PETROL PRODUCTS BY COUNTRY 1970-1983
Unit: million metric tonnes
NORTH and CENTRAL AMERICA

	1970	1975	1980	1981	1982	1983 (est.)	1981 kgs per capita
Canada	63.7	73.3	76.9	71.2	61.2	55.4	2.9
USA	612.0	673.9	673.4	636.3	610.6	599.3	2.8
Bahamas	0.5	1.1	0.8	0.8			3.2
Barbados	0.1	0.2	0.2	0.2			0.7
Belize	0.04	0.06	0.06	0.06			0.4
Costa Rica	0.4	0.6	0.7	0.7			0.3
Cuba	5.9	8.1	9.5	9.3			1.0
Dominican Republic	0.9	1.9	1.8	1.8			0.3
El Salvador	0.4	0.6	0.5	0.6			0.1
Greenland	0.1	0.2	0.2	0.2			3.3
Guatemala	0.7	1.0	1.3	1.1			0.1
Haiti	0.1	0.1	0.2	0.2			0.04
Honduras	0.4	0.5	0.5	0.6			0.1
Jamaica	1.6	2.6	2.7	2.5			1.1
Nicaragua	0.4	0.6	0.6	0.6			0.2
Panama	0.6	1.0	1.2	1.1			0.6
Puerto Rico	6.5	8.0	8.3	7.3			2.3
Trinidad & Tobago	1.4	1.1	1.7	1.6			1.4
Antigua & Barbuda	0.1	0.03	0.1	0.1			0.9
Bermuda	0.1	0.1	0.2	0.1			1.9
British Virgin Islands	0.01	0.01	0.01	0.01			0.9
Cayman Islands	0.01	0.02	0.03	0.04			2.0
Grenada	0.01	0.02	0.02	0.02			0.2
Guadeloupe	0.1	0.1	0.2	0.2			0.6
Martinique	0.1	0.2	0.2	0.2			0.6
Mexico	21.6	30.6	50.3	53.5	54.8		0.7
Netherlands Antilles	5.3	3.3	3.1	2.6			9.8
US Virgin Islands	1.9	3.1	3.6	3.5			31.1
TOTAL	725.1	812.7	838.1	796.4	748.2		2.1

Cont'd.

Table 38 continued

SOUTH AMERICA

	1970	1975	1980	1981	1982	1981 kgs per capita
Argentina	20.1	20.5	22.2	21.0	20.5	0.7
Bolivia	0.6	0.9	1.1	1.0		0.2
Brazil	22.2	38.3	46.8	45.1	44.8	0.4
Chile	4.5	4.3	5.0	4.6		0.4
Colombia	4.8	5.7	6.5	6.3		0.2
Ecuador	1.1	1.8	3.7	3.6	3.6	0.4
French Guiana	0.03	0.04	0.1	0.1		1.5
Guyana	0.5	0.6	0.5	0.5		0.6
Paraguay	0.2	0.2	0.3	0.3		0.1
Peru	4.5	5.7	6.1	5.9	6.3	0.3
Suriname	0.5	0.6	0.7	0.7		1.7
Uruguay	1.6	1.7	1.7	1.6		0.5
Venezuela	6.9	10.1	16.9	15.6	15.6	1.1
TOTAL	67.6	90.5	111.7	106.3	106.7	0.4

Cont'd.

Table 38 continued

NORTH AFRICA and MIDDLE EAST

	1970	1975	1980	1981	1982	1983	1981 kgs per capita
Bahrain	0.1	0.1	0.2	0.3			0.9
Cyprus	0.5	0.5	0.8	0.8			1.3
Yemen	0.3	0.5	0.7	0.8			0.4
Iran	8.4	17.8	18.6	17.6	19.2		0.4
Iraq	3.2	3.7	4.8	4.2	4.1		0.3
Israel	4.3	4.9	6.0	6.1			1.5
Jordan	0.4	0.7	1.4	1.7			0.5
Kuwait	0.6	0.9	1.9	1.9	2.6		1.3
Lebanon	1.1	1.8	1.7	1.6			0.6
Oman	0.1	0.2	0.5	0.5			0.5
Qatar	0.1	0.2	0.3	0.4	0.4		1.5
Saudi Arabia	1.8	4.8	9.6	9.8	9.6		1.1
Syria	1.7	3.3	5.0	5.1			0.5
Turkey	6.9	12.3	14.2	13.8	14.8	15.0	0.3
United Arab Emirates	0.2	0.8	2.9	3.0	2.8		3.9
Algeria	1.7	3.4	4.6	4.3	4.9		0.2
Egypt	5.2	7.6	11.8	11.8			0.3
Libya	0.9	1.7	3.5	4.1	3.6		1.3
Morocco	1.5	2.4	4.0	4.0	4.1		0.2
Tunisia	0.9	1.3	2.3	2.4	2.9		0.4
TOTAL	39.9	68.9	94.8	94.2	99.3		

Table 38 continued

AFRICA

	1970	1975	1980	1981	1982	1981 kgs per capita
Angola	0.5	0.6	0.6	0.5		0.1
Benin	0.1	0.1	0.1	0.1		0.03
Burundi	0.02	0.02	0.04	0.05		
Central African Republic	0.1	0.04	0.1	0.1		0.03
Chad	0.04	0.1	0.1	0.1		0.02
Congo	0.1	0.1	0.1	0.1	0.1	0.1
Djibouti	0.03	0.1	0.1	0.05		0.2
Equatorial Guinea	0.01	0.02	0.02	0.02		0.1
Ethiopia	0.4	0.3	0.6	0.5		0.02
Gabon	0.3	0.4	0.4	0.5	0.5	0.9
Gambia	0.02	0.03	0.05	0.05		0.1
Ghana	0.7	0.8	0.5	0.6		0.05
Guinea	0.2	0.3	0.3	0.3		0.1
Ivory Coast	0.7	1.0	0.8	0.7		0.1
Kenya	0.7	1.3	1.5	1.3		0.1
Liberia	0.4	0.5	0.5	0.5		0.2
Madagascar	0.3	0.5	0.4	0.4		0.04
Malawi	0.1	0.1	0.1	0.1		0.02
Mali	0.1	0.1	0.1	0.1		0.02
Mauritania	0.1	0.2	0.2	0.2		0.1
Mauritius	0.2	0.2	0.2	0.2		0.2
Mozambique	0.4	0.4	0.5	0.4		0.04
Niger	0.1	0.1	0.2	0.2		0.03
Nigeria	1.6	2.8	6.9	6.8	7.2	0.1
Rwanda	0.02	0.03	0.05	0.05		0.01
Senegal	0.3	0.8	0.8	0.8		0.1
Sierra Leone	0.2	0.2	0.2	0.2		0.1
Somalia	0.1	0.2	0.3	0.3		0.1
South Africa	6.8	11.2	10.4	12.3		0.3
Sudan	1.4	1.1	1.1	1.1		0.1
Togo	0.1	0.1	0.1	0.2		0.1
Uganda	0.4	0.3	0.2	0.2		0.01
Cameroon	0.2	0.3	0.5	0.4	0.3	0.04
Tanzania	0.5	0.6	0.6	0.6		0.03
Upper Volta	0.05	0.1	0.1	0.1		0.02
Zaire	0.5	0.7	0.8	0.8		0.03
Zambia	0.4	0.6	0.7	0.7		0.1
Zimbabwe	0.4	0.7	0.5	0.5		0.1
Reunion	0.1	0.1	0.2	0.2		0.5
TOTAL	19.0	27.2	31.0	30.5	35.0	0.1

Cont'd.

Table 38 continued

WESTERN EUROPE

	1970	1975	1980	1981	1982	1983	1981 kgs per capita
Austria	8.5	9.4	11.6	9.6	9.5	9.0	1.3
Belgium	20.2	19.1	19.1	15.7	14.8	13.8	1.6
Denmark	16.5	13.8	11.7	10.4	9.9	9.1	2.0
Finland	9.9	10.2	10.8	9.9	9.4	8.4	2.1
France	81.3	87.2	92.0	81.1	75.0	70.5	1.5
Federal Republic of Germany	107.8	110.4	108.6	103.8	100.4	97.1	1.7
Greece	5.1	7.9	10.3	9.6	9.8	9.2	1.0
Iceland	0.5	0.5	0.5	0.5	0.4	0.4	2.1
Ireland	3.4	4.4	5.2	4.7	4.1	3.6	1.4
Italy	74.1	80.8	92.5	87.4	83.4	77.8	1.5
Luxembourg	1.3	1.2	1.0	0.9	0.9	0.9	2.6
Netherlands	19.7	17.9	25.4	26.6	21.2	19.7	1.9
Norway	6.4	6.4	8.9	7.7	7.4	7.2	1.9
Portugal	3.1	5.3	7.1	7.5	7.8	7.8	0.8
Spain	22.4	37.5	43.2	41.3	42.4	40.7	1.1
Sweden	26.5	23.2	22.3	20.2	18.3	15.7	2.4
Switzerland	11.0	11.1	11.5	10.6	10.1	11.1	1.6
United Kingdom	81.9	77.8	68.5	63.6	64.8	61.7	1.1
EEC	411.6	420.5	434.3	403.8	384.3	363.4	
TOTAL	499.9	524.1	550.2	511.1	489.6	463.7	1.2

Cont'd.

Table 38 continued

EASTERN EUROPE

	1970	1975	1980	1981	1982	1981 kgs per capita
Albania	0.5	0.7	1.0	0.9		0.3
Bulgaria	8.2	11.7	13.6	13.3		1.5
Czechoslovakia	8.8	14.2	15.9	15.6		1.0
German Democratic Republic	10.2	13.9	16.3	15.9		0.9
Hungary	5.2	8.6	9.0	9.0		0.8
Poland	7.0	10.7	14.3	13.4		0.4
Romania	9.5	11.9	15.8	13.9		0.6
Yugoslavia	6.8	10.5	14.3	12.7		0.6
USSR	218.3	308.6	347.5	354.8		1.3
TOTAL	174.5	390.8	447.7	449.5	454.2	1.1

Cont'd.

Table 38 continued

FAR EAST

	1970	1975	1980	1981	1982	1983 (est.)	1981 kgs per capita
Afghanistan	0.2	0.3	0.3	0.3			0.02
Bangladesh	0	1.0	1.4	1.4	1.2		0.02
Bhutan	0.001	0.001	0.001	0.001			0.001
Brunei	0.1	0.1	0.2	0.2	0.5		0.8
Burma	0.9	0.7	1.0	1.0	1.1		0.03
China	21.8	54.1	73.1	67.8	68.6		0.1
Kampuchea	0.3	0.02	0.01	0.01			0.002
East Timor	0.004	0.01	0.01	0.02			0.02
Hong Kong	2.6	3.3	5.0	5.3	6.6		1.0
India	14.4	18.5	25.2	27.2	34.8		0.04
Indonesia	8.2	12.4	19.0	20.0	21.0		0.1
Japan	154.2	183.7	189.2	179.8	159.4	147.8	1.5
Korea, Democratic People's	0.8	1.2	2.2	2.3			0.1
Korea, Republic	8.4	13.8	22.8	23.4	23.3		0.6
Laos	0.2	0.1	0.1	0.2			0.04
Malaysia	3.9	4.7	7.6	7.2	7.2		0.5
Mongolia	0.3	0.4	0.6	0.6			0.4
Nepal	0.1	0.7	0.9	0.8			0.01
Pakistan	3.8	3.4	4.4	4.3	4.9		0.1
Philippines	6.5	8.3	10.8	10.9	9.5		0.2
Singapore	1.8	4.6	5.3	7.7	8.9		3.1
Sri Lanka	1.0	0.8	0.9	1.0	1.3		0.1
Thailand	4.4	7.3	9.9	10.0	10.0		0.2
Vietnam	6.7	3.0	1.2	1.3			0.02
TOTAL	240.6	322.4	381.1	372.9	378.0		0.2

Cont'd.

Table 38 continued

OCEANIA

	1970	1975	1980	1981	1982	1983 (est.)	1981 kgs per capita
Australia	20.5	23.9	25.8	25.2	24.9	23.7	1.7
Fiji	0.2	0.2	0.2	0.2			0.4
French Polynesia	0.1	0.1	0.1	0.1			0.5
Guam	0.4	0.6	0.6	0.7			6.6
New Caledonia	0.4	0.7	0.5	0.4			2.6
New Zealand	2.9	3.5	3.4	3.1	3.0	2.9	1.0
Papua New Guinea	0.2	0.5	0.6	0.6			0.2
TOTAL	24.9	29.7	31.6	30.6	29.7		1.3

Sources: UN Yearbook of World Energy Statistics
OECD/IEA Oil Statistics Quarterly
Statistiques de l'Industrie Petroliere
OPEC Statistical Bulletin
Own Calculations

Table 39

CONSUMPTION INTAKE OF REFINERY FEEDSTOCKS BY COUNTRY 1975-1983

	1975	1980	1981	1982	1983 (est.)
Austria	157	268	254	272	
Belgium	756	1,467	1,410	1,536	2,863
Denmark	33	249	459	845	761
Finland	20	177	211		
France		4,218	7,070	9,477	8,624
Federal Republic of Germany	1,789	13,274	15,420	16,951	20,356
Ireland	122	148	69	44	152
Italy	6,578	6,871	6,958	7,148	9,452
Netherlands	2,100	2,514	3,718	6,399	7,261
Norway	137	401	225	519	397
Portugal	9	70	132	308	211
Spain	19	514	339	324	596
Sweden	686	434	205	473	451
Switzerland	333	776	738	347	177
United Kingdom	2,408				
EEC	13,786	33,404	41,101	51,122	59,480
TOTAL—Western Europe	15,147	36,108	43,167	53,369	61,348
Canada			958	1,450	1,496
USA	971	6,110	17,065	17,827	13,360
Australia		1,139	1,111	1,914	1,932
New Zealand	748	632	473	305	235
Japan		20,681	16,575	5,155	5,296
Turkey	64	64	38	4	60
OECD	16,930	64,670	79,349	80,020	83,667

Source: OECD/IEA Oil Statistics Quarterly
   Own Estimates

271

Table 40

TOTAL LIQUEFIED PETROLEUM GAS CONSUMPTION 1970-1983
Unit: thousand metric tonnes

Region	1970	1975	1980	1981	1982	1983 (est.)
North/Central America	49,322	55,152	60,958	65,215		
South America	3,589	4,878	5,814	5,405		
North Africa/ Middle East	1,181	2,339	3,584	3,672		
Africa	233	231	251	254		
West Europe	10,261	12,029	18,466	16,728	15,501	15,665
East Europe	5,566	8,160	9,777	10,519		
Far East	9,755	14,763	19,146	19,940		
Oceania	177	422	531	1,187		
OPEC	1,203	2,167	2,700	2,621		
OECD	67,027	78,850	94,112	97,158	93,776	85,740
OPEC percentage share	2	2	2	2		
OECD percentage share	84	80	79	78		
WORLD TOTAL	80,139	98,420	119,223	124,428		

Source: Own Calculations

Table 41

CONSUMPTION OF LIQUEFIED PETROLEUM GAS 1970-1983
Unit: thousand metric tonnes
NORTH and CENTRAL AMERICA

	1970	1975	1980	1981	1982	1983 (est.)	1981 kgs per capita
Canada	1,763	2,417	3,232	3,019	771	602	125
USA	45,305	49,668	53,506	57,186	57,541	51,697	249
Bahamas	10	7	20	20			81
Barbados	3	9	8	9			34
Belize	1	2	2	2			12
Costa Rica	7	14	22	24			11
Cuba	57	83	106	103			11
Dominican Republic	29	47	85	77			14
El Salvador	8	12	26	30			6
Guatemala	24	33	45	45			6
Haiti	1	1	3	4			1
Honduras	4	8	6	8			2
Jamaica	18	42	30	30			14
Nicaragua	8	12	14	17			6
Panama	20	36	62	61			31
Puerto Rico	22	38	50	40			12
Trinidad & Tobago	14	20	33	25			21
Gaudeloupe	6	9	11	11			35
Mexico	1,896	2,553	3,566	4,382			59
Netherlands Antilles	77	72	75	70			268
US Virgin Islands	15	30	41	36			316
TOTAL	49,322	55,152	60,958	65,215			171

Cont'd.

Table 41 continued

SOUTH AMERICA

	1970	1975	1980	1981	1981 kgs per capita	
Argentina	926	1,088	1,094	1,142	41	
Bolivia	3	27	97	86	15	
Brazil	1,348	1,899	2,653	2,393	20	
Chile	297	407	485	503	45	
Colombia	292	320	310	300	11	
Ecuador	6	26	99	112	13	
French Guiana	1	1	2	2	30	
Guyana	5	5	6	6	7	
Paraguay	3	4	9	9	3	
Peru	27	119	109	100	5	
Suriname	5	0	10	10	25	
Uruguay	36	40	44	49	17	
Venezuela	640	942	896	693	48	
TOTAL	3,589	4,878	5,814	5,405	22	

Cont'd.

Table 41 continued

NORTH AFRICA and MIDDLE EAST

	1970	1975	1980	1981	1982	1983 (est.)	1981 kgs per capita
Bahrain	5	7	12	13			40
Cyprus	21	25	35	35			55
Yemen	2	3	8	8			4
Iran	292	449	300	500			13
Iraq	29	89	100	50			4
Israel	112	129	165	155			39
Jordan	12	25	32	60			18
Kuwait	0	0	212	200			137
Lebanon	58	70	70	60			22
Qatar	0	1	6	6			24
Saudi Arabia	41	106	285	290			31
Syria	21	45	83	90			10
Turkey	220	507	747	761	795	877	16
United Arab Emirates	0	0	12	14			18
Algeria	155	346	635	620			32
Egypt	103	169	383	300			7
Libya	15	157	93	80			26
Morocco	77	165	314	323			16
Tunisia	18	46	92	107			16
TOTAL	1,181	2,339	3,584	3,672			

Cont'd.

Table 41 continued

AFRICA

	1970	1975	1980	1981	1981 kgs per capita
Angola	14	25	19	23	3
Congo	1	2	2	2	1
Djibouti	1	1	1	1	3
Ethiopia	2	3	4	4	0
Gabon	3	3	4	4	7
Ghana	4	7	8	8	1
Guinea			1	1	2
Ivory Coast	9	8	15	18	2
Kenya	6	15	23	21	1
Liberia		1	1	1	0
Madagascar	8	9	5	5	1
Mauritania	1	1	3	3	2
Mauritius	0	1	1	1	1
Mozambique	8	7	7	7	1
Nigeria	18	15	23	23	0
Reunion	5	9	14	13	26
Senegal	3	4	5	5	1
South Africa	36	89	70	75	2
Sudan	3	6	8	8	0
Uganda	2	3	0	0	0
Cameroon	2	2	8	6	1
Tanzania	4	6	7	8	0
Zaire	1	2	1	1	0
Zambia	1	8	10	10	2
Zimbabwe	1	2	4	5	1
TOTAL	233	231	251	254	3

Cont'd.

Table 41 continued

WESTERN EUROPE

	1970	1975	1980	1981	1982	1983 (est.)	1981 kgs per capita
Austria	107	122	159	149	118	108	20
Belgium	466	559	525	459	433	401	47
Denmark	227	183	238	247	256	238	48
Finland	62	87	125	129	137	27	
France	2,254	2,575	3,388	3,324	2,605	2,525	62
Federal Republic of Germany		1,739	2,006	2,586	2,424	2,642	39
Greece	118	150	192	198	183	180	20
Iceland	1	1	1	1			4
Ireland	64	102	203	171	188	151	50
Italy	1,756	2,118	3,506	2,394	2,248	2,320	42
Luxembourg	26	22	25	25	26	25	69
Netherlands	363	541	2,077	2,674	2,341	2,260	188
Norway	20	28	875	797	100	80	194
Portugal	267	395	503	540	556	565	54
Spain	1,355	1,543	2,718	2,575	2,138	2,510	68
Sweden	158	185	192	173	179	174	21
Switzerland	47	75	98	98	129	126	15
United Kingdom	1,231	1,337	1,055	1,147	1,220	1,450	21
EEC	8,244	9,593	12,795	13,063	12,144	11,965	
TOTAL	10,261	12,029	18,466	16,728	15,501	15,665	39

Cont'd.

Table 41 continued

EASTERN EUROPE

	1970	1975	1980	1981	1981 kgs per capita
Bulgaria	6	40	43	42	5
Czechoslovakia	81	113	148	144	9
German Democratic Republic	222	253	272	263	16
Hungary	169	223	257	264	25
Poland	90	140	188	189	5
Romania	179	257	222	200	9
Yugoslavia	78	285	401	422	19
USSR	4,741	6,849	8,246	8,995	34
TOTAL	5,566	8,160	9,777	10,519	25

Cont'd.

Table 41 continued

FAR EAST

	1970	1975	1980	1981	1982	1983 (est.)	1981 kgs per capita
Bangladesh	0	0	3	5			0
Brunei	2	12	40	18			75
Burma	5	2	3	3			0
Hong Kong	24	73	104	114			22
India	161	320	421	473			1
Indonesia	4	33	35	29			0
Japan	9,310	13,825	17,653	18,336	17,886	15,628	156
Korea, Republic	47	138	376	408			11
Laos	1	1	4	6			2
Malaysia	53	33	28	33			2
Pakistan	0	15	30	32			0
Philippines	89	180	213	224			5
Singapore	4	7	25	23			9
Sri Lanka	8	1	7	9			1
Thailand	40	105	199	222			5
Vietnam	7	8	5	5			0
TOTAL	9,755	14,763	19,146	19,940			9

Cont'd.

Table 41 continued

OCEANIA

	1970	1975	1980	1981	1982	1983 (est.)	1981 kgs per capita
Australia	165	401	496	1,123	1,282	1,271	75
Fiji	1	2	3	3			5
French Polynesia	2	4	6	6			40
Guam	1	3	4	4			38
New Caledonia	3	5	5	5			35
New Zealand	3	3	12	41			13
Papua New Guinea	2	3	5	5			2
TOTAL	177	422	531	1,187			52

Sources: UN Yearbook of World Energy Statistics
OECD/IEA Oil Statistics Quarterly
Own Calculations

Table 42

TOTAL MOTOR GASOLINE CONSUMPTION 1970-1983
Unit: thousand metric tonnes

Region	1970	1975	1980	1981	1982	1983 (est.)
North/Central America	178,045	324,709	330,545	327,767	321,745	
South America	19,262	25,182	27,440	29,222		
North Africa/ Middle East	6,272	11,716	18,453	17,222		
Africa	5,528	7,224	9,807	9,791		
West Europe	72,538	88,796	100,931	100,069	104,230	103,249
East Europe	54,669	74,086	82,070	83,360		
Far East	27,358	39,057	47,201	47,535		
Oceania	9,157	11,336	12,901	12,953	13,429	
OPEC	7,984	14,104	24,560	24,415		
OECD	366,054	434,385	451,481	447,538	453,934	452,175
OPEC percentage share	2	2	4	4		
OECD percentage share	77	75	72	71	72	
WORLD TOTAL	473,382	583,257	630,171	629,672	627,500	

Source: Own Calculations

281

Table 43

CONSUMPTION OF MOTOR GASOLINE 1970-1983
Unit: thousand metric tonnes
NORTH and CENTRAL AMERICA

	1970	1975	1980	1981	1982	1983 (est.)	1981 kgs per capita
Canada	19,608	25,442	27,692	26,794	24,176	23,566	1,107
USA	248,413	285,874	283,311	280,135	284,655	284,996	1,219
Bahamas	64	72	85	85			343
Barbados	44	61	51	50			188
Belize	11	18	22	21			126
Costa Rica	76	120	129	130			57
Cuba	791	991	1,066	1,172			120
Dominican Republic	203	322	289	267			48
El Salvador	90	115	133	140			28
Greenland	13	14	8	11			216
Guatemala	165	263	254	232			31
Haiti	24	30	45	42			8
Honduras	82	88	93	100			26
Jamaica	328	270	205	180			81
Nicaragua	109	143	130	135			48
Panama	156	234	221	223			115
Puerto Rico	1,184	1,651	1,914	1,832			564
Trinidad & Tobago	178	211	765	742			626
Bermuda	22	21	22	27			443
Gaudeloupe	38	48	45	40			127
Martinique	41	84	77	77			237
Mexico	6,187	8,337	13,667	15,038	15,170		202
Netherlands Antilles		95	150	100			383
US Virgin Islands	76	117	104	125			1,096
TOTAL	278,045	324,709	330,545	327,767	321,745		860

Table 43 continued

SOUTH AMERICA

	1970	1975	1980	1981	1982	1981 kgs per capita
Argentina	4,063	3,926	5,445	5,604	5,434	200
Bolivia	225	351	413	469		79
Brazil	7,067	10,386	7,928	9,129	8,950	75
Chile	1,202	980	1,100	1,165		103
Colombia	1,843	2,493	2,735	3,000		108
Ecuador	365	668	1,264	1,265	1,282	146
French Guiana	10	13	25	17		254
Guyana	41	43	40	40		44
Paraguay	88	64	90	90		28
Peru	1,116	1,538	1,168	1,140		62
Suriname	25	30	44	40		101
Uruguay	239	202	222	233		80
Venezuela	2,978	4,488	6,966	7,029		492
TOTAL	19,262	25,182	27,440	29,222		120

Cont'd.

Table 43 continued

NORTH AFRICA and MIDDLE EAST

	1970	1975	1980	1981	1982	1983 (est.)	1981 kgs per capita
Bahrain	31	50	110	120			373
Cyprus	97	78	100	99			155
Yemen	75	90	171	175			92
Iran	1,119	2,751	2,500	1,800	1,957		46
Iraq	382	420	1,150	1,100			81
Israel	479	667	750	782			198
Jordan	100	154	270	280			83
Kuwait	314	449	810	815	887		557
Lebanon	362	478	500	480			179
Oman	30	81	200	216			235
Qatar	39	69	166	180	209		726
Saudi Arabia	500	1,144	3,240	3,420			367
Syria	196	399	600	610			65
Turkey	970	1,743	1,862	1,932	1,996	2,032	42
United Arab Emirates	60	210	585	610			801
Algeria	422	653	1,210	1,090	1,150		56
Egypt	494	1,336	2,007	2,050			47
Libya	196	450	818	850			275
Morocco	315	363	432	452			22
Tunisia	91	131	154	161			25
TOTAL	6,272	11,716	18,453	17,222			

Cont'd.

Table 43 continued

AFRICA

	1970	1975	1980	1981	1982	1981 kgs per capita
Angola	65	51	80	80		11
Benin	17	34	28	30		8
Burundi	8	9	16	19		4
Central African Republic	21	13	22	21		9
Chad	14	20	24	25		5
Congo	22	28	35	40		25
Djibouti	8	9	11	11		34
Equatorial Guinea	3	5	2	4		11
Ethiopia	87	70	113	110		3
Gabon	23	18	30	35		63
Gambia	8	11	22	22		36
Ghana	156	237	210	210		17
Guinea	34	38	46	46		9
Ivory Coast	236	189	317	225		27
Kenya	165	234	313	300		17
Liberia	55	62	69	69		34
Madagascar	84	82	95	95		11
Malawi	26	34	40	40		7
Mali	27	36	54	50		7
Mauritania	11	18	31	30		18
Mauritius	17	31	33	33		35
Mozambique	87	55	63	55		5
Niger	14	21	35	36		7
Nigeria	439	1,026	2,955	2,900	2,773	36
Rwanda	7	12	20	19		4
Senegal	78	111	130	125		22
Sierra Leone	43	39	20	45		13
Somalia	28	30	45	50		10
South Africa	2,719	3,698	3,680	3,880		114
Sudan	180	149	157	152		8
Togo	16	29	50	55		20
Uganda	112	121	88	70		5
Cameroon	88	109	170	146		17
Tanzania	129	98	87	84		5
Upper Volta	17	26	53	56		8
Zaire	119	182	188	180		7
Zambia	136	177	170	160		27
Zimbabwe	171	238	180	160		21
Reunion	44	60	97	95		189
TOTAL	5,528	7,224	9,807	9,791		30

Cont'd.

Table 43 continued

WESTERN EUROPE

	1970	1975	1980	1981	1982	1983 (est.)	1981 kgs per capita
Austria	1,937	2,231	2,523	2,395	2,560	2,655	319
Belgium	2,174	2,795	2,940	2,623	2,594	2,488	266
Denmark	1,497	1,566	1,523	1,479	1,467	1,514	289
Finland	1,059	1,341	1,370	1,460	1,491	1,554	304
France	12,455	15,883	16,321	17,421	17,473	15,961	323
Federal Republic of Germany	15,338	20,069	22,897	22,087	25,150	25,354	358
Greece	649	947	1,377	1,477	1,566	1,614	149
Iceland	51	86	89	92	93	92	398
Ireland	628	796	1,032	1,041	1,009	956	303
Italy	10,383	11,478	13,948	13,996	13,988	13,781	245
Luxembourg	103	181	286	311	310	303	854
Netherlands	3,035	3,473	3,849	3,687	3,624	3,577	259
Norway	1,001	1,143	1,446	1,423	1,439	1,473	347
Portugal	497	802	752	778	812	793	78
Spain	2,528	4,535	5,758	5,303	5,568	5,542	141
Sweden	2,761	3,242	3,516	3,462	3,494	3,595	416
Switzerland	2,113	2,444	2,751	2,860	2,908	3,001	440
United Kingdom	14,329	15,784	18,553	18,174	18,684	18,996	326
EEC	60,591	72,972	82,726	82,296	85,865	84,544	
TOTAL	72,538	88,796	100,931	100,069	104,230	103,249	238

Cont'd.

Table 43 continued

EASTERN EUROPE

	1970	1975	1980	1981	1981 kgs per capita
Albania	81	90	160	155	55
Bulgaria	1,379	1,680	1,850	1,800	199
Czechoslovakia	1,114	1,575	1,639	1,785	117
German Democratic Republic	1,772	2,417	3,109	3,242	194
Hungary	628	1,101	1,401	1,351	126
Poland	2,591	2,732	3,988	2,868	108
Romania	2,370	2,431	1,847	1,590	71
Yugoslavia	1,134	1,765	2,176	2,069	92
USSR	43,600	60,295	65,900	67,500	252
TOTAL	54,669	74,086	82,070	83,360	224

Cont'd.

Table 43 continued

FAR EAST

	1970	1975	1980	1981	1982	1983 (est.)	1981 kgs per capita
Afghanistan	108	83	102	89			5
Bangladesh	—	55	58	45			1
Bhutan	1	1	1	1			1
Brunei	26	32	72	68			283
Burma	162	179	241	246			7
China	3,005	7,950	9,920	8,820			9
Kampuchea	45	15	12	12			2
East Timor	1	2	4	5			7
Hong Kong	92	109	190	254			49
India	1,463	1,221	1,459	1,629			2
Indonesia	1,147	1,758	2,866	3,321	3,289		22
Japan	15,587	21,529	25,183	26,027	26,086	25,795	221
Korea, Democratic People's	340	490	700	750			41
Korea, Republic	640	491	769	802			21
Laos	44	39	55	62			16
Malaysia	432	464	841	790			55
Mongolia	133	150	220	230			134
Nepal	27	18	23	9			1
Pakistan	401	248	578	600			7
Philippines	1,770	1,795	1,552	1,375			28
Singapore	208	279	349	376			154
Sri Lanka	148	95	102	100			7
Thailand	645	1,304	1,654	1,674			35
Vietnam	933	750	250	250			5
TOTAL	17,358	39,057	47,201	47,535			24

Cont'd.

Table 43 continued

OCEANIA

	1970	1975	1980	1981	1982	1983 (est.)	1981 kgs per capita
Australia	7,532	9,333	10,796	10,918	11,064	10,844	731
Fiji	34	35	43	42			66
French Polynesia	21	29	31	34			227
Guam	40	91	90	75			721
New Caledonia	33	49	93	86			601
New Zealand	1,406	1,668	1,706	1,663	1,727	1,693	532
Papua New Guinea	60	83	86	77			25
TOTAL	9,157	11,336	12,901	12,953	13,429		562

Sources: UN Yearbook of World Energy Statistics
OECD/IEA Oil Statistics Quarterly
Statistiques de l'Industrie Petroliere
OPEC Statistical Bulletin
Own Calculations

Table 44

TOTAL AVIATION FUEL (GASOLINE) CONSUMPTION 1970-1982
Unit: thousand metric tonnes

Region	1970	1975	1980	1981	1982
North/Central America	2,640	1,949	1,880	1,750	
South America	192	170	163	141	
North Africa/ Middle East	58	46	102	50	
Africa	104	87	74	65	
West Europe	727	505	622	960	
East Europe	37	19	15	13	
Far East	1,289	369	116	109	
Oceania	142	153	128	132	
OPEC	106	76	127	67	
OECD	3,348	2,532	2,394	2,630	
OPEC percentage share	2	2	4	2	
OECD percentage share	64	76	77	81	
WORLD TOTAL	5,220	3,315	3,129	3,250	3,282 (est.)

Source: Own Calculations

290

Table 45

CONSUMPTION OF AVIATION FUEL (GASOLINE) 1970-1981
Unit: thousand metric tonnes
NORTH and CENTRAL AMERICA

	1970	1975	1980	1981	1981 kgs per capita
Canada	161	170	176	172	7
USA	2,307	1,632	1,480	1,385	6
Bahamas	7	3	6	4	16
Barbados	1	1	1	1	4
Belize	0	0	1	1	6
Costa Rica	12	9	9	8	4
Dominican Republic	2	2	4	5	1
El Salvador	4	2	2	2	0
Guatemala	4	4	9	7	1
Haiti	1	3	6	8	2
Honduras	9	5	3	5	1
Jamaica	5	3	2	3	1
Nicaragua	6	7	7	6	2
Panama	0	0	6	5	3
Puerto Rico	5	7	75	50	15
Trinidad & Tobago	5	2	4	3	3
Antigua & Barbuda	34	17	18	20	260
Mexico	45	58	67	61	1
US Virgin Islands	29	22	2	2	18
TOTAL	2,640	1,949	1,880	1,750	5

Cont'd.

Table 45 continued

SOUTH AMERICA

	1970	1975	1980	1981	1981 kgs per capita
Argentina	28	23	18	20	1
Bolivia	4	9	9	4	1
Brazil	44	40	49	50	0
Chile	23	14	8	7	1
Colombia	20	25	31	26	1
Ecuador	22	7	5	2	0
Guyana	2	2	2	2	2
Paraguay	1	0	4	3	1
Peru	20	14	8	4	0
Suriname	2	7	1	1	3
Uruguay	5	3	4	3	1
Venezuela	21	26	24	19	1
TOTAL	192	170	163	141	1

Cont'd.

Table 45 continued

NORTH AFRICA and MIDDLE EAST

	1970	1975	1980	1981	1981 kgs per capita
Cyprus	1	0	0	0	0
Yemen	0	0	0	0	0
Iran	4	15	65	15	0
Iraq	20	0	0	0	0
Israel	4	4	4	5	1
Jordan	8	14	17	15	4
Kuwait	0	1	1	1	1
Lebanon	1	0	0	0	0
Oman	2	2	1	1	1
Saudi Arabia	4	0	0	0	0
Syria	0	3	3	3	0
Turkey	0	0	3	2	0
United Arab Emirates	2	3	2	2	3
Algeria	5	1	1	1	0
Libya	2	2	2	2	1
Morocco	5	1	3	3	0
TOTAL	58	46	102	50	1

Cont'd.

Table 45 continued

AFRICA

	1970	1975	1980	1981	1981 kgs per capita
Benin	1	1	0	0	0
Burundi	0	0	4	4	1
Central African Republic	0	0	0	0	0
Chad	1	1	1	1	0
Congo	2	1	1	1	1
Djibouti	0	0	0	0	0
Ethiopia	0	15	7	6	0
Gabon	3	5	10	10	18
Ghana	0	0	0	0	0
Guinea	1	2	1	1	0
Ivory Coast	0	0	0	0	0
Kenya	5	8	6	0	0
Liberia	1	0	1	1	0
Madagascar	5	3	1	1	0
Malawi	1	1	1	1	0
Mali	1	0	0	0	0
Niger	3	2	1	1	0
Nigeria	13	5	5	5	0
Rwanda	0	0	0	0	0
Senegal	2	3	4	3	1
Sierra Leone	8	1	0	0	0
Somalia	0	2	7	7	1
South Africa	3	0	2	2	0
Sudan	14	17	4	4	0
Uganda	2	1	0	0	0
Cameroon	3	2	2	2	0
Tanzania	3	1	5	4	0
Zaire	28	11	8	8	0
Zambia	4	4	3	3	1
Zimbabwe	0	0	0	0	0
TOTAL	104	87	74	65	0

Cont'd.

Table 45 continued

WESTERN EUROPE

	1970	1975	1980	1981	1981 kgs per capita
Austria	0	1	0	0	0
Belgium	7	10	5	5	1
Denmark	4	5	4	5	1
Finland	12	9	8	4	1
France	58	25	29	30	1
Federal Republic of Germany	406	325	298	450	7
Greece	16	0	10	259	27
Iceland	0	0	1	1	4
Ireland	0	0	0	0	0
Italy	58	26	193	147	3
Luxembourg	1	0	0	0	0
Netherlands	8	8	5	3	0
Norway	5	1	7	6	1
Portugal	2	4	2	1	0
Spain	36	24	4	3	0
Sweden	20	12	11	12	1
Switzerland	0	0	1	0	0
United Kingdom	93	55	44	34	1
EEC	652	454	588	933	
TOTAL	727	505	622	960	2

Cont'd.

Table 45 continued

EASTERN EUROPE

	1970	1975	1980	1981	1981 kgs per capita
Romania	21	15	10	10	0
Yugoslavia	16	4	5	3	0
TOTAL	37	19	15	13	0

Cont'd.

Table 45 continued

FAR EAST

	1970	1975	1980	1981	1981 kgs per capita
Afghanistan	7	8	7	8	0
Bangladesh	—	2	3	2	0
Burma	3	1	1	1	0
Hong Kong	0	0	0	0	0
India	7	8	0	0	0
Indonesia	10	11	12	10	0
Japan	56	111	19	17	0
Korea, Republic	0	2	11	10	0
Laos	63	6	4	3	1
Malaysia	2	2	9	7	0
Pakistan	30	13	21	21	0
Philippines	12	47	10	12	0
Singapore	2	2	0	7	3
Sri Lanka	0	0	1	1	0
Thailand	5	6	18	10	0
Vietnam	1,092	150	—	—	0
TOTAL	1,289	369	116	109	0

Cont'd.

Table 45 continued

OCEANIA

	1970	1975	1980	1981	1981 kgs per capita
Australia	74	95	63	73	5
Fiji	0	0	0	1	2
French Polynesia	10	10	12	17	113
Guam	3	1	0	0	0
New Caledonia	4	1	1	1	7
New Zealand	23	19	31	21	7
Papua New Guinea	16	12	8	6	2
Wake Island	11	10	10	10	5,000
TOTAL	142	153	128	132	6

Source: UN Yearbook of World Energy Statistics 1981

Table 46

TOTAL AVIATION KEROSENE (JET FUEL) CONSUMPTION 1970-1983
Unit: thousand metric tonnes

Region	1970	1975	1980	1981	1982	1983 (est.)
North & Central America	40,999	44,519	48,820	45,570		
South America	1,437	2,651	3,925	3,648		
North Africa/ Middle East	648	1,817	1,545	1,437		
Africa	591	1,131	796	703		
West Europe	6,295	8,074	9,292	7,698	7,909	8,175
East Europe	121	283	311	284		
Far East	2,837	4,167	5,461	5,815		
Oceania	1,005	1,208	1,503	1,143		
OPEC	860	2,211	2,182	1,777		
OECD	48,327	54,328	60,364	55,318	53,843	54,536
OPEC percentage share	1.6	3.4	3.0	2.7		
OECD percentage share	89.3	84.6	83.7	83.0		
WORLD TOTAL	54,099	64,223	72,078	66,678	67,280 (est.)	

Source: Own Calculations

Table 47

CONSUMPTION OF AVIATION KEROSENE (JET FUEL) 1970-1983
Unit: thousand metric tonnes
NORTH and CENTRAL AMERICA

	1970	1975	1980	1981	1982	1983 (est.)	1981 kgs per capita
Canada	1,596	2,432	3,137	3,072	2,163	2,133	127
USA	38,305	40,795	43,903	40,744	40,583	41,491	177
Bahamas	0	0	0	0			0
Barbados	1	2	2	3			11
Belize	2	4	3	3			11
Costa Rica	28	11	22	24			11
Dominican Republic	58	34	42	45			8
El Salvador	14	14	8	9			2
Guatemala	20	27	22	19			3
Haiti	0	1	1	2			0
Honduras	6	11	22	24			6
Jamaica	13	29	34	30			14
Nicaragua	6	20	20	20			7
Panama	50	84	55	56			29
Puerto Rico	501	326	277	245			75
Trinidad & Tobago	6	8	57	52			44
Mexico	389	717	1,207	1,212			16
US Virgin Islands	4	4	8	10			88
TOTAL	40,999	44,519	48,820	45,570			120

Cont'd.

Table 47 continued

SOUTH AMERICA

	1970	1975	1980	1981	1981 kgs per capita
Argentina	301	463	732	769	27
Bolivia	13	43	92	91	15
Brazil	371	880	1,346	1,250	10
Chile	100	86	162	175	15
Colombia	167	315	453	490	18
Ecuador	79	60	129	133	15
Guyana	0	5	5	5	6
Paraguay	5	6	7	7	2
Peru	198	356	380	340	19
Suriname	0	14	17	15	38
Uruguay	9	7	16	18	6
Venezuela	194	416	586	355	25
TOTAL	1,437	2,651	3,925	3,648	15

Cont'd.

Table 47 continued

NORTH AFRICA and MIDDLE EAST

	1970	1975	1980	1981	1982	1983 (est.)	1981 kgs per capita
Bahrain	7	10	16	16			50
Cyprus	0	0	0	0			0
Yemen	15	25	36	30			16
Iran	241	1,074	400	300			8
Iraq	76	140	207	100			7
Israel	8	9	10	10			3
Jordan	0	0	0	0			0
Kuwait	36	33	198	200			137
Lebanon	5	14	5	10			4
Oman	0	1	3	3			3
Qatar	0	3	3	3			12
Saudi Arabia	20	60	127	140			15
Turkey	45	74	67	102	89	96	2
United Arab Emirates	1	2	1	1			1
Algeria	60	174	175	185			9
Egypt	21	65	60	50			1
Libya	63	72	165	165			53
Morocco	0	0	10	20			1
Tunisia	50	61	62	102			16
TOTAL	648	1,817	1,545	1,437			

Cont'd.

Table 47 continued

AFRICA

	1970	1975	1980	1981	1981 kgs per capita
Angola	16	28	15	12	2
Benin	0	0	0	0	0
Burundi	3	1	1	2	0
Central African Republic	3	4	7	7	3
Chad	2	3	5	5	1
Congo	0	1	5	5	3
Djibouti	0	0	0	0	0
Ethiopia	33	34	56	55	2
Gabon	3	3	0	0	0
Ghana	32	47	0	0	0
Guinea	0	0	0	0	0
Ivory Coast	39	64	74	68	8
Kenya	17	186	92	9	1
Liberia	0	1	1	1	0
Madagascar	11	10	5	5	1
Malawi	2	3	3	3	0
Mali	0	0	0	0	0
Mauritania	0	0	0	0	0
Mauritius	0	0	0	0	0
Mozambique	10	10	0	0	0
Niger	10	16	13	14	3
Nigeria	51	106	70	65	1
Rwanda	1	4	5	4	1
Senegal	1	6	6	5	1
Sierra Leone	3	12	6	5	1
Somalia	0	0	0	0	0
South Africa	126	370	280	285	8
Sudan	35	32	48	48	3
Uganda	78	36	0	0	0
Cameroon	3	1	10	10	1
Tanzania	29	50	35	25	1
Zaire	20	30	20	20	1
Zambia	36	6	5	5	1
Zimbabwe	20	55	5	5	1
Reunion	5	10	24	35	70
TOTAL	591	1,131	796	703	3

Cont'd.

Table 47 continued

WESTERN EUROPE

	1970	1975	1980	1981	1982	1983	1981 kgs per capita
Austria	58	45	136	148	137	160	20
Belgium	259	330	350	441	420	449	45
Denmark	185	228	231	234	239	246	46
Finland	45	97	92	93	181	190	19
France	1,606	1,876	2,314	2,032	2,045	2,084	38
Federal Republic of Germany	782	1,088	1,486	251	243	246	4
Greece	135	270	297	82	88	85	8
Iceland	0	0	0	0	1	1	0
Ireland	49	43	49	32	32	31	9
Italy	369	520	755	803	824	942	14
Luxembourg	0	0	0	0	1	1	0
Netherlands	143	155	25	36	52	37	3
Norway	5	5	5	5	5	5	1
Portugal	116	274	212	136	141	133	14
Spain	735	976	836	1,001	1,070	1,035	27
Sweden	152	170	196	200	205	192	24
Switzerland	54	67	97	82	81	89	13
United Kingdom	1,602	1,930	2,211	2,122	2,144	2,249	38
EEC	5,130	6,440	7,718	6,033	6,088	6,370	
TOTAL	6,295	8,074	9,292	7,698	7,909	8,175	16

Cont'd.

Table 47 continued

EASTERN EUROPE

	1970	1975	1980	1981	1981 kgs per capita
Yugoslavia	121	283	311	284	13
TOTAL	121	283	311	284	13

Cont'd.

Table 47 continued

FAR EAST

	1970	1975	1980	1981	1982	1983	1981 kgs per capita
Afghanistan	7	16	15	18			1
Bangladesh	—	20	45	54			1
Brunei	0	18	17	21			88
Burma	25	24	43	45			1
Kampuchea	24	0	0	0			0
Hong Kong	9	18	25	17			3
India	407	522	635	659			1
Indonesia	36	68	121	130			1
Japan	1,194	1,912	2,599	2,689	2,059	1,629	23
Korea, Republic	374	328	471	539			14
Laos	19	8	5	4			1
Malaysia	78	35	47	55			4
Pakistan	230	287	360	388			5
Philippines	149	218	265	250			5
Singapore	6	13	25	35			14
Sri Lanka	0	4	26	29			2
Thailand	279	676	762	847			18
TOTAL	2,837	4,167	5,461	5,815			3

Cont'd.

Table 47 continued

OCEANIA

	1970	1975	1980	1981	1982	1983	1981 kgs per capita
Australia	770	912	1,222	916	952	922	61
Fiji	12	13	0	0			0
Guam	70	120	90	90			865
New Zealand	122	129	144	97	88	90	31
Papua New Guinea	26	29	47	40			13
TOTAL	1,005	1,208	1,503	1,143			50

Sources: UN Yearbook of World Energy Statistics
OECD/IEA Oil Statistics Quarterly
Own Estimates

Table 48

TOTAL KEROSENE CONSUMPTION 1970-1983
Unit: thousand metric tonnes

Region	1970	1975	1980	1981	1982	1983 (est.)
North & Central America	18,093	13,607	14,231	12,001	11,168	
South America	3,670	3,962	3,830	3,400		
North Africa/ Middle East	4,588	6,737	8,556	8,330		
Africa	1,525	1,643	2,525	2,569		
West Europe	8,789	6,413	4,659	4,537	4,438	4,392
East Europe	24,148	30,774	35,389	35,697		
Far East	26,939	40,054	49,053	51,188		
Oceania	755	1,017	675	405		
OPEC	5,607	9,135	12,732	12,496		
OECD	38,048	35,963	35,485	33,874	30,548	28,154
OPEC percentage share	6.3	8.8	10.7	10.6		
OECD percentage share	42.9	34.5	29.8	28.6	25.9	
WORLD TOTAL	88,591	104,360	119,098	118,319	117,878	

Source: Own Calculations

Table 49

CONSUMPTION OF KEROSENE 1970-1983
Unit: thousand metric tonnes
NORTH and CENTRAL AMERICA

	1970	1975	1980	1981	1982	1983 (est.)	1981 kgs per capita
Canada	2,972	3,479	3,079	2,467	1,916	1,655	102
USA	12,354	7,262	7,466	5,879	5,845	5,892	26
Bahamas	9	50	60	50			202
Barbados	10	9	6	6			23
Belize	8	3	2	2			12
Costa Rica	18	18	20	18			8
Cuba	421	447	440	428			44
Dominican Republic	18	17	29	30			5
El Salvador	36	32	37	48			10
Greenland	1	2	4	7			137
Guatemala	52	55	62	72			10
Haiti	3	2	8	9			2
Honduras	30	32	42	46			12
Jamaica	23	56	84	81			36
Nicaragua	29	16	15	17			6
Panama	40	16	10	12			6
Puerto Rico	161	240	105	90			28
Trinidad & Tobago	95	133	79	76			64
Antigua & Barbuda	33	5	18	20			260
Gaudeloupe	9	8	18	20			63
Martinique	12	49	75	60			185
Mexico	1,675	1,614	2,393	2,403	2,343		32
Netherlands Antilles	47	27	40	40			153
US Virgin Islands	19	25	129	105			921
TOTAL	18,093	13,607	14,231	12,001	11,168		31

Cont'd.

Table 49 continued

SOUTH AMERICA

	1970	1975	1980	1981	1982	1981 kgs per capita
Argentina	886	834	662	568	491	20
Bolivia	103	143	132	93		16
Brazil	517	794	514	500		4
Chile	361	326	227	208		18
Colombia	457	432	352	340		12
Ecuador	65	184	343	284	290	33
French Guiana	1	0	17	19		284
Guyana	27	19	29	29		32
Paraguay	20	14	15	21		6
Peru	510	627	858	810		44
Suriname	8	5	2	2		5
Uruguay	180	163	119	101		35
Venezuela	535	421	560	425	431	30
TOTAL	3,670	3,962	3,830	3,400		14

Cont'd.

Table 49 continued

NORTH AFRICA and MIDDLE EAST

	1970	1975	1980	1981	1982	1983	1981 kgs per capita
Bahrain	9	8	9	10			31
Cyprus	14	9	10	8			13
Yemen	30	35	45	50			26
Iran	1,601	3,289	4,400	4,230	4,706		108
Iraq	578	414	450	400			30
Israel	129	9	75	80			20
Jordan	89	118	160	170			51
Kuwait	100	111	340	350	332		239
Lebanon	25	81	25	35			13
Oman	5	15	12	12			13
Qatar	4	5	4	5	5		20
Saudi Arabia	125	170	120	160			17
Syria	294	419	525	575			62
Turkey	462	458	433	286	294	260	6
United Arab Emirates	8	14	33	38			50
Algeria	65	123	50	55			3
Egypt	828	1,142	1,548	1,543			35
Libya	87	63	132	140			45
Morocco	73	73	75	70			3
Tunisia	62	91	110	113			17
TOTAL	4,588	6,737	8,556	8,330			

Cont'd.

Table 49 continued

AFRICA

	1970	1975	1980	1981	1982	1981 kgs per capita
Angola	24	13	21	21		3
Benin	16	25	18	20		5
Burundi	3	2	1	5		1
Central African Republic	15	0	3	3		1
Chad	4	4	5	5		1
Congo	14	10	2	2		1
Djibouti	8	11	12	12		37
Equatorial Guinea	4	5	4	6		16
Ethiopia	7	4	1	1		0
Gabon	15	24	18	20		36
Gambia	1	3	5	5		8
Ghana	74	97	103	105		9
Guinea	14	16	20	19		4
Ivory Coast	37	42	49	57		7
Kenya	10	20	16	73		4
Liberia	11	9	7	6		3
Madagascar	31	34	62	61		7
Malawi	11	11	9	9		1
Mali	8	12	11	10		1
Mauritania	1	1	3	3		2
Mauritius	11	20	15	19		20
Mozambique	27	48	55	40		4
Niger	3	3	5	4		1
Nigeria	217	373	1,200	1,185	796	15
Rwanda	3	2	7	6		1
Senegal	11	12	18	18		3
Sierra Leone	16	19	30	30		8
Somalia	7	9	29	26		5
South Africa	553	436	450	460		14
Sudan	122	56	40	40		2
Togo	13	6	0	0		0
Uganda	39	46	36	30		2
Cameroon	10	38	60	68		8
Tanzania	50	52	54	50		3
Upper Volta	8	10	10	11		2
Zaire	71	96	80	80		3
Zambia	4	12	25	20		3
Zimbabwe	30	38	12	10		1
Reunion	9	9	12	11		22
TOTAL	1,525	1,643	2,525	2,569		9

Cont'd.

Table 49 continued

WESTERN EUROPE

	1970	1975	1980	1981	1982	1983 (est.)	1981 kgs per capita
Austria	18	8	18	11	12	13	1
Belgium	51	98	103	60	57	59	6
Denmark	225	120	82	81	83	83	16
Finland	30	14	8	18	32	33	4
France	55	34	125	159	160	162	3
Federal Republic of Germany	83	63	105	106	103	104	2
Greece	75	38	36	29	31	30	3
Iceland	2	3	2	2	2	2	9
Ireland	97	92	72	79	83	78	23
Italy	3,193	2,382	977	1,234	1,153	1,152	22
Luxembourg	1	1	0	0	1	1	0
Netherlands	1,226	475	326	259	275	283	18
Norway	396	316	557	282	253	219	69
Portugal	80	69	68	66	68	64	7
Spain	282	121	77	75	86	75	2
Sweden	246	123	32	34	35	33	4
Switzerland	16	11	7	11	11	12	2
United Kingdom	2,713	2,445	2,064	2,031	1,994	1,989	36
EEC	7,719	5,748	3,890	4,038	3,939	3,941	
TOTAL	8,789	6,413	4,659	4,537	4,438	4,392	14

Cont'd.

Table 49 continued

EASTERN EUROPE

	1970	1975	1980	1981	1981 kgs per capita
Albania	15	40	70	65	23
Bulgaria	134	170	230	220	24
Czechoslovakia	389	397	578	510	33
German Democratic Republic	12	17	—	—	0
Hungary	90	131	198	224	21
Poland	169	203	265	227	6
Romania	825	1,014	868	770	34
Yugoslavia	34	10	30	31	1
USSR	22,480	28,792	33,150	33,650	126
TOTAL	24,148	30,774	35,389	35,697	109

Cont'd.

Table 49 continued

FAR EAST

	1970	1975	1980	1981	1982	1983 (est.)	1981 kgs per capita
Afghanistan	10	18	11	4			0
Bangladesh	—	326	389	452			5
Brunei	2	0	0	0			0
Burma	272	191	80	84			2
China	4,305	11,260	16,175	15,125			15
Kampuchea	19	0	0	0			0
East Timor	1	1	3	4			5
Hong Kong	170	190	248	169			33
India	3,239	2,988	4,277	4,861			7
Indonesia	2,207	3,944	5,082	5,204	5,244		35
Japan	12,773	17,393	19,249	20,371	17,720	15,628	173
Korea, Democratic People's	35	70	160	180			10
Korea, Republic	388	509	1,085	1,116			29
Laos	11	6	10	14			4
Malaysia	166	236	315	344			24
Mongolia	8	8	13	15			9
Nepal	28	22	27	29			2
Pakistan	739	597	694	620			7
Philippines	418	406	481	497			10
Singapore	167	741	209	1,467			600
Sri Lanka	319	211	182	186			12
Thailand	114	167	243	316			7
Vietnam	1,548	770	120	130			2
TOTAL	26,939	40,054	49,053	51,188			22

Cont'd.

Table 49 continued

OCEANIA

	1970	1975	1980	1981	1982	1983	1981 kgs per capita
Australia	667	914	575	314	317	309	21
Fiji	13	13	18	18			28
French Polynesia	16	14	8	3			20
New Caiedonia	4	4	4	4			28
New Zealand	31	44	24	20	18	18	6
Papua New Guinea	10	13	18	18			6
TOTAL	755	1,017	675	405			18

Sources: UN Yearbook of World Energy Statistics
OECD/IEA Oil Statistics Quarterly
Statistiques de l'Industrie Petroliere
OPEC Statistical Bulletin
Own Estimates

Table 50

TOTAL GAS/DIESEL OIL CONSUMPTION 1970-1983
Unit: thousand metric tonnes

Region	1970	1975	1980	1981	1982	1983 (est.)
North & Central America	155,064	175,004	184,531	176,413		
South America	16,187	23,401	34,879	32,725		
North Africa/ Middle East	10,159	19,698	26,338	25,853		
Africa	6,636	9,018	10,968	10,961		
West Europe	145,949	160,074	176,007	162,497	156,589	150,679
East Europe	75,077	108,387	126,048	125,616		
Far East	42,196	66,554	93,217	92,263		
Oceania	5,022	6,897	8,114	8,064		
OPEC	8,846	16,892	26,373	24,824		
OECD	317,234	358,078	387,996	365,351	348,186	341,697
OPEC percentage share	1.9	3.0	4.0	3.9		
OECD percentage share	69.4	62.8	58.5	57.3	56.8	
WORLD TOTAL	457,085	570,611	663,345	638,014	613,255	

Source: Own Calculations

Table 51

CONSUMPTION OF GAS/DIESEL OIL 1970-1983
Unit: thousand metric tonnes
NORTH and CENTRAL AMERICA

	1970	1975	1980	1981	1982	1983 (est.)	1981 kgs per capita
Canada	18,908	20,754	22,102	20,538	18,908	16,748	848
USA	127,217	139,374	142,840	135,675	126,406	128,398	590
Bahamas	280	738	340	340			1,371
Barbados	33	41	44	50			188
Belize	14	27	30	28			168
Costa Rica	151	279	330	355			157
Cuba	1,066	2,412	2,040	2,007			205
Dominican Republic	170	405	396	384			69
El Salvador	113	206	199	205			42
Greenland	85	148	168	152			2,980
Guatemala	171	261	387	334			45
Haiti	56	58	84	86			17
Honduras	159	198	246	255			67
Jamaica	240	347	229	220			99
Nicaragua	129	181	176	185			66
Panama	131	295	371	320			165
Puerto Rico	248	535	664	625			192
Trinidad & Tobago	72	95	194	149			126
Antigua & Barbuda	15	4	10	12			156
Bermuda	28	113	104	77			1,262
Cayman Islands	4	9	16	17		944	
Gaudeloupe	14	22	32	30			95
Martinque	15	29	8	9			28
Mexico	4,803	7,436	12,065	12,907	12,329		173
Netherlands Antilles	470	272	300	350			1,341
Saint Lucia	13	13	35	30			246
St. Pierre-Miguelon	8	9	10	11			1,833
US Virgin Islands	411	621	1,076	1,020			8,947
TOTAL	155,064	175,004	184,531	176,413			463

Cont'd.

Table 51 continued

SOUTH AMERICA

	1970	1975	1980	1981	1982	1981 kgs per capita
Argentina	5,245	5,928	7,180	6,853	6,968	224
Bolivia	89	159	220	175		30
Brazil	5,598	9,877	16,363	15,771	16,092	130
Chile	672	807	1,193	1,062		94
Colombia	721	846	1,227	1,285		46
Ecuador	250	444	918	908	913	105
French Guiana	19	22	75	65		970
Guyana	97	169	160	160		177
Paraguay	56	103	168	105		32
Peru	915	1,138	1,424	1,395		76
Suriname	85	152	162	144		363
Uruguay	338	402	477	488		167
Venezuela	2,101	3,353	5,311	4,312		302
TOTAL	16,187	23,401	34,879	32,725		134

Cont'd.

Table 51 continued

NORTH AFRICA and MIDDLE EAST

	1970	1975	1980	1981	1982	1983	1981 kgs per capita
Bahrain	22	65	94	130			404
Cyprus	131	98	153	151			237
Yemen	35	152	205	230			121
Iran	2,026	5,370	5,250	4,750	5,473		121
Iraq	692	1,043	1,379	1,200			89
Israel	682	902	1,022	968			245
Jordan	109	230	508	620			184
Kuwait	132	220	300	300	548		205
Lebanon	208	304	255	270			101
Oman	30	132	262	261			284
Qatar	30	72	163	169	183		681
Saudi Arabia	569	655	1,140	1,120			120
Syria	494	1,757	1,796	1,820			195
Turkey	1,751	3,233	4,193	4,221	4,773	5,078	91
United Arab Emirates	105	587	1,527	1,550			2,034
Algeria	765	1,449	1,974	1,810	2,092		92
Egypt	1,227	1,364	2,211	2,295			53
Libya	419	733	1,548	1,650			533
Morocco	484	808	1,420	1,385			67
Tunisia	248	524	938	953			146
TOTAL	10,159	19,698	26,338	25,853			

Cont'd

Table 51 continued

AFRICA

	1970	1975	1980	1981	1982	1981 kgs per capita
Angola	296	307	220	195		27
Benin	57	78	40	40		11
Burundi	4	6	13	14		3
Central African Republic	24	14	22	21		9
Chad	20	33	34	33		7
Congo	50	27	35	30		19
Djibouti	10	12	15	10		31
Equatorial Guinea	5	9	11	12		32
Ethiopia	182	94	313	306		10
Gabon	100	113	176	200		360
Gambia	6	17	25	24		39
Ghana	217	264	176	185		15
Guinea	40	41	45	44		9
Ivory Coast	214	264	272	285		34
Kenya	194	319	405	363		21
Liberia	133	150	120	115		56
Madagascar	118	148	130	131		15
Malawi	43	58	83	80		13
Mali	24	46	45	45		6
Mauritania	90	87	120	120		71
Mauritius	79	64	88	57		61
Mozambique	100	52	75	65		6
Niger	30	59	111	100		18
Nigeria	420	685	1,800	1,780	1,837	22
Rwanda	6	11	14	13		3
Senegal	64	264	255	240		41
Sierra Leone	26	21	39	44		12
Somalia	35	114	227	223		46
South Africa	2,391	3,450	3,870	3,985		117
Sudan	562	580	520	535		28
Togo	33	34	38	124		46
Uganda	98	81	60	50		4
Cameroon	76	142	230	60		7
Tanzania	103	253	225	205		11
Upper Volta	20	31	50	68		10
Zaire	266	282	330	360		13
Zambia	221	338	290	330		55
Zimbabwe	220	349	275	300		39
Reunion	21	32	65	65		129
TOTAL	6,636	9,018	10,968	10,961		40

Cont'd.

Table 51 continued

WESTERN EUROPE

	1970	1975	1980	1981	1982	1983 (est.)	1981 kgs per capita
Austria	1,657	2,220	2,817	2,434	2,491	2,433	324
Belgium	7,916	8,262	8,445	6,657	6,133	5,883	675
Denmark	6,112	5,804	5,348	4,853	4,685	4,332	947
Finland	4,096	4,378	4,571	4,256	4,142	3,544	886
France	8,961	7,801	12,709	12,583	11,843	11,307	233
Federal Republic of Germany	53,720	55,955	54,858	51,134	48,614	48,221	829
Greece	1,721	2,079	3,243	2,995	3,339	3,064	309
Iceland	315	335	231	218	196	182	944
Ireland	727	1,028	1,301	1,299	1,261	1,211	378
Italy	12,382	19,496	25,066	21,296	21,414	19,626	372
Luxembourg	491	509	519	496	488	484	1,363
Netherlands	6,631	6,186	7,599	7,956	6,521	6,437	558
Norway	2,605	2,910	3,695	3,140	2,985	2,800	766
Portugal	763	1,124	1,704	1,820	1,884	1,892	183
Spain	4,886	7,685	10,178	10,853	10,115	270	
Sweden	9,054	8,251	8,184	7,517	6,582	5,470	903
Switzerland	6,457	6,741	7,138	6,351	5,808	6,821	977
United Kingdom	17,455	19,310	18,113	17,314	17,350	16,857	310
EEC	116,116	126,430	137,201	126,583	121,648	117,422	
TOTAL	145,949	160,074	176,007	162,497	156,589	150,679	395

Cont'd.

Table 51 continued

EASTERN EUROPE

	1970	1975	1980	1981	1981 kgs per capita
Albania	140	185	300	250	89
Bulgaria	2,163	3,500	4,100	4,000	442
Czechoslovakia	2,671	3,571	4,051	3,980	260
German Democratic Republic	3,592	3,889	4,864	4,564	273
Hungary	2,014	3,662	4,431	4,279	399
Poland	2,539	4,551	5,856	5,188	145
Romania	2,501	3,487	4,834	4,130	184
Yugoslavia	2,351	3,109	3,512	3,125	139
USSR	57,106	82,433	94,100	96,100	359
TOTAL	75,077	108,387	126,048	125,616	335

Cont'd.

Table 51 continued

FAR EAST

	1970	1975	1980	1981	1982	1983 (est)	1981 kgs per capita
Afghanistan	63	122	147	150			9
Bangladesh	—	267	449	410			5
Brunei	27	29	59	80			333
Burma	274	258	385	390			11
China	6,360	14,030	17,525	16,025			16
Kampuchea	11	0	0	0			0
Hong Kong	486	682	1,002	786			191
India	4,714	7,610	11,330	12,189			18
Indonesia	1,237	2,168	4,887	5,075	5,469		34
Japan	18,792	28,296	35,476	35,148	34,123	33,868	299
Korea, Democratic People's	330	400	900	950			52
Korea, Republic	1,543	2,895	5,041	5,247			136
Laos	39	21	48	53			14
Malaysia	1,295	1,887	3,448	3,145			218
Mongolia	34	67	125	135			79
Nepal	40	32	38	43			3
Pakistan	1,211	1,407	2,142	2,093			25
Philippines	1,463	1,797	4,186	4,195			85
Singapore	309	985	1,522	1,425			583
Sri Lanka	240	291	348	370			25
Thailand	1,887	2,480	3,759	3,629			75
Vietnam	1,741	830	400	425			8
TOTAL	42,196	66,554	93,217	92,263			44

Cont'd.

Table 51 continued

OCEANIA

	1970	1975	1980	1981	1982	1983	1981 kgs per capita
Australia	3,891	5,508	6,440	6,377	6,507	6,055	427
Fiji	74	102	151	162			253
French Polynesia	11	12	15	15			100
Guam	43	73	105	145			1,394
New Caledonia	49	54	70	60			420
New Zealand	726	839	938	895	880	871	286
Papua New Guinea	110	167	232	240			78
TOTAL	5,022	6,897	8,114	8,064			350

Sources: UN Yearbook of World Energy Statistics
OECD/IEA Oil Statistics Quarterly
Statistiques de l'Industrie Petroliere
OPEC Statistical Bulletin
Own Estimates

Table 52

TOTAL RESIDUAL FUEL OIL CONSUMPTION 1970-1983
Unit: thousand metric tonnes

Region	1970	1975	1980	1981	1982	1983 (est.)
North & Central America	151,767	166,170	160,610	130,112	104,368	
South America	21,869	27,935	32,798	29,104		
North Africa/ Middle East	16,658	26,434	36,942	37,430		
Africa	4,494	9,422	6,557	6,152		
West Europe	238,240	231,049	222,576	202,357	177,681	141,517
East Europe	105,631	155,780	177,958	178,138		
Far East	128,313	152,171	160,095	149,244		
Oceania	7,259	6,808	5,788	4,895	4,308	
OPEC	9,761	15,628	23,224	24,484		
OECD	473,012	477,165	441,837	379,503	331,728	276,727
OPEC percentage share	1.5	2.0	2.9	3.3		
OECD percentage share	70.8	61.2	54.3	50.7	49.9	
WORLD TOTAL	667,740	779,637	813,841	748,852	664,403	

Source: Own Calculations

Table 53

CONSUMPTION OF RESIDUAL FUEL OILS
Unit: thousand metric tonnes
NORTH and CENTRAL AMERICA

	1970	1975	1980	1981	1982	1983	1981 kgs per capita
Canada	18,481	15,482	13,579	11,419	8,873	6,615	472
USA	113,352	122,760	111,048	84,073	72,839	61,781	366
Bahamas	109	215	250	300			1,210
Barbados	39	51	79	74			278
Belize	4	3	2	3			18
Costa Rica	87	147	138	180			79
Cuba	3,515	4,149	5,838	5,591			571
Dominican Republic	450	1,049	969	943			169
El Salvador	167	235	143	155			31
Guatemala	240	350	477	344			46
Haiti	31	37	58	60			12
Honduras	104	115	105	130			34
Jamaica	906	1,829	2,092	1,962			884
Nicaragua	137	211	194	195			69
Panama	238	311	480	450			232
Puerto Rico	3,752	4,539	4,465	3,808			1,172
Trinidad & Tobago	200	191	132	167			141
Gaudeloupe	30	41	58	75			237
Martinique	43	61	48	49			151
Mexico	5,829	9,063	15,709	15,867	17,112		213
Netherlands Antilles	4,615	2,805	2,500	2,000			7,663
US Virgin Islands	1,345	2,315	2,218	2,250			19,737
TOTAL	151,767	166,170	160,610	130,112	104,368		341

Cont'd.

Table 53 continued

SOUTH AMERICA

	1970	1975	1980	1981	1982	1981 kgs per capita
Argentina	8,181	7,685	6,565	5,506	5,249	196
Bolivia	143	191	115	115		19
Brazil	7,270	13,608	17,176	15,161	13,431	125
Chile	1,728	1,412	1,621	1,305		116
Colombia	1,067	1,150	1,157	755		27
Ecuador	297	430	942	888	849	103
Guyana	343	341	295	295		327
Paraguay	35	51	44	45		14
Peru	1,712	1,778	2,104	2,095		115
Suriname	373	412	479	450		1,134
Uruguay	775	864	812	667		228
Venezuela	− 55	13	1,488	1,821	1,821	127
TOTAL	21,869	27,935	32,798	29,104		120

Cont'd.

Table 53 continued

AFRICA

	1970	1975	1980	1981	1982	1981 kgs per capita
Angola	101	187	200	200		28
Benin	2	7	8	8		2
Burundi	1	1	3	3		1
Central African Republic	3	4	6	6		3
Congo	25	30	5	5		3
Djibouti	6	18	25	15		46
Ethiopia	132	122	75	52		2
Gabon	137	206	190	220		396
Ghana	237	120	42	50		4
Guinea	160	164	174	173		34
Ivory Coast	202	425	121	79		10
Kenya	301	495	630	560		33
Liberia	231	223	300	300		147
Madagascar	17	165	95	80		9
Malawi	4	4	4	3		0
Mali	6	9	25	22		3
Mauritania	25	48	35	35		21
Mauritius	83	70	52	79		84
Mozambique	120	213	270	230		21
Niger	7	5	5	5		1
Nigeria	462	596	800	890	2,367	11
Rwanda	2	1	3	3		1
Senegal	185	417	373	390		67
Sierra Leone	134	110	100	90		25
Somalia	0	2	5	5		1
South Africa	1,015	3,139	2,000	1,600		47
Sudan	517	307	325	325		17
Togo	23	24	44	46		17
Uganda	100	59	32	30		2
Cameroon	20	32	60	90		10
Tanzania	177	177	180	185		10
Upper Volta	2	4	15	—		0
Zaire	44	86	140	150		6
Zambia	3	84	180	200		34
Zimbabwe	0	0	1	1		0
Reunion	—	15	19	6		12
TOTAL	4,494	9,422	6,557	6,152		34

Cont'd.

Table 53 continued

NORTH AFRICA and MIDDLE EAST

	1970	1975	1980	1981	1982	1983	1981 kgs per capita
Bahrain	3	3	4	5			16
Cyprus	248	325	531	506			794
Yemen	110	165	210	300			157
Iran	3,080	4,900	5,700	6,000	6,004		153
Iraq	1,437	1,574	1,466	1,300			96
Israel	2,882	3,128	3,983	4,089			1,033
Jordan	98	138	398	569			169
Kuwait	18	69	80	80			55
Lebanon	408	871	810	790			294
Oman	10	4	4	3			3
Qatar	11	0	0	0			0
Saudi Arabia	578	2,629	4,680	4,690			503
Syria	659	638	2,005	2,020			217
Turkey	3,229	6,063	6,713	6,259	6,655	6,392	135
United Arab Emirates	0	0	760	760			997
Algeria	233	640	533	560	1,086		29
Egypt	2,530	3,535	5,588	5,570			128
Libya	134	246	791	1,200			388
Morocco	542	1,013	1,707	1,730			84
Tunisia	448	492	979	999			153
TOTAL	16,658	26,434	36,942	37,430			

Cont'd.

Table 53 continued

WESTERN EUROPE

	1970	1975	1980	1981	1982	1983	1981 kgs per capita
Austria	4,537	4,499	5,508	4,166	3,656	2,898	555
Belgium	9,169	6,953	6,714	5,351	5,305	3,947	543
Denmark	8,203	5,846	4,314	3,510	3,015	2,368	685
Finland	4,607	4,195	4,574	3,828	3,375	2,673	797
France	51,374	56,144	53,903	42,738	35,171	24,518	792
Federal Republic of Germany	30,837	26,858	21,457	23,451	20,812	16,670	380
Greece	2,337	4,321	4,981	4,414	4,002	3,498	455
Iceland	87	82	171	170	122	132	736
Ireland	1,835	2,323	2,493	2,043	1,457	1,170	594
Italy	43,125	42,316	45,653	45,485	40,916	37,972	795
Luxembourg	680	514	137	104	106	81	286
Netherlands	8,045	4,281	9,103	9,732	4,757	1,255	683
Norway	2,230	1,777	2,071	1,698	1,437	1,211	414
Portugal	1,248	2,484	3,795	4,110	4,281	4,358	414
Spain	12,004	21,582	21,918	20,747	19,193	16,117	551
Sweden	14,131	11,183	10,193	8,780	7,432	5,454	1,055
Switzerland	2,303	1,694	1,271	1,073	898	799	165
United Kingdom	41,488	33,997	24,320	20,957	21,746	16,396	375
EEC	197,093	183,553	173,075	157,785	137,287	107,875	
TOTAL	238,240	231,049	222,576	202,357	177,681	141,517	505

Cont'd.

Table 53 continued

EASTERN EUROPE

	1970	1975	1980	1981	1981 kgs per capita
Albania	303	351	500	450	161
Bulgaria	4,504	6,300	7,400	7,250	800
Czechoslovakia	4,523	8,257	9,115	8,840	577
German Democratic Republic	4,420	7,099	7,742	7,330	438
Hungary	2,251	3,362	2,554	2,790	260
Poland	1,625	3,078	3,997	3,961	110
Romania	2,717	3,731	6,812	6,080	271
Yugoslavia	2,825	4,583	7,338	6,437	286
USSR	82,463	119,019	132,500	135,000	504
TOTAL	105,631	155,780	177,958	178,138	477

Cont'd.

Table 53 continued

FAR EAST

	1970	1975	1980	1981	1982	1983 (est.)	1981 kgs per capita
Afghanistan	28	38	34	37			2
Bangladesh	—	355	496	431			5
Brunei	0	0	1	0			0
Burma	123	52	240	243			7
China	8,100	20,875	29,525	27,825			28
Kampuchea	76	0	0	0			0
East Timor	2	4	5	6			8
Hong Kong	1,832	2,230	3,469	3,795			736
India	4,374	5,794	7,032	7,404			11
Indonesia	3,429	4,325	5,794	6,075	6,924		40
Japan	95,076	96,186	83,084	71,325	62,472	57,549	606
Korea, Democratic People's	110	200	400	450			25
Korea, Republic	5,385	9,484	15,082	15,298			395
Laos	4	0	6	8			2
Malaysia	1,874	2,037	2,916	2,855			198
Mongolia	76	138	230	250			146
Pakistan	1,208	837	580	525			6
Philippines	2,518	3,809	4,091	4,303			87
Singapore	1,121	2,560	3,193	4,330			1,772
Sri Lanka	251	232	229	285			19
Thailand	1,386	2,515	3,288	3,349			70
Vietnam	1,340	500	400	450			8
TOTAL	128,313	152,171	160,095	149,244			72

Cont'd

Table 53 continued

OCEANIA

	1970	1975	1980	1981	1982	1983	1981 kgs per capita
Australia	6,010	4,834	4,325	3,732	2,958	2,652	250
Fiji	26	11	10	10			16
French Polynesia	0	0	0	0			0
Guam	225	327	330	350			3,365
New Caledonia	335	606	358	214			1,497
New Zealand	624	791	512	338	250	221	108
Papua New Guinea	4	177	199	196			64
TOTAL	7,259	6,808	5,788	4,895	4,308		213

Source: UN Yearbook of World Energy Statistics
   OPEC Statistical Bulletin
   OECD/IEA Oil Statistics Quarterly
   Statisques de l'Industrie Petroliere
   Own Estimates

Table 54

INTERNATIONAL PIPELINE CONSTRUCTION-BY COUNTRIES
(Underway, Planned, and Proposed—October 1982)

Country	Total Miles, Gas Pipelines			Total Miles, Crude Oil Pipelines			Total Miles, All Pipelines		
	Planned	Underway	Total	Planned	Underway	Total	Planned	Underway	Total
Abu Dhabi	275	185	460	64	0	64	339	185	524
Algeria	2,187	815	3,002	0	393	393	2,511	1,208	3,719
Argentina	4,136	0	4,136	435	0	435	5,063	0	5,063
Australia	2,541	84	2,625	157	0	157	3,123	496	3,619
Austria									
Bangladesh	150	170	320				150	170	320
Belgium	245	60	305				245	60	305
Bolivia	5,279	166	5,445				5,815	166	5,981
Brazil	2,228	264	2,492	1,169	51	1,220	3,637	315	3,952
Bulgaria									
Cameroon	125	0	125	2	0	2	127	0	127
Canada	10,267	529	10,796	1,525	0	1,525	12,709	529	13,238
Chad									
Chile	650	22	672	10	16	26	706	63	769
China	1,243	0	1,243	2,717	0	2,717	4,333	0	4,333
Colombia	961	15	976	1,159	50	1,209	2,694	165	2,859
Congo				14	0	14	14	0	14
Costa Rica							35	0	35
Czech'vakia	563		563				563	0	563
Denmark	7,730	377	8,107				7,730	377	8,107
Dubai				17	0	17	17	0	17
East Germany	250	0	250				250	0	250
Ecuador	494	0	494	231	0	231	725	0	725
Egypt	152	168	320	75	0	75	227	168	395
France	930	800	1,730	350	0	350	1,449	800	2,249
Greece	70	0	70	50	0	50	120	0	120
Guatemala				237	0	237	237	0	237
Honduras				160	0	160	160	0	160
Hungary	344		344	188	0	188	644	0	644
India	786	0	786	18	0	18	1,760	0	1,760
Indonesia	660	37	697	105	0	105	765	37	802
Iran	1,368	0	1,368	1,000		1,000	2,368	0	2,368
Iraq	400	250	650	2,066	0	2,066	3,477	250	3,727
Ireland	0	159	159				0	159	159
Italy	726	285	1,011				726	285	1,011
Ivory Coast				31	0	31	31	0	31
Japan	1,080	0	1,080				1,080		1,080
Kenya							248	0	248
Korea				450	0	450	450	0	450

Table 54 contd.

Country	Total Miles, Gas Pipelines			Total Miles, Crude Oil Pipelines			Total Miles, All Pipelines		
	Planned	Underway	Total	Planned	Underway	Total	Planned	Underway	Total
Kuwait	89	0	89	400	35	435	489	58	547
Libya	0	58	58	723	481	1,204	723	539	1,262
Malaysia	220	0	220				220	0	220
Mexico	0	68	68	399	0	399	1,559	219	1,778
Netherlands	366	0	366	124	0	124	490	0	490
New Zealand	879	65	944				879	170	1,049
Nigeria	2,496	120	2,616	346	0	346	2,842	120	2,962
Norway	0	531	531				0	531	531
North Yemen							228	0	228
Oman	132	203	335	472	271	743	604	474	1,078
Pakistan	483	0	483				483	0	483
Panama				0	78	78	0	78	78
Peru				1,017	0	1,017	1,017	0	1,017
Poland							342	0	342
Portugal							69	0	69
Qatar	0	40	40				0	40	40
Saudi Arabia	1,642	160	1,802	448	155	603	3,696	315	4,011
South Africa	225	0	225				225	0	225
Spain	586	40	626	159	37	196	813	7	890
Sudan				894	0	894	894	0	894
Syria									
Sweden	1,350	0	1,350				1,350	0	1,350
Switzerland	7	0	7				7	0	7
Taiwan				160	0	160	330	220	550
Tanzania							678	0	678
Thailand	503	62	565	286	0	286	1,117	62	1,179
Tunisia	179	329	508				179	329	508
Turkey	0	50	50	280	0	280	530	50	580
Trinidad	45	0	45				45	0	45
United Kingdom	392	292	684				392	430	822
United Arab Emirates	46	0	46				69	0	69
USSR	9,600	3,622	13,222				10,448	3,622	14,070
Uruguay	189	0	189				189	0	189
Venezuela	171	143	314	263	122	385	450	265	715
West Germany	266	22	288	81	0	81	548	22	570
Yugoslavia									
Zaire							380	0	380

Table 55

SHIPPING:
WORLD—SEABORNE TRADE 1971-1981
Unit: billion tonne-miles

	Crude Oil	Oil Products	Total Oil	All Trades Total	% Oil
1971	6,555	900	7,455	11,730	64
1975	8,885	845	9,730	15,366	63
1980	8,385	1,020	9,425	16,777	56
1981	7,371	1,000	8,371	15,840	53
1982	6,045	920	6,965	14,190	49
% change 1981/1982	− 18	− 8	− 17	− 10	

Source: Shipping Statistics and Economics

Table 56

TANKER FLEET
GENERAL DEVELOPMENT—TANKERS OF 10,000 dw OR MORE (1,000 tdw)

1970	132,137
1975	255,770
1980	327,882
1981	324,794
1982	320,236
1983	303,714

Source: Statistiques de l'Industrie Petroliere

Table 57

DEVELOPMENT OF FLEET BY FLAG

	Jan-1970		Jan-1981		Jan-1982		Jan-1983	
	1,000 tdw	%	1,000 tdw	%	1,000 tdw	%	1,000 tdw	%
Liberia	30,784	23.3	100,273	30.9	91,646	28.6	84,411	27.8
Japan	13,304	10.1	30,048	9.3	29,741	9.3	28,510	9.4
Greece	5,744	4.3	22,836	7.0	26,968	8.4	23,711	7.8
Norway	15,607	11.8	23,954	7.4	24,442	7.6	21,200	7.0
United Kingdom	18,690	14.1	24,793	7.6	21,218	6.6	17,624	5.8
USA	8,901	6.7	16,156	5.0	16,397	5.1	16,771	5.5
Panama	5,210	3.9	12,167	3.7	14,387	4.5	15,923	5.3
France	5,087	3.9	14,986	4.6	13,246	4.2	11,551	3.8
Spain	2,185	1.7	8,475	2.6	9,044	2.8	8,229	2.7
Italy	3,671	2.8	8,044	2.5	7,227	2.3	7,004	2.3
Russia	3,927	3.0	6,052	1.8	6,072	1.9		
Singapore	148	0.1	4,734	1.5	5,163	1.6	4,843	1.6
Denmark	2,312	1.7	4,877.	1.5	4,710	1.5	4,578	1.5
West Germany	3,120	2.4	5,072	1.6	4,765	1.5	4,287	1.4
Others	13,447	10.2	42,327	13.0	45,210	14.1	48,933	16.1
TOTAL	132,137	100.0	324,794	100.0	320,236	100.0	303,714	100.0

Source: Statistiques de l'Industrie Petroliere

338

Table 58

FLEET BY FLAG AT 1/1/1983

Flag	Total Number	1,000 tdw	Diesel Number	1,000 tdw
Liberia	649	84,411	416	29,208
Japan	208	28,510	134	10,090
Greece	307	23,711	239	14,505
Norway	158	21,200	118	10,218
United Kingdom	200	17,624	143	7,000
USA	311	16,771	36	1,400
Panama	198	15,923	160	8,553
France	64	11,551	26	1,932
Spain	65	8,229	49	4,245
Italy	80	7,004	63	4,103
Saudi Arabia	40	6,514	18	920
Russia	199	6,139	164	3,983
Singapore	64	4,843	60	4,525
Denmark	44	4,578	34	1,403
West Germany	31	4,287	17	937
Brazil	46	3,087	34	1,124
Kuwait	22	2,613	18	1,264
Sweden	22	2,525	16	505
India	38	2,230	36	2,199
Iraq	18	2,085	16	1,387
Finland	28	1,908	25	1,148
Netherlands	12	1,782	6	352
Holland	24	1,726	11	248
Iran	14	1,651	7	266
South Korea	16	1,639	10	453
Libya	13	1,478	12	1,166
China	49	1,460	49	1,460
Portugal	13	1,436	8	656
Turkey	17	1,411	14	931
Cyprus	22	1,371	18	593
Mexico	34	1,272	31	927
Argentina	39	1,119	29	807
Algeria	11	1,080	10	694
Philippines	10	997	8	534
Poland	9	924	7	654
Taiwan	10	888	10	888
Venezuela	19	768	14	584

Cont'd.

Table 58 contd.

Flag	Total Number	Total 1,000 tdw	Diesel Number	Diesel 1,000 tdw
Australia	16	763	15	746
Belgium	10	587	9	515
Romania	6	511	6	511
Bulgaria	12	500	12	500
Yugoslavia	5	351	5	351
North Korea	3	331	2	237
Morocco	8	316	8	316
Peru	8	284	6	167
Canada	13	275	9	210
Nigeria	1	273	1	273
Bahamas	4	270	4	270
Indonesia	19	264	19	264
Ecuador	7	233	7	233
Cameroon	1	222	—	—
Uruguay	2	159	1	29
Thailand	2	140	2	140
Gabon	1	138	1	138
Abu Dhabi	1	136	1	136
Qatar	1	136	1	136
Egypt	5	126	5	126
Lebanon	2	112	2	112
East Germany	2	95	2	95
Others	31	747	29	712
TOTAL	3,264	303,714	2,213	128,079

Source: Statistiques de l'Industrie Petroliere

340

Table 59

WORLD VEHICLE REGISTRATIONS in 1982

	Total	Cars	Trucks & Buses
North America	173,692,000	135,154,000	38,538,000
Western Europe	119,971,302	107,899,400	12,071,902
Eastern Europe	32,351,849	21,676,960	10,674,889
Africa	9,756,847	6,254,891	3,501,956
Middle East	8,354,687	5,254,238	3,100,449
Far East	49,392,370	29,883,252	19,509,118
Pacific	9,475,534	7,553,549	1,921,985
Caribbean	2,051,272	1,634,167	417,105
Central & South America	27,032,813	19,791,292	
TOTAL	432,078,674	335,101,749	96,976,925

Source: World Automotive Market

Table 60

CRUDE PETROLEUM PRICES

Country	API	1971	1974	1976	1979	1980	1981	1982	1983
		(US Dollars per Barrel)							
ALGERIA									
Blend	44°	2.72	16.19	12.85	14.81	30.00	40.00	35.50	30.50
CHINA									
Daqing	33°					32.13	37.00	34.90	28.70
ECUADOR									
Oriente	29.7°	2.50	13.70	11.45	13.03	36.41	40.07	33.05	28.00
GABON									
Mandji	28.9°	1.70	13.03	13.20	13.23	28.00	35.00	34.00	29.00
Gamba	31.8°	1.71	13.79	13.50	13.67	29.00	36.00	35.00	30.00
INDONESIA									
Minas	34.1°	2.21	10.80	12.80	13.90	27.50	35.00	34.85	29.53
Ardjuna	36.7°				14.40	28.95	36.45	35.20	30.20
Attaka	42.1°				14.95	30.25	37.75	36.25	30.95
Cinta	33.5°				13.50	27.50	34.50	33.30	28.25
Walio	34.3°				13.65	27.20	35.00	34.40	29.00
IRAN									
Light	34-34.9°	2.17	11.88	12.50	13.45	30.00	37.00	31.20	28.00
Heavy	31-31.9°	2.13	11.64	12.25	13.06	29.27	36.00	29.30	26.90
IRAQ									
Basrah	34°	2.16	11.67	11.43	13.29	27.96	35.96	34.21	29.21
Kirkuk	36°	3.21	7.10	11.65	13.52	18.18	36.18	34.83	29.43
KUWAIT									
Export	31-31.9°	2.09	11.55	11.30	12.83	27.50	35.50	32.30	27.30
Hout	35-35.9°	2.09	11.70	11.50	13.33	28.00	36.00	—	29.014
LIBYAN A.J.									
Brega	40°	3.42	12.32	12.62	14.69	34.67	41.00	37.00	30.50
Amna	36°		12.10	12.05	14.03	34.02	40.30	35.15	29.20
MEXICO									
Isthmus	34°					33.50	38.50	32.50	29.00
Maya	23°					28.00	34.50	25.00	25.00
NIGERIA									
Light	37°	3.21	14.69	13.10	14.81	29.97	40.02	35.52	30.00
Medium	26°	3.10	14.48	12.55	14.23	28.70	38.72	34.52	28.00
NORWAY									
Ekofisk	42°				15.85	34.50	40.00	34.25	30.25
OMAN	34°				13.94	32.50			
QATAR									
Dukhan	40°	2.28	12.41	12.74	14.04	29.42	37.42	35.45	28.49

...contd.

Table 60 contd.

Country	API	1971	1974	1976	1979	1980	1981	1982	1983
Marine	36°	2.20	12.01	12.54	13.78	29.23	37.23	34.30	29.30
SAUDI ARABIA									
Light	34°	2.18	11.65	11.51	13.34	26.00	32.00	34.00	29.00
Medium	31°	2.09	11.56	11.28	12.89	25.45	31.45	32.40	27.40
Heavy	27°	1.96	11.44	11.04	12.51	25.00	31.00	31.00	26.00
UNITED ARAB EMIRATES									
Murban	39°	2.24	12.64	11.92	14.10	29.56	36.56	34.56	29.56
Umm Shaif	37°	2.23	12.09	11.70	13.78	29.36	36.36	34.36	29.36
Zakum	40°		12.57	11.83	14.01	29.46	36.46	34.46	29.46
UNITED KINGDOM									
Forties	37°			12.45	15.50	33.75	39.25	33.50	30.00
VENEZUELA									
Tiajuana	26°	2.02	11.80	11.12	13.36	25.20	32.88	32.03	27.88
Tiajuana	31°	2.20	12.10	12.30	14.22	26.90	36.00	35.00	29.84
Oficina	34°	2.34	12.34	12.80	14.69	28.75	38.06	37.06	31.09
San Joaquin	42°	2.56	12.70	13.40	15.32	30.90	41.00	38.50	32.42

Sources: UN Yearbook of World Energy Statistics
Petroleum Economist
OPEC Statistical Bulletin

Table 61

RETAIL PRICES OF GASOLINE
(US Dollars per US Gallon)

	1971	1974	1976	1979	1980	1981	1982	1983
ARGENTINA (Buenos Aires)								
Regular gasolene 84/85 RON	0.37	1.28	0.55	1.14	1.57	1.29	1.09	0.90
AUSTRALIA (Sydney)								
Regular gasolene 89/92 RON	0.39	0.59		1.30	1.51		1.68	1.81
AUSTRIA (Vienna)								
Regular gasolene 88/89 RON	0.52	1.19	1.38	2.02	2.58	2.16	2.31	2.40
BELGIUM (Brussels)								
Regular gasolene 91/93 RON	0.70	1.43	1.43	2.27	3.17	2.71	2.64	2.53
BRAZIL (Rio de Janeiro)								
Regular gasolene 82 RON	0.40	0.61	1.51	1.55	3.23		2.83	2.47
CANADA (Ottawa)								
Regular gasolene 91/94 RON	0.42	0.57	0.63	0.71	0.81	1.16	1.29	
CHILE (Santiago)								
Regular gasolene 81 RON	0.28	0.64	0.87	1.53	1.74	1.91		
DENMARK (Copenhagen)								
Regular gasolene 92 RON	0.69	1.35	1.40	2.80	3.30	2.85	2.70	2.83
ECUADOR (Quito)								
Regular gasolene 80 RON	0.22	0.18		0.16	0.19		0.45	0.91
ETHIOPIA (Addis Ababa)								
Regular gasolene 79/82 RON	0.76	1.16	1.19	1.38	1.38	2.18		
FINLAND (Helsinki)								
Regular gasolene 92 RON	0.66	1.21	1.48	2.07	3.01	2.77	2.63	
FRANCE (Paris)								
Regular gasolene 90/92 RON	0.76	1.32	1.40	2.43	2.97		2.24	2.51
GERMANY, FED. REP. OF. (Bonn)								
Regular gasolene 91.5 RON	0.66	1.26	1.39	2.09	2.46		2.04	2.14
GREECE (Athens)								
Regular gasolene 90 RON	0.76	1.70	1.59	2.79	2.85	2.11	2.19	2.47
INDIA (New Delhi)								
Regular gasolene 81/83 RON	0.67	1.83	1.42	1.97	2.49	2.45	2.55	
INDONESIA (Djakarta)								
Regular gasolene 87 RON	0.32	0.41	0.64	0.61	0.91	0.91	1.40	1.76
IRAN (Tehran)								
Regular gasolene 90 RON	0.30	0.34	0.32	0.54	0.54	1.62		
IRELAND (Dublin)								
Regular gasolene 86 RON	0.68	0.97	1.28	2.47	2.73			
ISRAEL (Tel Aviv)								
Regular gasolene 83 RON	0.64	1.26	1.64	2.32	2.49	2.02		

...contd..

Table 61 contd.

	1971	1974	1976	1979	1980	1981	1982	1983
ITALY (Rome)								
Regular gasolene 86/87 RON	0.92	1.69	1.70	2.44	3.08	2.73	2.65	3.16
JAPAN (Tokyo)								
Regular gasolene 89/91 RON	0.56	1.36	1.07	2.27	2.72	2.55	2.46	2.70
LEBANON (Beirut)								
Regular gasolene	0.36	0.51						
MEXICO (Mexico City)								
Regular gasolene 81 RON	0.24	0.42	0.65	0.47	0.46	0.44	0.48	0.51
NETHERLANDS (The Hague)								
Regular gasolene 92/93 RON	0.69	1.37	1.52	2.18	2.89	2.49	2.47	2.36
NORWAY (Oslo)								
Regular gasolene 93 RON	0.82	1.47	1.51	2.10	2.87	2.68	2.67	2.67
PAKISTAN (Islamabad)								
Regular gasolene 80 RON	0.87	0.80	1.01		1.70	1.91		
PARAGUAY (Asuncion)								
Regular gasolene 85 RON	0.51	1.50	1.50	2.34	3.00	3.61		
PHILIPPINES (Manila)								
Regular gasolene 81 RON	0.18	0.61	0.71	1.06	2.18	2.40		
PORTUGAL (Lisbon)								
Regular gasolene 85/87 RON	0.75	1.41	1.81	2.17	3.13	2.75	2.64	2.98
SAUDI ARABIA (Jidda)								
Regular gasolene 83.5 RON	0.24	0.12	0.11	0.19	0.23	0.18	0.18	0.18
SOUTH AFRICA (Pretoria)								
Regular gasolene 87 RON	0.46	0.78	0.98	2.43	2.48	2.20	1.96	2.14
SPAIN (Madrid)								
Regular gasolene 90 RON	0.57	1.16	1.21	2.29	2.81	2.53	2.22	2.38
SWITZERLAND (Bern)								
Regular gasolene 90/91RON	0.58	1.12	1.50	2.60	2.79	2.47		2.29
THAILAND (Bangkok)								
Regular gasolene 83 RON	0.35	0.64	0.77	1.38	1.71	2.05		
TURKEY (Ankara)								
Regular gasolene 85 RON	0.39	0.75	0.63	1.32	2.09	1.56	1.88	
UNITED KINGDOM (London)								
Regular gasolene 90 RON	0.65	1.25	1.25	2.16	2.60	2.62	2.67	2.57
UNITED STATES (New York)								
Regular gasolene 90/94 RON	0.35	0.60	0.64	0.91	1.22	1.25	1.26	1.15
VENEZUELA (Caracas)								
Regular gasolene 74 RON	0.08	0.13	0.14	0.13	0.13	0.13	0.26	0.26
YUGOSLAVIA (Belgrade)								
Regular gasolene 87/88 RON	0.45	1.20	1.02	2.47	2.63		2.94	2.32

Source: UN Yearbook of World Energy Statistics
OPEC Statistical Bulletin

Table 62

WORLD AIR TRANSPORT OPERATIONS 1978-1982
Unit: millions
INTERNATIONAL & DOMESTIC

	1978	+ – Change	1979	+ – Change	1980	+ – Change	1981	+ – Change	1982	+ – Change
NON-SCHEDULED SERVICES—SCHEDULED AIRLINES										
Passenger-Kilometres Flown	50,900	–5.7%	43,900	–13.8%	49,300	12.3%	47,600	–3.4%	52,830	11.0%
Tonne-Kilometres Performed	6,400	–7.0%	5,790	–9.5%	6,530	12.8%	6,340	–2.9%	6,630	4.6%
Available Tonne-Kilometres	9,270	–10.3%	8,760	–5.5%	10,120	15.5%	9,720	–4.0%	10,300	6.0%
NON-SCHEDULED SERVICES—NON-SCHEDULED OPERATORS										
Passenger-Kilometres Flown	73,600	5.6%	75,000	1.9%	60,550	–19.3%	58,920	–2.7%	66,670	13.2%
Tonne-Kilometres Performed	8,650	11.0%	8,620	–0.3%	6,820	–20.9%	6,680	–2.1%	7,300	9.3%
Available Tonne-Kilometres	12,240	25.7%	12,000	–2.0%	8,990	–25.1%	9,280	3.2%	10,090	8.7%
SCHEDULED AND NON-SCHEDULED SERVICES COMBINED—ALL OPERATORS										
Passenger-Kilometres Flown	1,060,500	12.6%	1,178,900	11.2%	1,198,850	1.7%	1,223,520	2.1%	1,263,500	3.3%
Tonne-Kilometres Performed	128,590	11.7%	141,270	9.9%	144,060	2.0%	148,060	2.8%	152,150	2.8%
Available Tonne-Kilometres	214,560	6.9%	233,070	8.6%	244,720	5.0%	250,510	2.4%	256,250	2.3%
INTERNATIONAL										
NON-SCHEDULED SERVICES—SCHEDULED AIRLINES										
Passenger-Kilometres Flown	41,100	–0.2%	36,700	–10.7%	43,100	17.4%	41,100	–4.6%	47,800	16.3%
Tonne-Kilometres Performed	5,330	–3.3%	4,760	–10.7%	5,380	13.0%	5,140	–4.5%	5,780	12.5%
Available Tonne-Kilometres	7,810	–3.2%	7,260	–7.0%	8,250	13.6%	7,740	–6.2%	8,880	14.7%

Table 62 contd..

WORLD AIR TRANSPORT OPERATIONS 1978-1982

	1978	+−Change	1979	+−Change	1980	+−Change	1981	+−Change	1982	+−Change
NON-SCHEDULED SERVICES—NON-SCHEDULED OPERATORS										
Passenger-Kilometres Flown	70,600	5.8%	72,300	2.4%	59,600	−17.6%	57,600	−3.4%	64,200	11.5%
Tonne-Kilometres Performed	7,990	11.3%	8,000	0.1%	6,650	16.9%	6,420	−3.5%	7,020	9.3%
Available Tonne-Kilometres	11,400	28.7%	11,100	−2.6%	8,700	−21.6%	8,830	1.5%	9,720	10.1%
SCHEDULED AND NON-SCHEDULED SERVICES COMBINED—										
ALL OPERATORS										
Passenger-Kilometres Flown	496,700	12.9%	549,000	10.5%	568,700	3.6%	592,700	4.2%	611,000	3.1%
Tonne-Kilometres Performed	66,760	12.5%	73,320	9.8%	76,310	4.1%	80,030	4.9%	88,130	3.9%
Available Tonne-Kilometres	111,680	9.9%	121,110	8.4%	128,350	6.0%	133,310	3.9%	138,880	4.2%
DOMESTIC										
NON-SCHEDULED SERVICES—SCHEDULED AIRLINES										
Passenger-Kilometres Flown	9,800	−22.8%	7,200	−26.5%	6,200	−13.9%	6,500	4.8%	5,030	−22.6%
Tonne-Kilometres Performed	1,070	−21.3%	1,030	−3.7%	1,140	10.7%	1,190	4.4%	850	−28.6%
Available Tonne-Kilometres	1,450	−36.1%	1,510	4.1%	1,870	23.8%	1,980	5.9%	1,420	−28.3%
NON-SCHEDULED SERVICES—NON-SCHEDULED OPERATORS										
Passenger-Kilometres Flown	3,000	—	2,700	−10.0%	950	−64.8%	1,320	38.9%	2,470	87.1%
Tonne-Kilometres Performed	590	1.7%	620	5.1%	170	−72.6%	260	52.9%	280	7.7%
Available Tonne-Kilometres	840	−4.5%	900	7.1%	290	−67.8%	450	55.2%	370	−17.8%
SCHEDULED AND NON-SCHEDULED SERVICES COMBINED—										
ALL OPERATORS										
Passenger-Kilometres Flown	563,800	12.4%	629,900	11.7%	630,150	—	630,820	0.1%	652,500	3.4%
Tonne-Kilometres Performed	61,760	10.9%	67,950	10.0%	67,740	−0.3%	68,020	0.4%	69,020	1.5%
Available Tonne-Kilometres	102,870	3.8%	111,970	8.8%	116,370	3.9%	117,200	0.7%	117,370	0.1%

Source: ICAO/IATA

TOURISM ECONOMICS:
CONCEPTS AND PRACTICES

TOURISM ECONOMICS: CONCEPTS AND PRACTICES

OZAN BAHAR AND METIN KOZAK

Nova Science Publishers, Inc.
New York

LIBRARY OF CONGRESS CATALOGING-IN-PUBLICATION DATA

Kozak, M. (Metin), 1968-
  Tourism economics : concepts and practices / Metin Kozak and Ozan Bahar.
    p. cm.
  ISBN 978-1-60456-430-3 (hardcover)
  1. Tourism--Economic aspects. I. Bahar, Ozan. II. Title.
  G155.A1K683   2008
  338.4'791--dc22   2008008594

Published by Nova Science Publishers, Inc. ✦ New York

CONTENTS

INTRODUCTION

It is known that tourism is a practical field emerging as a result of its close relations with many scientific fields and industries. Foremost, among these fields, is perhaps economics because tourism is both a production and consumption event in economic terms. As a result of the consumption resulting from tourism, it is possible that income is transferred from one country/region to another. The one defined as a tourist or daily visitor spends the income s/he gains in the home country/region for consumption purposes in the place s/he visits. In this respect, it is possible to say that the reason for a tourist to travel is consumption, rather than production. This kind of consumption tendency has a stimulating role in awakening the national and international dynamics. Examining these dynamics in terms of supply is within the interest and field of practice of this subdiscipline defined as "*tourism economics*".

Since tourism is a production and consumption process, it is a natural result in economic terms for all tourist-receiving countries that some economic effects occur. It is possible to classify these effects as societal, environmental and physical as well as economic [1]. Because as stated above, tourism is a complicated and multisectional socio-economic event concerning various scientific areas and disciplines. In other words, tourism, which is a sector of industries, is an international economic movement based on the economic, social, political, legal, technological and environmental relationship of the various subsystems belonging to 41 different industries.

When the related literature is reviewed, it is known that in the global research carried out for development and prosperity until the 1950s, the tourism industry was underestimated [2]. By the end of World War II, it is seen that tourism activities have started to develop, especially in western countries, which are today's developed countries. Because its economic importance is understood, it has become the most rapidly developing industry in the global economy of the 21st century, together with telecommunication and information technologies [3]. According to the 2006 expectations of WWTC, it is foreseen that the tourism economy will contribute to the world GDP by 10.3 percent and to total exporting by 11.8 percent. Moreover, it is estimated that the people working in the tourism industry will reach up to 234,305,000 by 2006, forming 8.7 percent of total employment capacity [4].

Besides this, it is rather difficult to say that these kinds of activities generate completely positive effects on a country's national economy. More clearly, the development of tourism in a country/region carries some social and economic changes with their positive as well as negative aspects. It is possible to assess the effect of tourism on the balance of payments, its income-constituting effect, employment effect, effect on the balanced development among the regions, and effects on the other industries in the economy and the effect on the improvement of the superstructure as well as infrastructure within the scope of macroeconomic effects. Similarly, its effect on real wages and prices, opportunity cost, and internal and external economies are the selected examples of the micro effects of tourism. This book aims to examine both effects of tourism in a national economy in detail.

In order not to fall behind in the development race of the global world, gain a competitive power and develop new technologies, the developed and developing countries need new industrialization strategies and policies, new technologies and know-how. Nevertheless, for the countries with a certain tourism potential, there is no such situation. In other words, natural, cultural and historical assets that are peculiar to a country/region and not human-made but instead, made up within the natural structure of that country naturally, may create a competitive advantage over the other competitive tourism destinations. However, that the tourism regions (destinations) gain a competitive power and sustain a long-term market share permanently depend on compliance to customer expectations and accordingly their satisfaction level and the quality of products. Hence, it is possible to say that many countries with a sufficient capacity of tourism sources are in a competitive race among each other. Each tourism country develops various policies in order to have a greater share of international tourism within the effective market economy and increasing competition. However, while doing this, they should keep healthy data on hand. This work gives some clues on what kind of data are needed and how these should be analyzed in terms of economic effects.

The ones who are strong can compete with their rivals in terms of many factors such as price, cost, quality, customer satisfaction, marketing and distribution. However, due to the property of competition, as well as being important to be successful only in a certain year, month or period, it is also important to sustain this in the long term. When a short assessment is made in terms of the international tourism industry, it can be said that the destinations having the desirable share of tourism revenues and the number of tourists in the world and sustaining it in the further periods by increasing it will attain a higher competitive power in this industry. Therefore, particularly the positive economic effects of tourism on a country or region should be analyzed well.

It is good to see that writing a book has become a tradition regarding scientific or current issues in the fields deeply concerning the society. However, as one of our famous writers points out, it shouldn't be ignored that there are dangers occurring by "having an opinion before having knowledge". Considering the issue in terms of academic studies, it is not possible that the scientists create a work in a field about which they have insufficient ideas or have never done research personally. Indeed, there shouldn't be such implications! For this reason, the contribution of the results gathered by the various empirical field studies that the authors have done recently regarding the issue on this work cannot be denied. As it will be seen in the following chapters, the authors have preferred to refer to their own research directly so that some issues are made clearer and supported from theoretical and empirical aspects. In order for the issues to be clear and on the agenda not only the results of the work done by the authors but also the results of some recent outstanding empirical studies are reviewed. This point puts forward another feature of the work. Recalling that there are limited sources on tourism economics worldwide, the authors aim to make an important contribution with this study.

On the other hand, since when economics is mentioned, the mathematical equations or some definitions come to mind, there may be some problems at times on the effective assimilation and blending or the applicability of the theoretical knowledge to current life. While in order to eliminate such problems, the theoretical explanations of the issues in the study are simplified, attempts are made to visualize some theoretical knowledge by giving additional information in different tables, figures or notes in order to facilitate comprehensiveness. Additionally, each chapter is followed by the description of the related subject after a brief explanatory introduction and ends with a summary at the end of the chapter. The discussion questions prepared in order to test to what extent the subjects are utilised and synthesized are at the end of chapter endings. We hope that our efforts on this subject will succeed in proportion with the pleasure the readers will have and the level of knowledge they will gain on tourism economics.

To sum up, the authors are in charge of all information within this work which dwells on the explanations in a simple language by examples from daily life rather than from an artificial and virtual atmosphere in order for the information to be more permanent and the aim is to enable a pleasurable knowledge sharing. It is for this reason that we believe that the constructive criticism from readers will support not only us but also by means of the preparation of more comprehensive works, all instructors teaching in the related field for the development of more efficient methods on learning and teaching. No matter whether it is the lecturers, the students or the industry representatives, any kind of advice, criticism or contribution will serve as a guide for the future editions of the work.

Finally, it is necessary to emphasize this point as well. Creating a work is not only possible by the individual efforts of the authors. Mutual communication, sharing of knowledge, love and respect have had a role in the creation of this work. Therefore, we are grateful to our students whose direct and indirect contribution is felt during the lectures, our colleagues with whom we discuss matters and the representatives of the industry in determining the accuracy and comprehensibility of the subjects. Furthermore, we would like to thank the "tourist" groups, who make up the basis of most data and have visited Turkey from different countries and shared their experiences for this indirect contribution they have made. Lastly, we are thankful to our wives; Hesna and Tülay and our precious children; Mehmet Kayra, Çağdaş Miray and Deniz Giray who were always supportive and by our side all through the work.

We wish that this work would be beneficial to a blooming acceleration of tourism activities both at the national and international level.

Ozan Bahar
Metin Kozak

REFERENCES

[1]   See detailed information about positive and negative impacts of tourism on the social and natural environment see: N. Kozak et al. (2001), *Genel Turizm İlkeler-Kavramlar*, Ankara: Detay Yayıncılık, 5th edition, pp. 93–104.

[2]   G. I. Crouch and J. R. B. Ritchie (1999), "Tourism, Competitiveness and Societal Prosperity", *Journal of Business Research*, 44, pp. 137–152.

[3]   E. Heath (2003), "Towards a Model to Enhance Destination Competitiveness: a Southern African Perspective", *Journal of Hospitality and Tourism Management*, 10 (2), p. 124.

[4]   WTTC (2006), *The 2006 Travel & Tourism Economic Research,* London.

Chapter 1

INTRODUCTION TO ECONOMICS
AND BASIC CONCEPTS

As an introduction to the subject of tourism economics, first of all, it is better to have a general review on the basic concepts, theories and systems. In this respect, the first chapter aims at providing brief information about the purpose of the topic "economics", its parts and basic concepts. Afterwards, the meaning of the economic system as well as its classification are the additional subjects to be examined within this chapter. Hence, for the readers with an intention of gaining detailed knowledge on tourism economics to understand the further issues well, this chapter as an introduction has a great importance. Understanding these basic concepts here clearly contributes to a more pleasurable and useful course of learning. As a conclusion being a significant sub-sector of economics, tourism has a structural property that cannot be thought apart from it.

I. INTRODUCTION TO ECONOMICS

Today, tourism which has had a mark on the 21st century after the concepts of telecommunication and information technology, is one of the leading service industries of the world and has been developing each day, and appears as a sub-sector of economics. As known, the three basic sectors in economics are agriculture, manufacturing, and services (see Figure 1.1). Services are divided into two categories: information and traditional services. The first group includes banking, insurance, information technologies, consultancy, building and technical services, advertisements and distribution, health, education, public services etc. Undoubtedly, the ones working in this services group have a big amount of human capital. In the other group in which tourism exists as well, there are more conventional methods such as transportation and social services [1].

Tourism is a leading activity among the conventional services as the subject matter of international trade. Therefore, it is possible to accept tourism as a sub-sector of economics. In order to understand the economic structure of tourism and maintain a better cause and effect relationship among the events, a basic knowleddege of economics is required. Indeed, the purpose of the topic of economics dealing with the usage of scarce resources is explaining the cause and effect relationship regarding the events and activities under the scope of it and putting forward scientific rules.

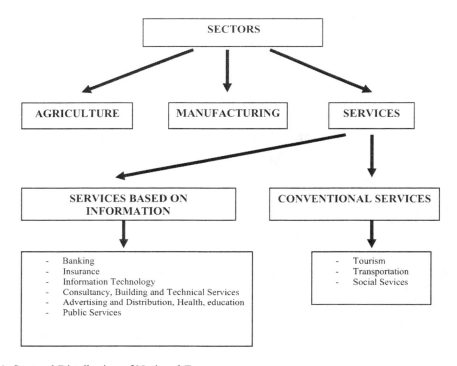

Figure 1.1. Sectoral Distribution of National Economy.

However, it should never be forgotten that these established rules are the general rules of tendency because human behaviors form the essence of economics is a social science. Due to the fact that human beings react distinctively and show different attitudes to events, the above-mentioned rules stress the general tendency of a community or a group.

The laws and principles of economics are not as certain as the laws and theories of applied sciences because economics make examinations on human and social behaviors as it is in the context of other social sciences. These behaviors are subject to change depending on the society, individual and time [2]. Indeed, the behavioral patterns of people / societies, conditions for production and consumption, manufacturing methods, their preferences and necessities differ according to the changing time and technological developments. It is for this reason that, as explained earlier, the economical laws and theories have a quality of expressing the general tendency, apart from exactly reflecting the truths.

As the fundemental problems of economics are scarcity and preference, it will be a more proper approach to assume the history of economic events equal to the history of hmankind. From this point of view, the reason for the branch of economics to emerge is that the resources to meet the unlimited necessities are scarce. According to the generally accepted definition, the subject of economics is to manage the scarce resources that will fulfill the human needs. Indeed, increasing the human prosperity by reducing the imbalance between necessities and resources lies on the basis of all economic activities [3]. In other words, that the necessities are abundant, but in contrast, the resources are scarce lead to an imbalance in economics, and this is defined as *"scarcity"* [4]. However, the scarcity concept used here should not be confused with the scarcity concept used in daily life. That is, the latter refers to the fact that a product is less obtainable in the market or does not exist at all. On the other hand, in economics literature, every product that is incapable of meeting all the demand of people is considered scarce whether it is acquired easily or not. This is enough for now as this subject will be reviewed again while the definition of products is given in the following chapters.

In light of the above, human beings, no matter which economic system is chosen, ought to find a solution for the three main problems of coping with scarcity in economic terms. As emphasized previously, the purpose of economics is also towards this perspective. These problems [5] are indicated as: a) *absolute usage of resources*; b) *using resources efficiently*; and c) *economic growth and development*. What goods and services will be produced and in what quantities? How will the various goods and services be produced? For whom will the various goods and services be produced? Economists call these three questions brief economic questions (the problem of resource allocation). Every society faces these economic problems and seeks solutions for the problems by the economic policy and system it has chosen.

A. Absolute Usage of Resources

The first and perhaps the most important problem that people should solve in economic terms is the absolute usage of resources. What does it mean? By this, it is meant that all factors of production (labor, land, capital and entrepreneurship) that a country, a region or a firm have are used in the production activities. Using all four above-mentioned factors of production in the production process is called *"perfect employment"* or *"absolute usage"* [6]. In the struggle with scarcity it is aimed to produce the maximum amount of output with the available resources. However, in the real world, it is seen that not all the available resources are employed in the production process. Some of them have to be wasted, and that is called *"imperfect employment"* or *"imperfect usage"*. It is better to explain this situation by means of the production possibilities frontier. The production possibilities frontier is defined as the geometrical place of the maximum amount of the combination of goods and services when all resources are used in production with the available technology on hand [7]. The production possibilities frontier, also called the transformation curve, is shown in Figure 1.2.

According to this approach, on the horizontal axis there is textile production and on the vertical axis there is automobile production. In Figure 1.2, point F where BF production possibilities frontier intersects the x-axis shows the maximum amount of textile when all production resources are used in textile production; that is 20 units. Similarly, point B where BF production possibilities frontier intersects the y-axis shows the maximum amount of automobiles that can be produced when all resources are used in automobile production (200 units). On the curve, there are *n* points such as B, C, D, E, F. Each of these points shows the combinations of goods by using the available resources and technology in the economy.

A point which is not on the production possibililities curve, say point L, shows the production point which is not possible to produce with the available technology and resources. On the other hand, each point below the production possibilities frontier shows the production level occuring as a result of not using all of the production resources.

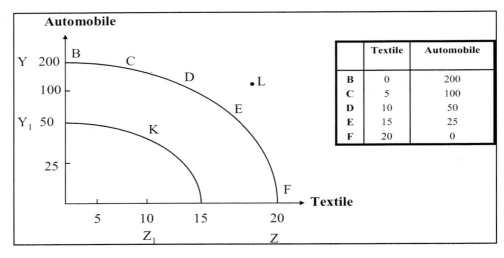

Figure 1.2. Production Possibilities Frontier.

In other words, some resources remain unused and the condition of imperfect employment occurs. When point K is examined, it is seen that some factors of production are wasted (the work force with a desire to work but cannot find a job or not making use of nature as desired or sufficiently enough) and economy produces lesser amounts. In case of such an imperfect employment when resources are not used efficiently, textile is produced in OZ_1 amounts and automobile in OY_1 amounts in the economy. However, if the wasted resources can be used in the production process somehow, it is possible for the textile production to increase by Z_1Z and automobile production by Y_1Y. Thereby, when some resources are not used in production, the loss in economical terms is Z_1Z. in textile and Y_1Y in automobile.

B. Efficient Utilization of Resources

The second important question considered in economics is whether resources are used efficiently or not. The point to be solved here is supposed to be how scarce resources are used carefully, wisely and without causing waste. In other words, the problem of efficient usage disappears when the resources are used in a way that they sastisfy the necessities. In this respect, in the war against scarcity, there are alternative domains of usage for the scarce resources requires that proper decisions to be taken during their production process (see Figure 1.3).

At this point, the concept of opportunity cost or the cost forgone arises, and this is explained in the following chapters more elaborately. In this situation, the actors responsible for economics will need to choose among the alternative usages of the scarce resources and answers to the following questions will be searched as well [8]. What goods will be produced in what quantities? (efficient allocation of resources); With what methods will the production be made and how? (production efficiency); For whom will the goods and services be produced? (efficiency in sharing).

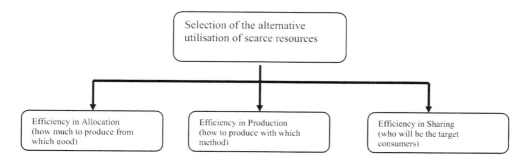

Figure 1.3. Choice of Alternative Usage of Scarce Resources.

In order for the efficiency in allocation to be attained, the goods and services that are determined to be produced should be in a way that they satisfy society's needs perfectly [9]. In other words, the indicator that efficiency in production is maintained is the condition of the goods produced with scarce resources are not distinct from the goods people want to buy. The question of "what to produce" is already answered by this. For instance, while a society needs 100 units of bread and 250 units of clothing, if the resources of the country are allocated for the production of 50 units of bread, 100 units of clothing and 300 units of glasses, it is seen that the scarce resources in the economy are not used efficiently. The increase in the capacity of a tourism area up to 25,000 while the need is only 10,000 in total can be an example given for the tourism industry. Unless the economic choice emphasized below is made in accordance with society's necessities, it will be difficult to meet some of these necessities, or the manufactured goods will remain unused and resources will have been used in vain. As a result, the questions of what goods and services the society needs and how much to produce are the most significant ones to be answered in economic terms. Choosing between the alternative resources is very important in terms of their efficient usage.

To maintain efficiency in production, it is impossible to produce more of a good without producing less of another good when the allocation of resources among goods is changed [10]. More clearly, efficiency in production means maintaining the maximum production level with the resources existing in a society. In the event that the production efficiency is maintained, there is no possibility to increase the production by changing the alternative usage domains of resources. The production possibilities frontier is based on the assumption that the most suitable production technology has been chosen. Here, we face the question of how the production will be made. Will the production be made as capital intensive labor? Or, will both of the methods be used together? To illustrate, considering the automobile production, it is a type of capital intensive production. On the other hand, textile production or production of a good or service in the tourism industry is a type of labor intensive production. In the capital intensive production method more capital and less labor are used whereas, in the labor intensive production method, more labor and and less capital are used. Unless a production technology that is suitable to the structure of society is selected, as stated earlier, some scarce resources remain wasted and the imperfect employment condition occurs.

Box 1.1. Maintaining Efficiency in Production

Since the economies in western developed countries are strong, these countries completed their industrialization process long ago. The capital accumulation in these countries is at high levels, and should a good that is to be capital intensively produced is produced with the labor force again, it is possible to indicate that capital remains wasted and, hence, causing a loss in the wealth of the nation. Similarly, in a less-developed country, which is based on an agricultural population, changing the production technique and using highly technological machines will cause several people to be left outside the agricultural sector, and thus, not being able to find jobs, this workforce will be unemployed. This situation is, again, an imperfect employment condition. As some factors of production are wasted, the level of goods and services to be produced in a perfect employment condition cannot be reached, causing a loss in the prosperity of the society in general.

Efficiency in sharing means the usage of the goods and services are produced by the people who take the best advantage of them. In an economy where efficiency in sharing is maintained, it is not possible that sharing in the society allows anyone to become richer without allowing even one person to become poorer with a new arrangement [11]. The answer to the question of efficiency in sharing can be answered by for whom the production is going to be made. This is another crucial question to be answered because as a result of the sharing decision, some people in the society will become wealthy while some others will become poor. Who is going to be richer or poorer will appear clearly as a result of sharing the produced goods and services. In the event that production and sharing can be done at the same time, the highest level of prosperity that can be reached in the economy is attained. The target that today's societies want to reach is that this common efficiency is achieved together.

C. Economic Growth and Development

Growth and development in economical terms are the concepts with different contents. *Growth* expresses the changes occuring on account of the quantity not the quality. *Economic growth* is a concept used for expressing the increase in a country's financial wealth [12]. In this regard, growth of an economy does not cover its structural change. Only the increase in production and income per person is considered sufficient for growth. However, when *development* is pronounced, particularly structural changes of a society in terms of economic and socio-cultural as well as changes in production and income per person is meant [13]. Examples to the structural change include a decrease in the share of the agricultural sector, the development of the services sector as well as industries, and an increase in the quantity and quality of factors of production.

The developed countries problem of increasing production capacity is called *economic growth* and the same problem in less-developed countries is called *economic development* [14]. However, although the growth process comprises the developed and less-developed countries, the development process is a concept used only for the less-developed ones. Indeed, the less-developed country can maintain growth in an economical aspect but it may not be able to develop. Nevertheless, what is important for such economies is that the development process is realized successfully [15].

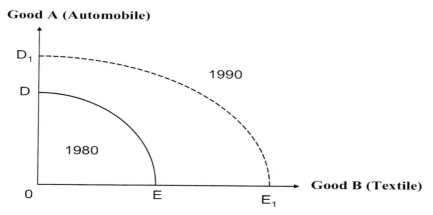

Figure 1.4. Economic growth.

As stated previously, DE is the production possibilities frontier. In this economy, change in the production technology over time (as it is in passing froma labor intensive economy to a capital intensive one) and the increase in the quantity and quality of the factors of production cause the production possibilities curve to change. Figure 1.4 shows the growth process of an economy over time. In 1980, the production possibilities frontier was DE. It becomes D_1E_1 in 1990. In so doing, the curve has shifted to the right position and a growth in the economy has appeared. This case indicates that when comparing to 1980, more goods and services are produced in that economy.

It cannot be asserted that every country in the world experiences such a growth and development process. While some experience a rapid growth and development process, some others may have no development at all. The World Bank and other outstanding international economics institutions treat the nations as less-developed, developing and developed in regards to the consideration of GDP per person [16]. The reason for this is, as explained above, the developmental variation among the nations. As a result, the main purpose of all countries is to reach this development process and increase the economic prosperity of itscitizens.

II. DIVISIONS OF ECONOMICS

The subject of economics is divided into two parts according to the field of examination of the economic activities and actors. The part that analyzes the behavior of small units such as individual, family, consumer, firm or organization is called *micro economics*. On the other hand, the part analyzing the behavior of a society such as national income, gross production – consumption, gross investment, growth and development is called *macro economics* [17]. As the names imply micro means "small" and macro means "big".

Economics = Micro Economics + Macro Economics

Micro economics deals with issues such as setting up individual economic decisions, resource allocation, prices, production and identification of income distribution, and its tools of analysis resemble a microscope. *Macro economics* deals with macro subjects such as national income, employment, money and banking, growth and investment, etc. and the tools of analysis are similar to a telescope. For instance, while examination on a single good is made in micro economics, the general level of price is pointed out in macro economics. The difference in between is that micro is interested in the demand and supply of small units and certain goods and services, whereas macro is interested in the demand and supply of total goods and services intended for the general society.

Another distinction in economics is the distinction of positive and normative economics. Being the theory laden part of the economics, the *positive economics* approaches the events objectively [18]. The preferences, beliefs and values are out of the question. Putting forward a phenomenon scientifically and analyzing it to prove whether the assumptions are true are the subjects of interest for positive economics. Such assessments as how it should have been do not exist in positive economics. The economic events are examined within the cause and effect relationship, being tested by the actual data from reality. *Normative economics* surveys what should be done to foster social prosperity. In other words, it covers the values and subjective targets are under discussion [19]. To illustrate, to search for an answer to the question of what inflation is and how it emerges, and to explain it by basing it on a theory is the subject of positive economics; to assess whether the income distribution in a country is fair or to search for a solution for the reduction of poverty and deriving some conclusions is the subject of normative economics.

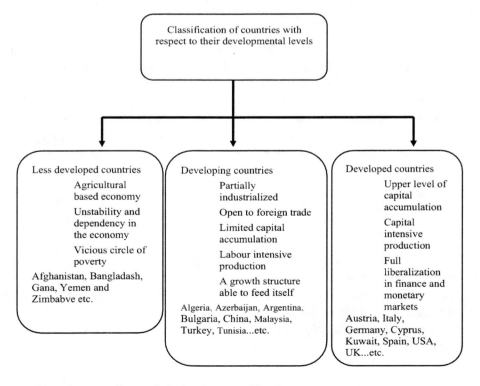

Figure 1.5. Countries according to their developmental levels.

III. Basic Economic Concepts

Explaining some economic concepts that are being used by the subject of economics will be benefical for a clear understanding of the following issues. For this reason, in this part, some basic concepts will be dwelled on briefly. Acknowledgement of what these economic concepts represent will be helpful in understanding the other issues in the following parts more clearly.

A. Factors of Production

Labor, land, capital and entrepreneurship are the factors of production in economics. Below, there are brief explanations about each of these factors of production [20]. It is not possible for production in an economy to occur without the existence of the factors of production. However, as it will be explained in the following chapter, the conditions of perfect or imperfect employment emerges in accordance to the partial or complete use of the factors of production.

Labor

Any kind of human effort towards the production of a useful good or service is under the scope of the labor factor. In other words, labor is the human physical and mental ability used in the production of a good or service. As labor is a type of factor of production supplied by the human capital, it differentiates from the others in this respect. Additionally, labor is assessed as the time a person spends in working for a certain time and it is charged accordingly. The labor factor can be educated or uneducated, qualified or unqualified, decision maker or implementator. The cook responsible for cooking in the kitchen, a tourist guide in charge of guiding around or a waiter responsible for serving can all be assessed within this scope.

Land

It is the physical surface on which production is made. The natural resources under and on the land, such as petrol, coal, and trees are also viewed within the scope of this factor. The fertile land needed for a competitive-based and healthy agricultural production, the mining quarries requested for energy production or nature (forests, beaches and so forth) needed for a tourism focused production can all be presented as examples on this issue.

Capital

These are the long-term devices used by human beings for the production of goods and services. Or any type of good that used to be produced in the past but not any more represents capital. The physical tools such as a building, machine, and equipment are called physical capital and know-how and skills of the working person is called human capital. In the case of agriculture, the tractor a farmer uses or various machinery used in a factory for production are the examples of physical capital.

Entrepreneurship

Entrepreneurship is the factor that generates the production of the desired goods and service by organizing other factors of production with regards to the expectations of the society. Labor, land and capital, together some particular skills that people have for the development of new job opportunities as well as methods for the current job are defined as entrepreneurship. Entrepreneur can be an individual; it can also be a firm, company or any kind of state instution. A tour operator for the production of package tours, and a hotel established for the accommodation needs of the people are the economical units that exist as a result of entrepreneurship.

B. Goods

Everything attributable to fulfilling the needs of people is called "goods". Generally speaking, while making a distinction between goods and services, the term good is used for both. In fact, the physical assets with the quality of satisfying the needs are called goods. Services, while being under the scope of this definition of goods, are the ones satisfying the needs but do not have a tangible (physical) quality [21]. For instance, tourism, banking, insurance and so forth are included in the services class. However, the essential point here is the service the goods make for human beings.

Although the goods in the economy are mainly divided into two goods, *scarce* and *free*, as seen in Figure 1.6, we have three separate classifications here [22]. The goods that can be attained without effort, payment and whenever requested are free goods (such as air and water, besides a fountain). Free goods can be found in nature whenever needed and can be attained without effort. The goods that are attained with effort and that should be paid for in order to be aqcuired are scarce goods, such as bread, tea, petrol, sugar, etc. Goods are divided into two as to their level of endurance; *durable* (TV, refrigarator) or *feeble* (food, beverage) goods. Durable goods are the ones which have a benefit even a long time after the acquisition. Feeble goods are the ones which end immediately after they provide benefits.

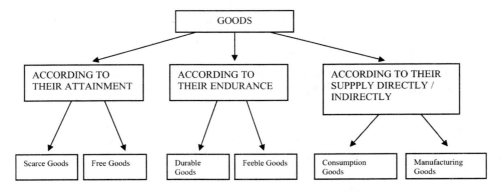

Figure 1.6. Classification of goods.

Finally, goods are divided as *consumption* or *manufacturing* (investment) goods according to whether they fulfill human needs directly or indirectly. While goods such as food, beverage, and lodging are intended for consumption, capital goods, semi-product goods, machinery, raw materials, petrol, mineral, electrictiy are intended for manufacturing.

C. Utility

There are various types of goods and services. They satisfy people when used in different places and at different times, and when consumed. Among the main concepts to be learned in economics is *utility*. In brief, it is possible to define utility as of the ability of the goods in fulfilling human needs [23]. Goods and services provide benefits by satisfying the human needs. For this reason, utility is a subjective and relative concept. In other words, the utility of a good may change from person to person. Moreover, even if laws and regulations forbid the usage of some types of goods, that good is considered benefical in economic terms when there is someone in need of it [24]. For instance, drug usage is both forbidden and disapproved by society. However, for an addict using it, this good may provide an economic utility for him. From this point of view, there is a discrepancy between the utility concept used in daily language and the one used in economics science. In order for a good to be beneficial in economic terms, it doesn't need to be compatible with laws or traditions/customs. What is important here is whether the item providing utility fulfills a human need and satisfies the individual.

D. Need

Need is the feeling that gives pleasure, joy and contentment when fulfilled, but causes grief, sorrow and pain when unfullfilled [25]. Human needs consist of eating, drinking, shelter, health, entertainment and education. Considered in economica aspects, economists divide the needs into two parts: mandatory needs and optional needs. For instance, needs such as eating, drinking, accommodation and health are mandatory and should be fulfilled for people to survive. On the other hand, entertainment and needs with cultural aspects are not mandatory and may change from person to person or society to society. While smoking is mandatory for the person who has the habit and is an addict, it may not be mandatory for a non-smoker. Similarly, staying at facilities such as Magic Life or Robinson Club Select for 10 days, having a holiday in Hawaii or the Caribbean Islands every year, or having a luxurious car may be a mandatory need for some people, whereas for others, these may not be mandatory at all. However, in time, with the development of technology and economic opportunities, the definition of some needs have been changed. About 30 or 40 years ago it was not mandatory to have a television or computer while it seems to be mandatory nowadays.

As indicated in Figure 1.7, needs have five properties [26]. As stated earlier, the first one is that human needs are of various types and unlimited. The second one is that the desire and the intensity to fullfill these are different. While the need for food turns out to be in the first rank for most people, the need for entertainment may be second or sixth or even the last rank and may not even have any importance at all. The third one is that as the needs are fullfilled the desire and want towards them decrease and another one appears.

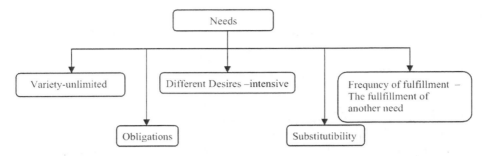

Figure 1.7. Properties of needs.

In other words, the more they are satisfied the less they are requested. As a person who is very hungry or thirsty fulfills his needs towards food and water, their consumption do not satisfy him/her any more after a certain point and begins to cause pain or suffering. The fourth feature is that some needs may turn out to be mandatory while they were not in the beginning, such as smoking. The last property of needs is that they can be substitutable. For instance, a person with a need for tea may substitute it with coffee when s/he cannot find tea. In this case, coffee becomes the substitute of tea.

E. Value and Price

There has been an ongoing debate on the issue of what the real value of a good is and how it should be determined since Adam Smith (1776) because several different definitions are being made about the value of a good and which one defines the "value" in a real sense is continuously being discussed by the economists. For example, according to an approach, *value* is the relative importance attributed to goods and services and given to them. The view that is put forward by some economists that the value of any good may be determined by its cost is not a generally accepted approach. According to another definition, value is the rate that a particular good or service may be charged for other goods or services; that is also called the value of barter (exchange) [27]. According to another approach, if a good is demanded and is found less available in the market it is more valuable than other goods, such as the caviar served in a restaurant. Or when the issue of value is analyzed in a different perspective, as the number and quantity of the factors of production that are used in the production of the goods increase, it is asserted that the value of that good will increase as well.

Here, the important point to be taken into consideration is that a good or service with a value has the property of satisfying the human need. Besides this, it may not be possible to state that any type of good satisfying the human need has a value. That is, while air is a very valuable need for people, it doesn't have a value in economic terms. Price is the cost paid in order to attain a certain good or service. Or the value of a good or service measured with a standard monetary unit is its price.

F. Opportunity Cost

As explained, the existence of an economic problem in a country requires choosing among the alternative uses of the scarce resources. However, choosing among various alternatives (resources) means sacrificing the unselected alternatives or leaving them unused. Demanding to have something more requires less of another thing, which is called *opportunity cost* or *alternative cost* in economics literature [28]. In brief, the opportunity cost of a good is giving up the production of another good for releasing the required resources in order to increase the production of that good by one unit [29].

As an example, let's assume a student spending his/her time by going to a match instead of studying. For this student spending time in a match rather than studying, the foregone activity of studying becomes the alternative cost of going to a match. Or a government's assignment of its investment resource to establish a factory rather than a highway construction means sacrificing the highway that could be constructed with the mentioned resources. Hence, the sacrificed highway here is the alternative cost of the establishment of the factory. Thereby, the citizens of the country face an alternative cost in social terms. If the economic utility of establishing a factory is greater than that of the highway construction, there isn't any loss in the welfare for the society. However, when the country needs the construction of a highway, spending the resources in a factory establishment causes both inefficient usage of resources and a big loss of welfare for the citizens of the given country. For this reason, as indicated prior, it is important that the resources be allocated in the best way that increases the welfare of the people of that country.

G. Employment

Another phenomeneon that is considered important for economics that we try to measure is the general level of employment and developments thereof. The reason why the issue of employment is important for economics and hence economists is clear. An increase in the welfare of human beings and society economically is an incident that can be possible by the effective and intensive usage of factors of production. In this respect, how the efficient usage of employment and as a result the factors of production should be achieved is among the questions that economics has to answer and investigate [30].

In the dictionary, employment is explained as "have someone work or use somebody for a service". From the aspect of economics, it is using all factors of production in a broad sense and only the labor factor in a narrower sense in production of goods and services in order to obtain profit [31]. Employing all the factors of production and not wasting a single resource in an economy is the most ideal, desirable and wanted situation. As emphasized before, economists express the condition of assigning four of the factors of production into the production process as "perfect employment" and some of them remaining outside the production as "imperfect employment" (see Figure 1.8). The imperfect employment of the labor factor is called *"unemployment"* in economics [32]. The most common indicator used in order to show the general level of employment is the rate of unemployment. *The unemployment rate can be defined as the rate of the ones who cannot find a job although they are ready to work at a rate of current salary* [33]. In economics literature various kinds of employment types are indicated and defined. Here, for the purpose of limiting the subject to only the subjects of unemployment and rates of unemployment are presented [34].

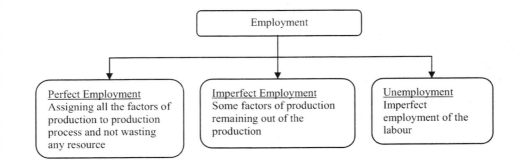

Figure 1.8. Types of employment.

All of the stock of labor, capital, land and entrepreneurship in a country's economy may not be used in the production process actively every time. Indeed, in the case of the factors of production is not being used completely, it can be said that there is "unemployed (unused/wasted) capacity" or "unemployed resources". When considered generally, the concept of employment, although covering four factors of production, means the participation of more labor into the production. The reason for this is the assumption that when labor is employed the others will be employed automatically. Hence according to the settled tendency, when employment in economics is pronounced the employment of labor is meant by that.

H. Balance of Payments

The balance of payments is important in terms of its showing the foreign economic and financial relations of a country in a certain period of time. The balance of payments is a chart that makes the accounting of the economic relations that the people or institutions in a country have with the outside world [35]. In other words, the balance of payments is the account that shows the result of the economic affairs of a citizen of a country with the other part of the world. When examined economically, in order for the balance of payments to occur the income a country aqcuires from abroad should be equal to the payments it made to them. The case when there is no equality or there is a surplus or deficit in the balance of payments is assessed as the indicator of the mentioned country's economic and financial reputation in the international arena as it will reflect the improvement or the destruction of its international power of payment [36].

Having a surplus or a deficit in the balance of payments is very important as it creates comprehensive effects on the country's economy. Because several economic variables such as national income, level of employment, economic growth and development rate, exchange rates, inflation rates, salary rises, income distribution, interest rates and foreign debts in a country are affected by the balance of payments positively or negatively. Hence, whether the country's foreign economic affairs are in order, necessary precautions are taken should there be a problem, and reconsidering the economical policies are regarded as obligatory.

As seen in Table 1.1, there are two main items in the balance of payments: *Current operations account* and *capital investments account*. Since the formal reserves account and net mistakes and missings generate equalizing operation, the important items here are the "current operations account" as well as the "capital investments account" due to their ability to generate a surplus or deficit in economic terms [37]. Being the most significant part of the balance of payments, the current operations account shows the balance of current operations occuring as a result of the goods and services the country imports and exports. Current operations account constitutes the parts as trade of the goods, international services and one-way transfers. Trade of the goods also called as visible trade gives the difference between the total imports and exports that a country does with the outside world, and this is called balance of foreign trade. There is a point that should not be confused: The balance of payments is used in order to express all the international income and expenditures of a country, whereas the balance of foreign trade is used only to indicate the import and export of the goods.

In international services part, there is the payment flow from tourism as well as importing and exporting of services by the country. This item is also called *invisible trade*. The difference in income and expenditures emerging from the total international services importing and exporting is called the balance of services.

Table 1.1. Balance of Payments

A) Current Operations
  1. Trade of Goods
    • Imports
    • Exports
  2. International Services
    • Foreign Tourism
    • Foreign Capital Income and Expenditures
    • International Banking and Insurance Services
    • International Transportation Sevices
    • Workers' Income
    • Private Services
  3. One Way Transfer Accounts
    • Balance of Current Operations
B) Capital Investments
    • Long Term Capital Importing
    • Long Term Capital Exporting
    • Short Term Capital Importing
    • Short Term Capital Exporting
C) Official Reserves Accounts
    • Exchange Operations
    • Gold Operations
    • IMF Reserve Positions
D) Net Mistake and Missing

Box 1.2. Foreign Trade Deficit and Tourism Income

> While the proportion of tourism in closing the foreign trade deficit in Turkey was 6.5% in 1980, it rises dramatically in 2001 with a highest level of 170.2%, and although falling down in 2003, it increases approximately 11 times as much as the amount in 1980 and reaches 79.2% [38]. Here, it can be commented that merely the tourism income is sufficient enough to close the foreign trade deficit in Turkey. Tourism is the second industry following exporting, by which the biggest foreign exchange income is acquired.

Tourism, transportation, banking and insurance, the gains from foreign capital investment, income of the workers abroad, cost of the licenses, rents, commissions and official services abroad are under the scope of this item. As seen, tourism abroad is assessed within the international services item. When considered from the aspect of tourism economics, the most important item in the balance of payments is the international services because this item shows the expenditures that foreign tourists make in the country and the expenditure which the citizens make abroad, and as a result the net contribution that tourism makes to the economy. While the expenditures that the tourists make in another country have an effect of importing, the expenditures that the citizens of the country make abroad have the effect of importing. Thus, the importance of tourism revenues for a country should be measured by its place in the balance of payments.

If the foreign exchange income aqcuired from the tourism activities into a country is higher than the expenditures that the citizens of that country make abroad, namely the expenses, the effect of tourism on the balance of current operations will be positive. Indeed, as international tourism contributes greatly to the nation's economy due to this feature, it resembles an industry without a chimney. That is, under normal conditions, by the sales of some goods and services manufactured in domestic markets but in such a quality that they cannot be exported in external markets (souvenirs, accessories, food and beverages, etc.), an income of foreign exchange is acquired. Moreover, goods produced in a tourism area may be sold to tourists at a price higher than the exporting price by bearing lower transportation and insurance costs. Similarly, an increase in the foreign exchange income of the country is achieved by enabling the country to have an ecnomical acquisition by means of selling the natural, cultural, historical and social assets of a country (natural sight, waterfall, sea, sun, sand or museum visits, etc.) to the tourists, which have solely no meaning and no economic value at all. As a result, tourism income has a big share in the international services income taking place in the country's balance of payments.

In order to limit the issue, here we have reviewed only the balance of current operations covering tourism; nevertheless, it will be useful to present a brief explanation on the capital activities account as well. The capital activities account is composed of the physical investments (production facilities, building, land, etc.) that the individuals and the institutions inhabiting in a country make in a foreign country and the financial funds (purchasing and selling of foreign bonds, stocks, promissory notes, etc.) transferred overseas. In summary, the capital activities to and from the country are examined in this account. As it is in exporting of goods and services, the capital inflow to the country bring foreign exchange while capital outflow causes foreign exchange to leave the country as it is in the goods importing [39].

I. Full Competition Market

The full competition market is a virtual type of a market which cannot be encountered in the real world. This is a theoretically assumed market to clarify the issue of markets. In this respect, full competition market indicates the atmosphere where buyers and sellers barter without an obstacle under certain conditions. As seen in Figure 1.9, there should exist four conditions of the full competition market [40]. This is the ideal market type where competition occurs perfectly and the economical resources are used completely [41]. The market type where one of the four conditions forming a full competitive market is missing is called *imperfect market* [42]. Begining from this point it is essential to briefly define, the two most apparent market types of the imperfect market that is commonly used in the tourism industry and shows the market type the firm in the industry is involved in. These are monopolistic competitive markets and oligopolistic markets [43].

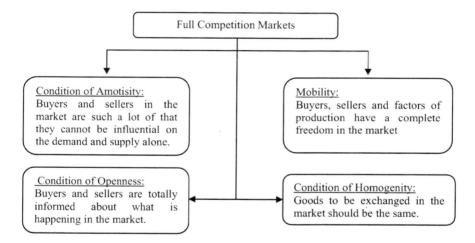

Figure 1.9. Conditions making up the full competition market.

Monopolistic Market: The market composed by the group of firms producing distinguished products (different packaging, appearence, brand name, color, etc.) and having a harsh competition among each other is called monopolistic market (chewing gum, detergent, soap, clothing, beverage, etc.). When a general assessment from the aspect of the tourism industry is made, there are various kinds of markets in the industry from food to accommodation, transportation to entertainment. However, generally speaking, while it is seen that the firms and organizations operating in the industry keep on their activities under imperfect market conditions, different market properties in which conditions of oligopolistic and monopolistic competition are dominant comes across. For instance, while the conditions of oligopolistic market are valid for a four or five star hotel, the conditions of monopolistic competition market are valid for a group of hotels with less stars.

Oligopolistic Market: There are a limited number of firms in an oligopolistic market. If any firm's decision on sales policy is to affect other firms, this means there are a limited number of firms in this market. Hence, oligopol is the market where a few sellers that can influence each other meet with an unlimited number of buyers (automobile, banking, refrigarator, TV, steel, etc).

J. Gross Domestic Product (GDP)

In an economy the sum of the values of the gross amount of the finished goods and services produced in a year with regard to the market price is called GDP.

National Income = Total Production = Total Expenditure

The expression of these values is monetary. In this respect, the national income becomes the sum of all the manufactured goods and services in a certain time period (generally in a year). This expression makes the definition of GDP in terms of production. As this production is divided among the ones participating in the production process, it is equal to the sum of total income and total income is equal to expenditures. Hence, GDP is also equal to total expenditure [44].

IV. ECONOMICAL SYSTEMS

The three main problems have been previously explained and people should find solutions for the common problems to be solved for all societies. Nevertheless, the solutions and methods for these three main problems may differ from society to society. Every society tries to solve these problems by adapting an economic system that it finds suitable for itself. Hence, when *economic system* is pronounced, the societal framework and type of organization formed in line with the solution of these three main economic problems [45]. In other words, the economic system can be perceived as the principles as a whole that lead to the economic activities. From this point, it is possible to say that the best economic system is the one providing people with the goods and services they desire as much as they can or as much as possible [46]. The purpose of all economic systems is to increase the welfare of the people economically. However, the economical tools used for this purpose differ. Economic systems are divided into three as capitalist, socialist and complex systems (Figure 1.10). These are examined below, respectively.

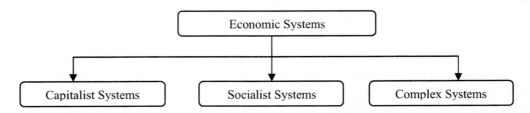

Figure 1.10. Economic systems.

A. Capitalist System

The idea that is called liberal thought, which means people behaving freely and having complete freedom in economic issues, lies in the core of capitalist system. When each individual thinks about his own interest, it is stated that this gives results suitable for the general interest of the society. In capitalism, there is a private possession and freedom of contracts. The solution of three main economic problems in this system reach a solution within the price and market mechanism as a result of the manufacturers' and consumers' activities [47]. In the capitalist system, the generally accepted view is that there is as if "*an invisible hand*" prepares the results suitable for the whole society and, hence, no intervention should be made to the system from outside, and the market price mechanism will provide the resource allocation in the best way automatically. Besides this, the essential part of the system is competition.

The problem of which goods to produce is solved by the firms producing the most profitable goods. The problem of how they will be produced is solved by using the production techniques that will enable production with the lowest cost. For whom these goods will be produced is solved by the indivudals decision on how to spend his salariy and other incomes occuring as a result of the usage of the factors of production. The producers will produce the goods and services from which they will get the highest profit in line with the consumers' tastes with the lowest cost and the owners of the factors will have to share to the extent that they contribute to production and hence, the income distribution (problem of sharing) will be provided in the best way naturally.

B. Socialist System

The socialist system seems to be totally opposite from the capitalist system. In this system, other decisions concerning production, distribution and economy are made by the government itself. Therefore, the state does not allow private ownership, the production process is collective and is done in factories as a group. In socialism, there is collective ownership and central planning. The system tries to solve the three main problems by the government identifying what people will produce, how they will produce them, who will have the share and in what amounts and how they should be living [48].

This central plan that the government has in a socialist system serves as the market and price mechanism in capitalist system. However, doubtless to say that the success of this system is only possible by the neccessities, preferences and production method being well planned well and as a result implemented. It is not achieved, production and consumption do not overlap and production in line with people's desires and demands is not achieved. The socialist system implemented in the past by the Soviet Union, China and particularly Eastern Bloc Countries has been abondened recently, and private ownership on production tools is now allowed and market economy is prefered.

C. Complex System

It may be difficult to see the above mentioned two systems in practice in the proper sense. It is for this reason that by claiming that the society's welfare is achieved by the complex system that is the mixture of the two systems, this type of system is brought forward. It is possible to see the properties of both capitalist and socialist systems in this system. In the complex economic system, the thought of people living freely and providing them with equal opportunities, having an eye on society's interest in the event of a dispute of the individual and society, restricting the ownership of production tools and social security issues stand in the forefront [49].

The two most important properties of the system are restricted and widespread ownership, as well as democratic plan and order of social state. In the complex economic system, as it is in the capitalist system, the market and price mechanism is considered. However, upon a negativity countering society's interest, the governmnet may allow leading and encouraging plans as well. Hence, for the purpose of performing economic developmet and growth, the developmental situation of the private sector is considered, and according to this the public sector may play an identifying role in the economy. In this system the government makes the production of some goods and services itself, and it can also intervene in the market from time to time in order to make the resource distribution efficiently, prevent unemployment and inflation or protect the consumer rights. The system implemented in Turkey particularly between the years of 1930 and 1980 can be given as an example of complex system.

V. SUMMARY

In this chapter, there is a short analysis of some basic concepts as an introduction to the course of tourism economics. Knowing these concepts as the basis of the issue may help to understand the following chapters more clearly. Consequently, what sectors the economy is composed of, the problem of scarcity, concepts of growth and development, the parts of economics discipline, basic concepts and systems are among subject matters underlined in this chapter. The second chapter presents the basic concepts regarding tourism and examines in detail the definition of tourism and its concepts such as tourism products.

DISCUSSION QUESTIONS

1. What is scarcity in economic terms? Explain briefly.
2. Explain the three main problems that the topic of economics should solve.
3. What are the factors of production? Explain briefly.
4. Give examples from the tourism industry for each factor of production.
5. Is tourism an obligatory necessity? If not, in what kind of situations can it become an obligatory necessity? Discuss.
6. Explain the concept of opportunity cost in terms of tourism economics and give proper examples regarding the issue.
7. What does the balance of payments represent in economic terms? Give brief information on the place and importance of tourism in the balance of payments.
8. What is an economic system? How many types of economical systems are there? Explain briefly.
9. Comment briefly on the relationship of each economic system with the tourism industry.

REFERENCES

[1]   H. Seyidoğlu (2001), *Uluslararası İktisat*, İstanbul: Güzem Yayınları, 14th edition, p. 732.

[2]   M. Eğilmez, E. Kumcu (2002), *Ekonomi Politikası Teori ve Türkiye Uygulaması*, İstanbul: Om Yayınevi, p. 14.

[3]   O. Bahar, M. Kozak (2005), "Türkiye'nin Turizm Sektöründeki Uluslararası Rekabet Gücü: Uluslar Bazında Bir Kıyaslama", *Uluslararası Rekabet Sürecinde Türkiye*, M. Faysal Gökalp and C. Yenal Kesbiç (Eds.), İstanbul: Beyaz Yayınları, p. 427.

[4]   E. M. Ünsal (2003), *Mikro İktisada Giriş*, Ankara: Turhan Kitabevi, p. 10.

[5]   Z. Dinler (2004), *İktisada Giriş*, Bursa: Ekin Kitabevi, 10. Basım, p. 5.

[6]   İ. Parasız (1998), *Para Politikası*, Bursa: Ezgi Kitabevi Yayınları, p. 1.

[7]   M. İ. Parasız (1996), *İktisadın A B C'si*, Bursa: Ezgi Kitabevi Yayınları, p. 15.

[8]   Dinler, pp. 7–10.

[9]   Ünsal, s. 14.

[10] Ünsal, s. 14.

[11] Dinler, s. 10.

[12] A. Eren (2002), *Türkiye'nin Ekonomik Yapısı ve Güncel Sorunlar*, Muğla: Muğla Üniversitesi No:11, p. 43.

[13] E. Han, A. A. Kaya (2002), *Kalkınma Ekonomisi Teori ve politika*, Eskişehir: Etam A.Ş. Matbaa Tesisleri, p. 2.

[14] M. Kar, S. Taban (2005), "İktisadi Gelişmenin Temel Dinamikleri ve Kaynakları", *İktisadi Kalkınmada Sosyal, Kültürel ve Siyasi Faktörlerin Rolü*, M. Kar, S. Taban (Eds.), Bursa: Ekin Kitabevi, p. 8.

[15] Han, Kaya, p. 2.

[16] World Bank (2000), *World Development Indicators*, CD.

[17] Y. Büyükerşen et al. (1999), *İktisada Giriş*, Eskişehir: Anadolu Üniversitesi İktisat Fakültesi Yayın No: 31, p. 8.

[18] B. Üstünel (1978), *Ekonominin Temelleri*, Ankara: Doğan Yayınevi, p. 21.

[19] Üstünel, pp. 21–22.

[20] Ö. Gürkan (1997), *İktisada Giriş*, Ankara: Attila Kitabevi, p. 10.

[21] Büyükerşen et al., p. 23.

[22] Büyükerşen et al., pp. 24–25.

[23] Gürkan, p. 8.

[24] Büyükerşen et al., p. 26.

[25] Gürkan, p. 6.

[26] Gürkan, pp. 6–7.

[27] Dinler, p. 42.

[28] R. G. Lipsey, P. O. Steiner and D. D. Purvis (1984), *İktisat*, Eskişehir: Bilim Teknik Yayınevi, p. 5.

[29] A. Eren (1998), *Mikroekonomi*, Muğla: Esin Ofset Matbaacılık, p. 94.

[30] M. Merih Paya (1997), *Makro İktisat*, İstanbul: Filiz Yayınevi, p. 27.

[31] N. Güran (1999), *Makro Ekonomik Analiz*, İzmir: Anadolu Matbaacılık, 2nd edition, p. 35.

[32] K. Lordoğlu, N. Özkaplan and M. Törüner (1999), *Çalışma İktisadı*, İstanbul: Beta Basım Yayım Dağıtım A. Ş., 3rd edition, pp. 48–49.

[33] Lordoğlu et al., pp. 53–57.

[34] For unemployment and its types see: M. Merih Paya (1997), *Makro İktisat*, İstanbul: Filiz Yayınevi, pp. 29–32.

[35] Seyidoğlu, p. 384.

[36] K. Yıldırım, D. Karaman (2001), *Makro Ekonomi*, Eskişehir: Eğitim, Sağlık ve Biliel Araştırma Çalışmaları Vakfı, Yayın No: 145, 2nd edition, p. 70.

[37] Seyidoğlu, pp. 388–399.

[38] O. Bahar (2006), "Turizm Sektörünün Türkiye'nin Ekonomik Büyümesi Üzerindeki Etkisi: Var Analizi Yaklaşımı", *Celal Bayar Üniversitesi İ.İ.B.F. Yönetim ve Ekonomi Dergisi,* 13(2), pp. 137–150.

[39] For detailed information about the balance of payments see: H. Seyidoğlu (2001), *Uluslararası İktisat*, İstanbul: Güzem Yayınları, 14th edition.

[40] E. Alkin (1984), *İktisat*, İstanbul: Filiz Kitabevi, pp. 14–15.

[41] Eren, *Mikro Ekonomi*, p. 122.

[42] Dinler, p. 199.

[43] Eren, *Mikro Ekonomi,* p. 177; Dinler, pp. 312–327.

[44] Gürkan, p. 216.

[45] H. Karakayalı, M. H. Yalçınkaya and Ç. Çılbant (2005), "Modern Çağın Kavramları: Küreselleşme ve Rekabet", *Uluslararası Rekabet Sürecinde Türkiye*, M. Faysal Gökalp and C. Yenal Kesbiç (Eds.), İstanbul: Beyaz Yayınları, p. 67.

[46] J. K. Galbraith (1988), *Ekonomi Kimden Yana*, İstanbul: Altın Kitaplar Yayınevi, p. 23.

[47] B. Ataç et al. (2000), *Kamu Maliyesi*, Eskişehir: Anadolu Üniversitesi İktisat Fakültesi Ders Kitapları No: 18, p. 4.

[48] Ataç et al., pp. 4–5.

[49] Gürkan, pp. 26–27.

Chapter 2

BASIC CONCEPTS ON TOURISM

It is a known fact that globalization has increased rapidly, the image of quality goods has been adopted by the people, and thanks to trademarking, attempts at creating customer happiness and addiction in world economies have been made. Therefore, it is very important for the tourism industry, as it is for every industry that the destinations of tourism market, that the firms and organizations analyze themselves and their rivals, decide on which marketing strategy to choose and identify a vision accordingly. In this rapidly growing tourism industry [1], to attain the competitive power, sustain it in the long run, and increase the market share, the income generated will allow the people living in a tourism destination to have a healthier life in terms of economic, cultural and social aspects. Indeed, as it is in other activities in economics, the most important tool for all people in a tourism destination is utility maximization and raising economic welfare to the highest level. This chapter aims at presenting detailed information on the definition of tourism, its properties and varieties as well as the concept of the tourism product, its properties and resources.

1. DEFINITION OF TOURISM

Tourism economics is a discipline that surveys the reasons of national and international tourism activities, their scope, development conditions, results, the cause and effect relationships among these events with scientific methods in economic terms [2]. In this respect, in this chapter, which is an introduction to the course of tourism economics, some basic concepts concerning tourism are clarified. For the issues in the following chapters to be understood more clearly and definitively, it is important to know these concepts. Otherwise. there may be a conceptual conflict and understanding the issues may become harder. From this point of view, first of all, concepts such as tourism, tourist, daily visitor, destination and tourism product are going to be explained, and who the daily visitors are will be put forward.

Particularly in the last century there have been very rapid technological developments in the world. Societies have given more and more importance to industry. As a result, problems such as noise, air pollution, traffic, extreme stress, etc. occuring due to rapid urbanization have adverse effects on the societal and psychological structure of human beings. As time passes by, an increase in such problems occurs due to people's needs for more relaxation, enjoyment, travelling and spending time for themselves. Moreover, the technological facilities and the technical level reached in transportation create major differences in the distance concept and hence in the concept of time and place. When there is a decrease in time spent at work due to technical improvements, most of the people are leaving their homes temporarily in order to satisfy their pschological, social and cultural needs such as travel, entertainment, relaxation and learning. Therefore, this movement called tourism which leads to broad effects in the societal and economic arena has become an essential necessity for today's developed (industrialized) societies.

The origin of the term "tourism" that dates back to the 17th century, is the word "tour" , which is the word "torah" meaning "learning, surveying" in the Hebrew language. Hebrew people call the people they send to far places in order to learn and investigate about the economic and social conditions of the people inhabiting there as *tourist* and their actions as *touring* [3]. It is known that the definitions or studies made concerning tourism dates back to the end of 19th century. While several authors and researchers have very different definitions and explanations regarding the issue, the first definition of tourism was made in 1905 by Guyer – Feuler. They define tourism as "an event, peculiar to the 'modern' age, based on the belief that nature gives happinness to the people and the event that enables gradual climatic changes and needs for relaxation, the desire to know the fabulous beauties enriched by the nature and art, and the nations and groups to approach each other especially as a result of the development of trade and industry and the transportation means becoming perfect [4]".

When the literature is looked up, it is seen that the contemporary tourism concept was first introduced by the Swedish economist Walter Hunziker and Kurt Krapf during the World War II period. According to this, tourism is defined as "all of the events and relations emerging from people's travel and accommodation outside the places that they continuously work or live" [5]. This definition was later accepted by the Association of International Experts on Scientific Tourism (AIEST). Likewise, convened in 1963, the World Tourism Organization also agreed with this definition.

There are also opinions and definitions that examine tourism with its complex and multi-disciplined aspect within the socio-economical scope. Casper defines the event tourism as the "global system based on tourist, tourism organizations, the subsystems formed by tourism organizations and the relations of systems with the economic, social, political, legal, technological and ecological compound of the services for transportation, accommodation, eating and drinking, entertainment and relaxation needs of the tourists [6]. For the countries, tourism is a sector that contributes to the national income, increases employment, has a positive effect on the balance of payments, improves the transfer of technology and knowledge, encourages the foreign investors and serves as a locomotive in the country's development [7].

As indicated above, the activity of tourism covers moving and travelling from one place to another, entertainment, relaxation, accommodation and transportation. However, in order for all these activities to be done, a person should have the sufficient income and put aside some monetary resources for travel purposes. For a person travelling for tourism purposes, the factor having priority is the accommodation factor. The tourist will stop off in a hotel, holiday resort, apart hotel or in boarding houses in return for its price. In addition, the tourist needs to spend money for sightseeing, eating and drinking, shopping or entertainment. This activity has a feature of a service and consumption while definitions regarding tourism are made. Based on this, definitions on tourism include these features:

- First of all, tourism mainly involves people's travel activities to various places and accommodation activities.
- People leave the places they permanently live, work and inhabit for travelling purposes with the tourism activity.
- Accommodation is obligatory in tourism activity; however, it covers a temporary period. This means that people return back to their inhabited place at the end of the journey.
- The tourist is a consumer at the place s/he visits. In other words, demand for tourism products in the place visited shows the tourist's consumption with tourism purposes.
- Travel should not be made for the purposes of working, doing business and yielding revenues.

Hence, in order to redefine tourism briefly in a way that it covers the points mentioned here: *Tourism is a travel and accommodation event that is made as a consumer in order to meet tourists' needs in relation to the relaxation, entertainment, and culture in a place outside their permanent inhabitation.* Therefore, the expenditure made for every kind of goods, services and experiences that a tourist has to buy commencing from his/her departure from the home country to the period of heading back to the hometown are within the field of interest of "tourism economics" at the national and international level.

A. Properties of Tourism

Tourism has a relationship with various industries and several branches of science such as primarily economics and geography, physics, pschology, sociology, political science, politics, law, medicine, history and so on. In other words, tourism is a compound of different activities, services and industries. Hence, tourism has several definitions so far. The reason for this is that, as recently explained, tourism is versatile and has a complex interaction structure. However, despite all these points indicated, tourism also has some outstanding properties as it is in every activity or work. While it is known that there are classifications of different qualities concerning the features of tourism, here the seven most basic and general features will be examined. These are:

- Tourism is a socio- cultural activity composed of social and cultural values; any kind of accumulation presented in the literary and figurative sense make up tourism products.
- Tourism is a very comprehensive sub-economical sector or discipline formed with transportation, accommodation, relaxation (entertainment, drinking) eating, health, shopping, vehicles, facilities and activities.
- Tourism has financial transaction properties due to the consumption expenditure of the people, namely the tourists in the place they visit.
- Tourism is a multi-way economic activity requiring the prevention of the local, societal, economic, ecological and environmental relations in the place where the people visit.

- Tourism, being a social phenomenon, is an important tool in protecting peace, friendship and brotherhood among different people, cultures, religions and nations.
- Finally, tourism, with its benefits, is a comprehensive micro- and macroactivity concerning producers, consumers, societies and national economies in a country.

Having many componenets makes it difficult to explain tourism in severe lines and definitions. In summary, for the tourism industry established wholly on the natural resources and beauties of a country or a destination, as long as there are various factors (work, curiosity, religion, culture and education, relaxation and entertainment, sports, medication and so on) allowing people to consume, this economical activity improves and grows day by day. As civilisation and technology develop and people's desires and expectations change accordingly, tourism becomes widespread around the world and it is expected to become an obligatory necessity for the mankind in the future.

B. Definitions of "Tourist" and "Daily Visitor"

The share of tourism in the economies of both less developed/developing and developed countries is increasing day by day. Hence, in tourism, which is an economical activity concerning a wide range of people, defining the concepts as tourist and daily visitor will enable this economic activity to be expressed more correctly and accurately, because unless it is known what some concepts cover, it will be difficult to make realistic comments and sensible predictions about them. This situation is particularly important for the representatives of the public and private sector who make political decisions for the purposes of tourism. Indeed, today, the developed countries analyze their activities and events with figures in a detailed manner and use this statistical knowledge in their future plans, projections or estimations. In terms of the tourism industry, it is perhaps more important for such data to be recorded, compiled and operated daily than the other industries in an economy. As explained later, the current demand must be known and immediate investment accordingly must be made in terms of tourism demand and supply. Hence, it is beneficial to take the necessary precautions for now by predicting the level of demand in the future.

There are a number of definitions concerning the subject in tourism literature. However, here we will dwell on only some of these in order to limit the issue. According to the definition made in 1937 by the statistics experts of Leaugue of Nations, tourist is accepted as *"the person who visits the country other than the one s/he inhabits for at least 24 hours"* [8]. Today's dictionaries, which is similar to the definition in the American government's dictionary, defines tourist as the one spending 24 hours or a night outside his/her home. In Turkey, on the other hand, according to the Travel Agencies Regulations the term "tourist" was defined in 1996 as: *"a person who temporarily goes outside the place s/he inhabits and travels as the consumer for a certain time without having an aim of earning money, for relaxing and entertainment or for cultural, scientific, sportive, administrative, diplomatic, religious, health and similar reasons and then turns back to his/her residence again"* [9]. According to this definition, it is possible to regard the below as a "tourist":

- People travelling for pleasure and for reasons such as family, health and so on.
- People travelling in order to attend the meetings (scientific, administrative, religious, commercial, diplomatic, sportive and so on) for various reasons.
- People cruising even if they stay less than 24 hours.

There are some exceptions for these definitions and not all individuals travelling are considered tourists. Accordingly, the below are not considered as a "tourist":

- People visiting a foreign country in order to work or have a commercial activity, whether s/he has a business contract or not.
- People going to a country to settle down and stay permanently.
- People going to a country as soldier, student, diplomat, immigrant or with the purpose of working.
- People passing by a country without stopping even if it takes longer than 24 hours.

The concept of daily visitor mainly resembles the definition of "tourist". The most notable difference between them results from the staying duration in the visited place. According to this, *daily visitor is the person who travels for pleasure for a time less than 24 hours* [10]. As understood from the definition, the daily visitor stays in the country that is visited less than 24 hours. This concept which is used in many countries means the person is travelling by a vehicle. Daily visitors cover the cruise passengers, as well. For example, the people who arrive in the Palma Harbour by boat and after visiting the city returning to their boat on the same day are not tourists but should be counted as daily visitors in the official statistics. The visitors not staying in an accommodation not even for a day although they stay in the country they visit for a day or more are also in this group [11]. These kind of visitors use their own boats or trains for accommodation and leave the country on the same day. Therefore, the person going to a country for visiting purposes and staying there for more than 24 hours is called tourist whereas the one staying less than 24 hours and not staying for the night is called the daily visitor [12].

Box 2.1. Tourism Statistics

Since the daily visitors do not obtain accommodations in the country they visit not even for a night, those countries do not include such visitors in the international tourism statistics. Indeed, from a legal perspective, tourism statistics do not cover the ones not entering the country; such as the ones not leaving the transit region at the airport. However, in Turkey, until the period of 1950 - 1965, there was no difference between tourists and daily visitors in statsitical terms. After 1965, the State Institute of Statistics started to make this difference complying with the definitions of World Tourism Association. Since then, the number of tourists coming to the country and statistics regarding other tourism data towards the properties of tourists have begun to be recorded in a healthier and more trustworthy way. However, with a decision taken in 2004, the fact that the Turkish workers living abroad and entering and exiting Turkey are considered within the scope of tourists brought some new discussions. The international tourism and travel statistics first started to be collected by the International Union of Official Tourism Organizations (IUOTO) founded in 1947 with the participation of 52 countries.

C. Concept of Tourist Destinations

The tourist destination is a commonly used concept in tourism literature. Explaining this concept that will often be used within this book is important for the subjects to be understood clearly. When the destination is mentioned, first of all the geograpical regions that are formed by the unification of the tourism products and that presents integrated services to the consumer group called "tourist" should come to mind [13]. Being a mixture of tourism products, destinations present an integrated experience to consumers. These products are experienced under the name of the related tourism region [14]. Traditionally, tourism regions are very well defined geograpical areas. Hence, a place considered as destination may be a a city, a town, an island as well as a country [15]. Or the tourism region called as destination can be paraphrased by consumers as a perceptual concept based on their travel plans, cultural backgrounds, travel purposes, educational levels or past experiences [16]. To illustrate, while London is a destination for German businnessmen, Europe is accepted as a destination for the Japanese tourists with a desire to travel to the six European countries with a one-week tour [17].

In another source, the tourism region is a national area with a certain image in people's minds and it is smaller than the whole country, larger than many cities in that country. It is defined as the region having important tourism attractions, attractive centers and activities such as festivals, carnivals, a good transportation web established in the country, a potential to develop, transportation facilities countrywise and among the regions connected with an internal transportation web, and a sufficient geographical region for the development of the touristic facilities [18]. As understood from this definition, some properties exist distinguishing a tourism region from other regions. Brief information regarding these properties, which have been developed in order to indicate the differences among the regions, are in Table 2.1.

Table 2.1. General Properties of Tourism Regions [19]

No	Properties	Contents
1	Attractiveness	All the natural specific events and works made by human beings for a specific purpose, which were able to remain until today from past civilizations as part of their heritage.
2	Attainability	The transportation system formed by the unification of all public transportation vehicles, rotas, terminals and services.
3	Facilities	Accommodation, nutrition, purchasing and other services for tourists.
4	Suitable packages	The composition of touring or travelling organized in advance by the mediators and tourism authorities.
5	Activities	All the activities that consumers may attend during their visits.
6	Supporting services	The general name for the services such as communication, postal service, newspaper agency, hospitals that tourists need at any time.

In summary, the *tourism regions* are the composition of the tourism attraction centers emerged by gathering of tourism services produced individually (e.g., accommodation, transportation, food and beverage consumption, entertainment and so on) and the variety of public goods (e.g., sea, lakes, sights, socio-cultural environment and so on) [20]. In other words, in order to mention a tourism region, there should be a geographical area with a specific brand name and image formed by the tourism attraction centers where public services are presented, having the same cultural, climatic and natural conditions, natural and cultural beauties, activities peculiar to those places that will be served for consumers, such as facilities like accommodation, transportation and communication. The positive and adverse effects of tourism activities in a region are under the responsibility domain of the branch of "tourism economics".

D. Types of Tourism

It is known that there have been many different types of classsifications in various sources regarding the major types of tourism so far. Nonetheless, it is possible to compile them in two major headings in terms of their economic importance. The reason for this is the close relationship of major tourism types. Although, below in Figure 2.1. major types of tourism are shown in major headings, as indicated above in this book, two fundamental types will be emphasized, which are domestic and international tourism. This classification is made according to the place the tourist goes and the region where the events and relations occur in economic terms.

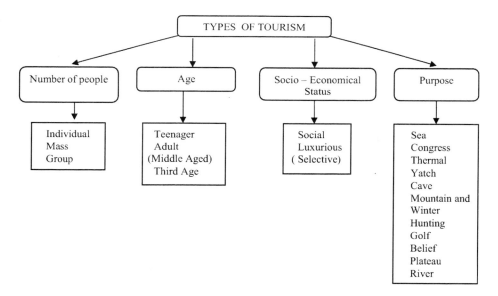

Figure 2.1. Major types of tourism.

1. Domestic Tourism

The tourism travels in a country realized by its own citizens is called domestic tourism [21]. A person living in Nice and going to Paris is an example of domestic tourism in France. It doesn't require a passport or a visa, foreign language or foreign exchange and nor does it have any effect that may provide input for the economy as a foreign exchange. However, the development of domestic tourism is significant for the firms and organizations in the country to compete with the similar firms and organizations at the international stage, because tourism organizations furnishing services domestically by means of tourism find opportunities of self development in time, on the subjects such as service quality, consumer satisfaction, speed and flexibility. It is easier for the organizations competing with each other in the country to compete with their rivals outside the country. Moreover, tourism firms and agencies within the country may use the income they get through domestic tourism in order to encourage international tourism.

By an increase in domestic demand for tourism and the improvement in domestic tourism, the tourism economy and accordingly the country's economy will be affected by this in a positive way [22]. For instance, by means of this, a more well-balanced distribution of the national income among the regions will be ensured, poverty will decrease and employment will increase by creating new job opportunities. Nevertheless, the consumption and production volumes and patterns among the regions change by means of domestic travel income and it is guaranteed that a more well-balanced developmental policy among the countries is applied.

Hence, the active participation of the people living abroad to tourism activities results in their learning natural and cultural values, an increase in the sense of tourism, and physcologically, people getting rid of the exhaustion and stress in their life and resting and having fun, causing increased production in their work life.

In summary, the development of international tourism in a country depends on the size of the demand for domestic tourism that exists in that country. Indeed, the existence and potential of the demand for domestic tourism in a destination country and the realization of that demand decrease the industry's dependency on the foreign markets and firms. The movements of domestic tourism in a country make up the dynamics of international tourism development [23]. Hence, the higher the demand for domestic tourism, the faster and more effective the development of foreign travel, accordingly.

Box 2.2. Development of Domestic Tourism

The development of domestic tourism in Turkey is still not at the desired levels although Turkey is among the first in the world ranking of the number of tourists. Besides economic reasons like a citizen's low purchasing power and an organization's different price policies for domestic and international tourists, there are also some societal reasons that hinder domestic tourism to develop; such as the lack of an established culture of holiday or the perception that holiday is a luxurious thing. Yet, when countries like the USA, England, Spain and France are considered, it is seen that domestic tourism has developed as much as foreign tourism and that they support each other from time to time. However in Turkey, while the national tourism industry discusses tending to the market of domestic tourism in case of crisis, there have been no serious development so far. Such policies observed in the early and late 1990s have appeared again during 2006. Some tourism associations convened and searched for ways to refresh domestic tourism by preparing suitable vacation packages and organizing exhibitions in some cities of Turkey within the scope of a campaign called "Anatolian Holiday Days". Considering the tourism's economic impacts such as the multiplier effect or added value, it will be seen that the development of domestic tourism in a country is at least as significant as the development of international tourism to that country (inbound tourism) as well.

However, doubtless to say, the necessary and sufficient condition for domestic tourism to be formed and the demand to be created is that there is an average expendable income level that will enable the citizens of that country to participate in tourism activities. As the income per capita and accordingly the wealth in a country increase, the demand for domestic tourism and the foreign travel increase as well. Evaluating from the aspect of tourism economics, the turning of the demand for tourism into a real demand is parallel to the existence of a sufficient, expendable income and spare time for a vacation.

2. *International Travel*

The tourism-related movements of the people living in a country to the places outside their country are called international tourism [24]. International tourism is an activity that is made up by tourists coming to a country from other countries and going to other countries from a country. Indeed, the ones having these trips are called "international tourist", like the person who is going to France from Japan in order to see the Eiffel Tower. The most important property of foreign travel is that it supplies the input or output of foreign exchange for the national economy according to the entrance or exit status from one country to another. Hence, it is possible to divide international travel into two categories as *active international travel* (income tourism) and *passive international travel* (expense tourism) in terms of their values [25].

a. *Active International Travel*: These are the tourism-purposed travels to a destination country by the other country's citizens. A trip from Austria to Turkey is an active international travel to the latter country. In other words, foreign tourists coming to a country form the basic source of the active international travel. This kind of tourism is important since it supplies an input of foreign exchange for the country and has an effect of importing in economical terms. It also plays a crucial role in closing the foreign trade deficit by having a positive effect on the balance of payments. Indeed, the economical significance of tourism is originated from its being an invisible type of import and export. Increasing the foreign exchange income is a necessary source for the economic development and growth of the countries. From this perspective, the contribution of the active international travel on the national economies may be mentioned.

b. *Passive International Travel*: This is the traveling of a country's citizens abroad. The trip from the UK to Eygpt is a passive international travel for the former country. In passive travel, there is no foreign exchange input to the country; in contrast, there is an outflow of foreign exchange. Since there is an outflow of foreign exchange from the country to abroad, the effect of this tourism on the balance of payments is negative and it has an export effect. In summary, it is a subject to be dwelled on carefully that the countries develop policies and strategies to improve active foreign travel, encourage this kind of tourism and support this with plans and projects oriented towards the future since it supplies an inflow of foreign exchange to the country and increases the national income.

Box 2.3. Active and Passive Foreign Travel Movements

Attaching importance to the active side of the international travel, some countries, like Turkey, with a great need of foreign exchange have the aim to attract more tourists. Governmental policies such as the implementation of Housing Fund repress the development of the passive tourism side even for a temporary period as dissuasive factors for the citizens to go abroad. It is known that similar policies were implemented in some other countries in the previous years. For example, England, for some economic reasons, started to implement policies that encourage its citizens to have holidays within the country in 1990s. It is said that a similar implementation is valid for Germany in 2006. Nevertheless, countries like the USA, England and Canada can have activities in both active and passive international tourism. In other words, while such countries welcome a great number of tourists, they also encourage a great number of their citizens to travel to other countries.

International travel is a tourism that has positive contributions on the national economy since it increases the foreign capital inflow to the country, assisting the lower and upper structure to develop, refreshing the other industries that it is connected to due to its income-raising effect. Yet, it should be kept in mind that foreign travel has also positive societal and psychological effects because of its reinforcing the peaceful environment among the nations, recognizing different cultures, people, groups and religions and the creation of an awareness of respect.

3. TOURISM PRODUCT

It is necessary to explain the concept of tourism products is the reason of the existence of tourism economics and the main target of demand and supply. In order to explain the meaning of tourism product, it is better to examine the tourism regions within the tourism system as well as the resources they present. From this perspective, it is required to make the definition of two main regions and connecting factors within the system [26]. (Figure 2.2.)

i.   *Tourist generating countries*: This region covers the places where the journey starts and finishes and where active or potential tourist groups permanently live. The motivation of tourists before the travel and holiday appear here and the related plans and decisions are prepared here as well.

ii. *Transportation:* The transportation systems that enable the tourist to visit another region and turn back and his/her experiences acquired during the transportation are within this process. Any type of highway, seaway, railway and airway vehicle has the function of a bridge between the region generating tourists (or the country) and the region accepting tourist (or the country), e.g., automobile, train, plane, boat, etc.

iii. *Tourist attracting countries:* Being the last phase, it covers the region where the tourist gains the holiday experience and accepted as the "tourist" by local people and recorded as "tourist" in statistics. All the plans and studies are to be made on the experiences to be acquired in this region.

Figure 2.2. Tourism System.

Box 2.4. Properties of Tourist Generating and Attracting Regions

Considering the countrywise tourism activities, it is seen that the countries sending tourists are the ones that completed their industrialization and have a high income per capita such as the western countries like the USA, Canada, UK, Germany, Finland, France as well as Far Eastern countries such as Japan and Australia. Similarly, some of these countries are on the top of the ranking of tourist accepting countries. On the other hand, there are some tourist attracting countries whose national economies are mostly dependent on the foreign travel activities although they rank in rather lower levels of tourist generating countries. Some island countries such as Jamaica, Maldivs, Seashell Islands and Republic of Dominique are among the important destinations being talked about in the international arena as a result of the high rates of the tourism income they have in GDP.

It is possible to examine the sources presented by tourism regions in different ways. For example, according to an approach, the main headings of tourism regions are evaluated under two headings [27] (Figure 2.3). These are called the primary and the secondary properties. These two properties, together, formed the attractiveness of a tourism region. It is necessary to clarify the concept of attractiveness. *The main factors that attract the tourists to visit a country, region or a place are the attractivenesses of that place* [28].

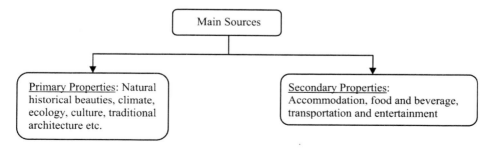

Figure 2.3. Main sources of tourism regions.

As an example, Niagara Falls, Eiffel Tower, Selimiye Mosque, Great Walls, etc., it is emphasized by the experts that the secondary properties are more effective and important than the primary properties on the development of tourism regions. That is, however, much the country's natural, historical and cultural beauties are, unless they are supplied in an organized way suitable for the tourism upper and lower structure, it will be seen that none of these are important, because nutrition and accommodation are the basic and mandatory necessities of human beings. A place where there is an attractive view, a bay, the sun and sand but where people cannot accommodate may have no value in terms of tourism.

Figure 2.4. shows the main tourism sources (tourism products). As seen a tourist visiting a country is not only interested in objective (physical) goods but also has an interaction with the subjective goods in accordance with his/her motivation [29]. Therefore, when defining the tourism product, it is necessary to consider both subjective and objective factors. As a conclusion, it is possible to define the tourism product as *the aggregation of subjective and objective values served by the factors constituting the tourism supply and that a tourist obtained with the purpose of fulfilling his/her necessities within the period from the beginning of his/her travel until the end.*" [30]

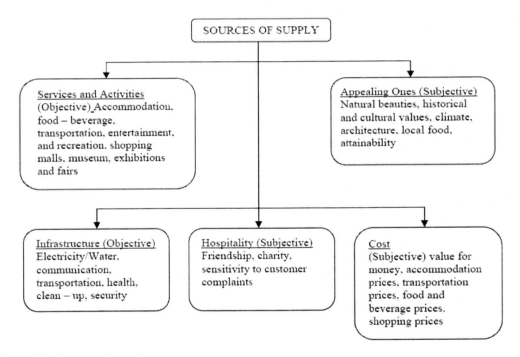

Figure 2.4. Sources of supply.

It is understood from this definition that the tourism product does not include a production process based on a unique product unlike other goods and services in the economy and does not have a homogeneous structure [31]. According to the structural property of tourism, a tourist does not benefit from a single factor; in contrast, s/he buys the "final product" that emerges from the combination of many geographical, economic and social factors and this product may appear as an important factor in the formation of holiday experience (tourism demand) and determination of the price policies of the organizations (tourism supply) [32].

A. Properties of Tourism Products

When the tourism economics literature is considered, it is seen that there are four main properties of a tourism product (Figure 2.5.) [33].

The first one is that the tourism product is something mostly with a service content rather than being a good with a physical existence unlike other industries in the economy. The product produced by firms and organizations operating in the tourism industry is a service product. Therefore, consumers do not have the chance to test the product without trying, as it is in the other goods in the economy. Indeed, the marketing chance of the tourism product is not as easy as it is in other product categories. For example, a person who is going to buy a refrigarator, a pair of shoes or a car may examine the product, try it or compare it with the similar products prior to the purchasing stage. However, a tourist does not have the opportunity to understand the package tour or the tourism product that s/he has just bought or have an idea about it without experiencing it [34]. Considering that the subject matter package holiday covers the flight, accommodation, consuming food and beverage, sunbathing on a beach, being served by a waiter or a taxi driver and benefiting from the country's historical and geographical beauties, it may be understood how complex a structure the tourism product has and that it differentiates from the other products in the economy with this aspect. Additionally, the tourism product (such as the package holiday) is produced and sold in large numbers. The volume of goods being produced in large numbers depends on its consumption, namely the demand.

The second property of the tourism product, the tourist who is the consumer of the service must come to the place where the product is sold and service is served. In other words, the produced tourism product must be consumed at the moment of production simultaneously. Marketing and sales also require the product to be consumed at the moment of production.

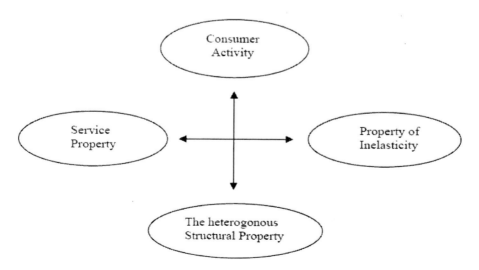

Figure 2.5. Properties of tourism products.

In other economic goods, after being produced in a place such as a factory, a product is delivered to the retailer and wholesaler and then to the shopping center such as a store where it will be sold to the consumer. This is not the way for marketing tourism products. Tourism, with this property, causes an international economic activity.

Another property of the tourism product is that the served goods resemble each other but they are not exactly the same. The tourism product is not homogeneous (similar) but has a heterogeneous (different, discrete, not the same type) structure and appearance. What the tourism product covers has been mentioned in the above paragraph. Considered from this perspective, it is a fact that each package tour that seems similar but made up of combinations of different products may not be of the same quality, standard and characteristics. For example, it is not possible for a five star hotel or a flight to be of the same standard every time even if they belong to the same or different firms. In a hotel where the sea, the sun, the beach and the facilities are very good, the negative attitudes and behaviors of employees to tourists are an obstacle for the tourism product to be similar. Hence, the influence of each package holiday product over the level of the tourists' satisfaction, their image perceptions or service quality perceptions may involve dissimilarities and this makes it harder for the served product to be homogeneous.

The fourth property of the tourism product is that tourism supply is constant or limited (inelastic) whereas tourism demand has a flexible (elastic) and sensitive structure. Generally in the tourism industry, the short term tourism supply is constant and there is no way to change this. For instance, if a region's hotel and hence bed capacity is at a certain level, the adaption of this capacity to the increased demand depends on the capacity of tourism investments. In other words, despite the seasonal or periodical increase, it is not possible to increase the number of hotels or beds immediately. At least a one- or two-year construction and waiting period is needed for such a change. At the end of this waiting period, new hotels may be built and current facilities may be enlarged. Therefore, while supply cannot comply with the short term increase in demand, it is possible to increase supply in certain amounts in the long run.

The policy makers with tourism purposes in a country may have to increase tourism supply in a way that they use the resources efficiently and by considering the rapidly changing structure of demand. Otherwise, each hotel room or airline seat unsold causes a big income loss, and as indicated before a waste of resources, first, for the tourism organization and then for the country. Meanwhile, as a result of the unnecessary usage of the scarce resources of the destination country, there will be a loss in the citizens' wealth as well. In situations when supply exceeds demand, the price of the served tourism products will also decrease, and the desired profit cannot be achieved and will cause trouble for the firms/organizations, causing them to lose money. This subject will be elaborated later.

B. Tourism Industry and Tourism Products

The tourism industry is composed of numerous organizations and enterprises, and supply services and goods to the people called tourists and daily visitors. These cover many small, medium or large enterprises serving travel agencies, tour operators, transportation, accommodation, food and beverage, entertainment and other assisting services. An important part of these enterprises is interested in travelling, accommodation and various tourism attractions in the industry. Besides, some enterprises and organizations providing special and supporting services such as tourism information centers, gift and souvenir producers and retail shops also take place within the industry. Due to this wide domain and distribution of the industry, it is necessary to make a classification for some subsectors. A classification by Holloway is shown in Table 2.2 [35].

Table 2.2. Organizations, Institutions and Attractions Supporting Tourism

No.	Supporting Groups	Services
1	Means of Transportation	Any kind of transportation tool that is necessary for a tourist's travel (e.g., automobile, aircraft, boat, train, etc.).
2	Accommodation Facilities	Facilities needed for a tourist accommodation (e.g., hotel, apart hotel, holiday village, boarding house, etc.).
3	Manmade attractions	Natural attraction areas made for public use (e.g., natural parks, aqualands, amusement parks, etc.).
4	Private Sector Supporting Services	Ones supplying goods and services to tourism organizations (e.g,. food, construction, medicine, architecture, etc.).
5	Public Sector Supporting Services	Institutions supplying support as information to tourists visiting a destination (e.g,. tourism information desks, etc.).
6	Travel Mediators	Businesses organizing and selling tours through tourist expectations marketing (e.g., travel agencies, tour operators, tour guides, etc.).

An important part of organizations and enterprises in the tourism industry primarily highlights the private goods that they have produced. Asked what their products are, most of these organizations reply as "accommodation – entertain" and most of the agencies reply as "travel services". Because of this reason, according to their own approach, while admitting that they take place within the tourism industry, most of them do not accept that they serve a comprehensive and a complete product that is called the tourism product.

This situation constitutes a significant problem for the economic analysis. This problem can be explained in a way that consumers do not perceive or demand the good produced by the industry as the same good. Tourists assess tourism products not as the product that they take from individual organizations subsequently but as a whole. Moreover, the facilities that most tourists want to benefit from may not have a commercial aspect. For instance, a tourist may want to buy an experience of lying on a sunny beach for a week. The purpose here, is to relax, be healthy and benefit from the sun. Another tourist travelling for business purposes may demand an opportunity of attending a business meeting in order to make a selling agreement for his/her own firm. In some other examples, tourists may request an actual commercial product rather than a total tourism and travel experience, activity and a job opportunity.

That the firms, organizations and the enterprises in the tourism and travel industry have developed rapidly and the marketing studies have become more complicated led the organizations to serve as a whole a package tour instead of serving their own private products. For this reason, the tourism product is accepted as the gross of goods and services in economical terms and it is demanded by tourists. Previously, it was indicated that there is difficulty in identifying the tourism product. As known many tourists demand products without a commercial value. For instance, tourists with a cultural purpose may request that they have a consumption with tourism purposes in order to see the livestyle and history of a different society. Doubtless to say, most of the consumers demand commercial goods such as transportation and accommodation so as to go to another destination and these goods are the components of the bunch of goods which will be made up in order to gain a tourism experience. This situation clarifies the issue on the grounds that the foregoing goods are assessed as tourism products [36].

Primarily, the tourism product is thought to be an absolute holiday package or a set of complementing goods, although perceived as unique by the consumer. A tourist makes a general evaluation in terms of the total price of his/her trip and the effect of a change in an item, such as the price of the accommodation, on the holiday decision will depend on the share of this item's holiday package on the total cost. In this respect, each item in the holiday package may be substituted with the other. Each item in the total holiday budget of the tourist will be in competing with one another. Second, the tourism product is accepted as the set of individual goods complementing each other. However, these products are regarded as separate goods while tourists make the purchasing decisions. Generally, while the first approach affects the purchasing decisions of tourists at the stage before purchasing, the second approach affects the decision such as a regular service of rent-a–car or food purchasing in a business trip. In practice, every travel is seen as a good by the tourist and every item constituting the travel contributes to this good according to its own characteristics.

C. Sources of Tourism Products

As examined in the first part, the sources in an economy are generally assessed in two groups. The first group is the "free" sources obtained without an effort and the second one is the "scarce" sources with a limited supply, which are attempted to be increased by means of human efforts. Some economic analysts are interested in the production of these scarce sources and their supply in required fields. The decisions regarding the supply of these scarce sources or their being served for the benefit of people are taken by various organizations in the society. Countries adopt different political systems on the issue of source usage and supply. As explained in the previous part in detail, while the free market mechanism in capitalist systems make these decisions the socialist systems adopt central planning in the supply and usage of sources. The national or regional decision organizations in these systems supply the sources within different industries and within their efficient or inefficient economic use. For instance, a central administration, industries or the suppression groups may haggle regarding the use of a land that may be considered for mining, agriculture or tourism (for alternative industries) or the land's remaining in its natural condition (economically inefficient).

Individually speaking, firms also compete with each other with regards to the sources at the industry level. For example, hotel keepers and owners of restaurants may compete for recruiting an expert cook with an attractive salary base. There may be significant competition on the scarce resources and the identification of the supply of these resources even in an efficient organization or enterprise. Moreover, as explained eloborately on the issue of opportunity cost, will a national tourism organization spend its limited budget on the introductory video products for the country or destination or instead on booklets and posters or on another similar issue? How and according to what is s/he going to make his/her choice? These decisions are identified by the authorities in the center of economy-related studies. Hence, the fundamental workspace of tourism economics ought to be related to the replies of the following questions [37]:

- What kind of a source supply mechanism will be applied in regions originating and attracting tourists?
- How will the tourism industry contend with the other industries and activities in competition for the scarce resources in the economy and what will be the most suitable distribution of these sources among the industries?
- Similarly, what kind of competition will be discussed for using the resources among the firms/enterprises supplying tourism products and operating in the tourism industry?
- Is there an important opportunity cost regarding the usage of private resources with tourism purposes? Does the usage of these resources oriented towards tourism give rise to a loss of welfare in terms of society?

Certainly, all of the questions presented above are positive (based on reality) questions rather than having a normative (based on personal judgements) characteristic. In other words, these questions related with what should be actually or what needs to be done about resource using. The above mentioned main discussion domains in economics call for two questions in tourism domain; the first of which is that there are usually significant discrepencies between the perceptions of consumers and producers regarding what makes up the tourism product. The second is that most of the products demanded for tourism are public or free goods.

Today, it is argued that there is certainly a very limited amount of free resources and as indicated above, the tourism industry is based mostly on these free and public goods. The tourism industry generally uses resources with an opportunity cost that is almost zero. Opportunity cost may arise not only for the resources with a certain economic value but also for making use of time and facilities easily. To illustrate, should a family go camping in an off-season, s/he uses highways, national parks and beaches in the region and sightseeing and taking pictures s/he "consumes landscape" without hindering or disturbing anyone and without being disturbed. On the other hand, if the same family traveled in a peak season, there might be several reasons that can influence the opportunity cost such as traffic congestion or overcrowded region and most probably, its cost might be high morally and financially.

The ones supplying goods and services in the tourism industry will make use of the compound or mixture of the same scarce resources that are used by the producers operating in other industries of the economy. Several lists with statistical importance have been made regarding the usage and combination of resources in tourism. This list may be summarized as in Figure 2.6 [38]. The figure shows that tourism mostly benefits from free resources and scarce resources are used as a combination of the public and private sector. This situation puts forth how resources are gathered in tourism, which goods and services do the producers produce and how tourists perceive the goods and services they consume as a "product". While the classifications are seen to have almost the same content, three kinds of classifications are made in terms of the intense usage of the tourism resources. According to this, tourism sources are classified as:

1. Natural resources
2. Artifical resources
3. Socio-cultural resources.

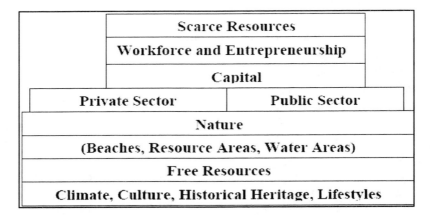

Figure 2.6. Significant resources for commerce and tourism.

These resources are the products of a country's natural, cultural and historical prosperity. In fact, these resources with no economic value and meaning are accepted as tourism resources since they are used by the tourists intensively and an income is earned in return. The countries with this prosperity gain an advantage of having an effective competitive edge in terms of tourism products and resources over the ones lacking them. The long term existence and sustainability of their economical importance are based on tourism and hence their being used consciously. While using the natural, cultural and historical resources for tourism with an expectation of high foreign exchange income, it is important for the countries with high expectation from the tourism industry that these resources be supported with future oriented sustainable tourism policies. Otherwise, the extinction of the natural, historical and cultural environment, which are the reasons of existence of tourism, may be faced. The environment is an obligatory condition for tourism and the future of the countries operating in tourism.

As it will be seen later regarding the tourism resources, there is also a resource distribution based on resource and usage. The common property in these two classifications is that the ones that are not manmade and exist in nature are based on resources. The primary properties of tourism resources indicated previously can be assessed within the group. The resources that are manmade and reflect the properties of the people of that region or that are made later for the benefit of tourists are expressed as the resources for usage. Similarly, it is possible to examine the secondary properties of tourism resources and manmade socio-cultural resources in this group.

SUMMARY

Some fundamental knowledge with an intention of forming a base for the field of tourism economics and a general introduction to the tourism concept are presented in this chapter. First, considering the definition of tourism concept, its main properties are given. Moreover, the main differences between the concept of a tourist and a daily visitor are dwelled on as an important subject in tourism economics. The concept of tourist destinations where the majority of tourism experiences occur and most of the tourist spending are made and the varieties of tourism are the additional topics outlined in this chapter. Since there are several factors affecting people's involvement in tourism activities, handling the issue in terms of domestic tourism or international tourism becomes an important part of the agenda. Finally, following this part that deals with the tourism product that constitutes the basis for economic activities, there is the third part in which the positive and adverse effects of tourism are examined in detail.

DISCUSSION QUESTIONS

1. Define tourism from different perspectives and indicate of which main properties such definitions generally consist.
2. State the properties of tourism briefly.
3. Define the concepts of tourists and daily visitors and indicate the main difference between them.
4. What does the concept destination refer to? In this respect, what may be the general characteristics of destinations?
5. What are the major types of tourism? Explain the concepts of domestic tourism and international tourism in terms of their economic importance.
6. What is a tourism product and which main factors does it constitute?
7. Describe the relationship between tourist generating countries and tourist attracting countries and the transportation.
8. What are the sources of a tourism product? What type of economic benefits do these sources have for a tourism destination?
9. Discuss the similar and different ways of tourism and industrial products.

REFERENCES

[1] C. R. de Freitas (2003), "Tourism Climatology: Evaluating Environmental Information for Decision Making and Business Planning in the Recreation and Tourism Sector," *International Journal of Biometeorol*, 48, p. 45.

[2] H. Olalı, A. Timur (1986), *Turizmin Türk Ekonomisindeki Yeri*, İzmir: Ofis Ticaret Matbaacılık, p. 195.

[3] H. Erdoğan (1995), *Uluslararası Turizm*, Bursa: Uludağ Üniversitesi Yayını, p. 8.

[4] N. Kozak et al. (2001), *Genel Turizm İlkeler-Kavramlar*, Ankara: Detay Yayıncılık, 5th edition, p. 1.

[5] W. Hunziker, K. Krapf, *Allgemeine Fremdenverkehrslehre*, Zürih 1942. Cited in Sait Evliyaoğlu (1989), *Genel Turizm Bilgileri*, Ankara, p. 53.

[6] Olalı, Timur, pp. 5–6.

[7]   X. Han, B. Fang (1997), "Measuring the Size of Tourism and Its Impact in an Economy," *Statistical Journal of The United Nations Economic Commission For Europe*, 14(4), p. 357.

[8]   M. Z. Dinçer (1993), *Turizm Ekonomisi ve Türkiye Ekonomisinde Turizm*, İstanbul: Filiz Kitabevi, pp. 8–9.

[9]   Kozak et al., p. 7.

[10] C. Y. Gee, E. Fayos-Sola (1997), *International Tourism: A Global Perspective,* Madrid: World Tourism Organization, p. 5.

[11] Erdoğan, p. 29.

[12] W. F. Theobald (1994), "The Context, Meaning and Scope of Tourism," *Global Tourism*, W. F. Theobald (Ed.), Oxford: Butterworth & Heinemann, 1st edition, p. 10.

[13] D. Buhalis (2000), "Marketing the Competitive Destination of the Future," *Tourism Management*, 21, p. 97.

[14] R. Haris, N. Leiper (1995), *Sustainable Tourism: An Australian Perspective,* Chatswood: Butterworth-Heinemann.

[15] C. M. Hall (1994), *Tourism and Politics: Policy, Power and Place,* Chichester: Wiley.

[16] Buhalis, p. 97.

[17] R. Davidson, R. Maitland (1997), *Tourism Destinations,* London: Hodder and Stoughton.

[18] C. Tosun, C. L. Jenkins (1996), "Regional Planning Approaches to Tourism Development: the Case of Turkey," *Tourism Management*, 17(7), pp. 519–531.

[19] O. Bahar, M. Kozak (2005), Uluslararası Turizm ve Rekabet Edebilirlik, Ankara: Detay Yayıncılık, p. 78.

[20] D. Dredge (1999), "Destination Place Planning and Design," *Annals of Tourism Research,* 26(4), pp. 772–791.

[21] Gee, Fayos-Sola, pp. 7-8; Kozak et al., p. 14.

[22] J. Jafari (1986), "On Domestic Tourism," *Annals of Tourism Research*, 13(3), pp. 491–496.

[23] Jafari (1986).

[24] Gee, Fayos-Sola, pp. 7-8; Kozak et al., p. 15.

[25] Dinçer, p. 11.

[26] O. İçöz, M. Kozak (2002), *Turizm Ekonomisi*, Ankara: Turhan Kitabevi, 2nd edition, p. 2.

[27] O. İçöz, M. Kozak (2002), *Turizm Ekonomisi*, Ankara: Turhan Kitabevi, 2nd edition, p. 2.

[28] C. E. Gearing, W. W. Swart and T. Var (1974), "Establishing a Measure of Touristic Attractiveness," *Journal of Travel Research*, 12(4), pp. 1–8; T. Var, R. A. D. Beck and P. Loftus (1977), "Determination of Touristic Attractiveness of the Touristic Areas in British Columbia," *Journal of Travel Research*, 15(3), pp. 23–29.

[29] İçöz, Kozak, p. 17.

[30] İçöz, Kozak, p. 15.

[31] S. I. J. Smith (1994), "The Tourism Product," *Annals of Tourism Research*, 21(3), p. 583.

[32] G. McIntyre (1993), *Sustainable Tourism Development: Guide for Local Planners,* Madrid: World Tourism Organization.

[33] Erdoğan, pp. 42–44.

[34] D. G. Taylor (1980), "How to Match Plant with Demand: A Matrix for Marketing," *International Journal of Tourism Management*, 1(1), pp. 56–60.

[35] J. C. Holloway (1994), *The Business of Tourism*, London: Pitman, 4th edition, p. 126.

[36] Bull (1995), *The Economics of Travel and Tourism*, Melbourne: Longman, 2nd edition, p. 26.

[37] Bull, p. 5.

[38] İçöz, Kozak, p. 20.

Chapter 3

RELATIONSHIP BETWEEN ECONOMICS AND TOURISM

Tourism is accepted as an important activity for many developed and developing countries. The economic significance of tourism and its potential from this aspect were understood in the mid-20th century and the tourism industry was generally underestimated until the 1950s in the research of development, growth and prosperity. Tourism movements have begun developing since the end of World War II, particularly in the western societies that are today's developed countries. After the 1960s, the economical importance of tourism has been understood and it has become one of the most rapidly developing industries. In this chapter, the economic importance of tourism, its industrial properties and its close relationship with economics are explained. Additionally, how to measure the economic effects of tourism and the methods used in the primary and secondary economic effects are among the issues to be examined in this chapter.

I. IMPORTANCE OF TOURISM

Today, tourism is regarded as one of the most important sources of economic growth and development both in developed and developing countries [1]. According to WTO, the number of international tourists became 763 million and international tourism income reached 623 billion US$ in 2005 [2]. When domestic tourism movements and spending are considered, it is better understood that tourism is an important economic activity. At present, the tourism industry equals 5% of the world's GDP and 8% of the world's export revenues [3]. Over 255 million people, which equals 11% of the world's total employment capacity, have links with the tourism industry [4].

International tourism affects and is in the interest of many countries due to its property of supplying foreign exchange. Indeed, tourism has a great contribution on employment, income level, alleviation of the internal and external debts, balance of payments and hence, increase in the citizens' wealth [5]. It especially affects the economies of the developing nations because tourism, which is directly related with the level of economic development and growth, is an important source of income for these countries with economies that are not widespread. As a result, as tourism deveops in a country, the foreign exchange hardship diminishes, the competitive power of the domestic firms rivals those outside the country. Hence, their productivity increases, the scale economies are utilized, the balance of foreign trade is positively affected, an area of employment is created, and by increasing the national income entirely it may have a positive effect on the economic growth.

A. Industrial Properties of Tourism

Tourism has some peculiar structural properties as it is in every industry in an economy. In order to put forward the economic structure of tourism as well as the cause and effect relationship among the events more clearly, it is better to clarify what these properties are. As indicated before, the economic system chosen by each society to solve economic problems differs. This differentiation affects the structure of the tourism industry indirectly. In other words, a country with an open economy to outside approaches the economical problems differently from that of a country which does not have an open economy. Therefore, in tourism which is a sub-sector of economics, the point of view for the economic events may depend on the country. However, besides all of the above indicated points, tourism has some properties that may be common in all countries in industrial terms. It is possible to list them as follows [6];

1. Tourism is an invisible export industry because the tourism product served in this industry has such a different appearance from other industries that it does not have a quality of a good with a physical existence and can be moved from one place to another. In this respect, since the product as a subject of international trade is service, tourism is also called an invisible export. The spending on eating and drinking, which a tourist does for consumption purposes, is operated as if a product is exported abroad due to its effect of supplying foreign exchange. A similar situation exists for the items such as gifts that tourists buy upon returning to their own countries. In such transactions, the responsibility belongs completely to the consumer.

2. As known, the tourism industry takes place among the services industries and it is involved in many other sub-sectors in accordance with its area of activity. The goods and services supplied for both domestic and foreign tourists are manufactured by numerous branches of activities. There are several industries in charge of producing the goods and services that a tourist may need from the moment s/he arrives in a foreign country until departure. While some of these are industries serving directly to tourists such as eating, drinking, transportation, health care, entertainment, some others have a direct contribution such as banking, construction and so on.

3. The reason for tourism to exist is the natural, historical, cultural surrounding and traditional values. These mentioned prosperities can only gain the quality of an *"economic product"* thanks to tourism. Due to this property tourism's dependence on outside decreases and it follows a development strategy based on national/regional sources with the production style utilizing these assets with the quality of a free good. This situation has an importance especially for the economy of a less developed country because for the development and industrialization of a less developed country, importing intermediate and capital goods in industries other than tourism occurs. If one assumes that tourism develops in a less developed country, there would be no need for a foreign exchange outflow for the import of intermediate

and capital goods; in contrast, a foreign exchange inflow would have been maintained for the country.

4. There are various market types in the tourism industry such as food, accommodation, transportation, entertainment. However, generally examined, although it is seen that the firms and businesses operating in the industry are performing their activities not in full market conditions, there are different market types with oligopolistic and monopolistic competition market conditions. For example, while the conditions of an oligopolistic market are valid for four or five stars hotels in the hospitality industry, monopolistic competition conditions are valid for those hotels with less stars. The reason for this is that the tourism product has a heterogeneous feature rather than a homogeneous one.

5. Tourism is an industry that is quickly affected by terrorism and war events and natural dissasters as well as some other events such as political or economic instabilities and governmental crises. Whereas the production, consumption or exporting of the other goods and services continue upon such events, a tourism product is highly affected by this. From this perspective, tourism appears as an industry which becomes riskier than the others. Especially, it is known that several terrorism incidents, epidemic diseases and natural disasters that have occurred recently have had adverse effects both on the regional tourism and international tourism movements; just as it is in the attack to the Twin Towers in New York, the SARS disease broke out in China, tsunami disaster in Thailand, the earthquakes in Turkey and the recently occurred bird flu in 2005-2006.

6. Consumption with tourism purposes is a necessity that is not obligatory, as explained before. With this type of consumption, the tourism product is involved in the group of necessities that are not obligatory. Considering Maslow's Needs Hierarchy, it is seen that in the bottom level there are physical necessities (food, shelter and security) that appear to be mandatory for human life and the pyschological necessities (resting, entertainment, relaxation and self-fullfillment) existing in the moral side of human life take place in upper levels. According to this theory, human needs are associated with each other. It is theoretically an incorrect approach to expect that an upper level need is met before a lower one is satisfied. Therefore, it is not right for the people who cannot satisfy their lower level needs to anticipate that their tourism needs in upper level are met.

7. The technological change worldwide and the differentiation in consumption and preferences of consumers have required the public and private sector representatives to act faster and be dynamic in this industry requiring high competitive power. Otherwise, the chance to compete with other international firms and businesses that act quickly in today's global environment may disappear. Briefly, the tourism industry has a dynamic quality. So, tourism is among the industries that are affected by the technological developments quickly. The countries of firms that aim to be one step further in the market by meeting the changing consumer expectations are searching for the ways to market new products. As well as covering the construction of more comfortable accommodation enterprises, these new products may cover faster flights that may have more passengers, completing the booking transactions in a shorter time thanks to technology or establishing the new aquaparks, recreation or entertainment centers.

8. The supply and demand of a product in other industries is affected by some economic factors such as the price of that good, the price of other complementary or supplementary goods, the level of technology and consumer preferences as well as the income. Nevertheless, in supply and demand of tourism general customs and traditions and elements such as social, psychological, political and legal factors and fashion have an important role. Therefore, consumption with tourism purposes differs mainly from all other consumption types in the economy. For example, although the demand for a country is very high, the decisions taken by the country's government regarding the visa restrictions may hinder more tourists to visit that country, such as the decisions taken by countries like the USA and the UK regarding the restriction of the entrance of foreign tourists. Or while a good is fashionable at a certain period, it may face another good in other periods. For instance, whereas in 1970s "apart hotel" type enterprises were more attractive in Mediterranean countries, today there is mostly a demand towards the "all inclusive" type of tourism products.

9. As seen in several scientific studies with a content of tourism, it is an important obstacle for identifying tourism's exact place in the economy that an exact and complete measurement of tourism spending cannot be achieved, that the statistical data are mostly based on the primary sources attained by the surveys carried out among the tourists because of the inadequacy of the secondary data regarding tourism and that tourism consists of sections of industries. This situation makes it difficult to attain healthy and adequate data regarding tourism as well as to measure or examine all the effects of tourism in economic terms. Indeed, scholars are developing some new measurement methods about the spending related with tourism every other day and the discussions on the issue continue in different dimensions.

It is possible to deal with the tourism activities as the sales of a tourism product developed in economic terms to the consumer, that is the tourist, and the benefit attained as a result of this sales. Hence, the goods and services product in the industry generates the supply side of the event while their being bought by the tourists makes up the demand side of it. Thereby, like other goods and services produced in the economy, from its production to consumption, a tourism product economically undergoes stages, e.g., production, consumption, marketing, distribution, sales. Nevertheless, the tourism industry differs from other industries in the economy due to its structure. The production of a product in tourism is much more complex than the goods and services production in other industries because tourism has a relationship with many industries such as transportation, communication, banking, accommodation, eating and drinking, entertainment, health [7]. The tourism industry serves the product as a whole to the consumer by buying goods and services from other industries in the economy. Therefore, each stage of the production and consumption of the tourism product is a separate economic activity and this chain of activities may not be regarded as separate from the economy. However, according to the economical system of each country and the system they choose, the economic effects of tourism may change from country to country.

II. TOURISM AS AN ECONOMICAL ACTIVITY

Tourism is associated with economics and has got a structure that cannot be thought as separate than the economics because tourism economics is a labor-intensive field that tries to meet the human needs of tourism with the tourism sources. Additionally, it is possible to mention the following in terms of the relation of tourism and economics: Tourism is a socio-economic event beginning with an economic decision on how to use spare time and savings and that has economic aspects such as investment, consumption, employment, exporting and public income [8]. Tourism also has a structure of serving passengers and meeting their needs, constituting any kind of spending that may arise during a tourist's journey. Besides, it has a quality of an umbrella that harbors hundreds of subsidiary companies, some of which are big and others are small, e.g., airline companies, sea lines, train and car rental companies, tour operators, and people serving tourists, motels, restaurants and convention centers [9]. From this, it is easy to explain the close relationship of tourism and economics by means of the properties listed below [10]:

1. Tourism is an economic activity affecting several people directly or indirectly as a producer or a consumer. Consequently, today many people travel for various reasons and contribute to the national economy in some extent by their spending which is a source of income for the countries, regions, firms, and enterprises, and as a result the local people that satisfy tourists' needs. As the necessities with regards to tourism increase, the production of goods and services in the subject matter country will increase leading to an increase in investment, production, employment and income in that region. From this aspect, tourism may be considered as a stimulating power for the economic development, growth and progress.

2. Tourism is not only an event of travelling or being accommodated. Its main purpose is the production and marketing of goods and services that the consumer (tourist) demands and satisfying the consumer needs. In this respect, in order to make the necessary infrastructure and superstructure investments and create a tourism supply, it is mandatory to have a certain amount of capital in economic terms. Thereby, by means of tourism it is maintained that a development in the physical structure in the country (physical investments such as marina, airport, roads, water, electricity, etc.). However, all these facilities and institutions are not just for tourists but they also have an important role in satisfying the local people's needs.

3. In the tourism industry, the desired tourism product is provided by bringing together nature, capital and labor by the economic units (accommodation and travel firms, tour operators and guides, etc.) forming the industry. These units are also among the main factors identifying the ability of a region or a country to compete internationally in tourism. Therefore, while a country is attractive naturally or culturally, another region may be important for the quality of its facilities and the varieties of its cuisine. Today, whereas Antalya, a Turkish tourist resort, has a good reputation for the quality of its facilities and natural frame, some capital cities such as Athens, Budapest and Prague can supply more different products with their historical, cultural and cuisine structure. Similarly, while London is rich in terms of

cultural tourism, Paris may have a different feature with the nightlife, entertainment oppportunities and shopping centers.

4. The foreign exchange left by the tourists as a result of the spending they made at the destinations visited contributes to the country's balance of payments in economic terms. As pointed out earlier, the calculations regarding tourism is made under a subtitle which is in the current accounts in the balance of payments called the balance of invisible operations. In this respect, tourism has a property of rectifying and balancing the balance of payments and the foreign trade deficit. For example, the spending tha tourists coming to Turkey from any European country make has a direct positive contribution to the country's balance of payments.

5. Since there is a firm communication and collabaration among all subsidiary and support units operating with regards to tourism, in other words, among units of transportation, accommodation, entertainment and travel, tourism provides new job opportunties directly and indirectly in the economy. For this reason, other industries supplying goods and services to tourism develop and an economic income is attained. Considering the regions where the tourism industry develops, it will be seen that the other subsidiary enterprises are operating in order to support tourism. These are both public and private enterprises. The importance of enterprises such as the banking services that are getting more and more in the industry, activities such as communication, food, fashion, souvenirs, furniture, transportation is increasing day by day.

6. Since tourism is a sub-branch of the services industry, the production of goods and services is not one hundred percent appropriate for automation and mechanization. The demand for labor is much more than it is in other industries. In order to meet the increased demand with the development in tourism, mostly labor will be needed and this will have a positive contribution to the development in employment and a decrease in unemployment in the region. In short, tourism causes new branches of activity and work. Hence, it becomes possible that more people work and are employed. With an increase in the demand towards a region, the supply of the additional staff for the new hotels, restaurants, bars, travel agencies and transportation companies planned to be established will be obtained from the mentioned tourism region or nearby regions or the other places in the country.

7. Tourism is used as the most important tool in attracting foreign capital investment to the country. By attracting the foreign capital to the country, the development of employment is supported, and adopting the international management techniques are enabled. Thereby, the transfer of new technologies will be the case. It is known that the reflection of the foreign capital investments and management approach started in the 1950s, made a peak in the 1980s to the present time, and has an important role in Turkey's acceleration in international tourism activities and becoming a competitive country. Thus, the service quality increases and the new generation works with a more professional understanding.

8. Briefly in many countries with tourism potential, the effectiveness of the industry on the national economies is increasing rapidly everyday. The properties that indicate the economic importance and value of tourism are that it affects foreign exchange supply positively due to its exporting rather than importing effect, the return of investments is high and profitable causing new employment areas, it creates a new

financial source in terms of the country's payments of internal and external debts, provides a value added and has an important role in the interregional balanced development. It is seen that many local public and private institutions have started to make an effort on bringing their region into foreground in domestic and international tourism. The brochures published locally, designed web sites, attending national and international tourism fairs are among the marketing and advertising methods that are utilized.

Today, tourism, the economic properties of which are mentioned above, is one of the three main services industry of the world that make a mark on the 21st century following telecommunication (communication) and information technology. The breaking out of tourism and its development differs by societies, regions and time. Tourism has different qualities according to each society's structure, lifestyle and economic level. The world population increases rapidly. Moreover, the income per capita increases although not with the same rate in every country. Parallel to these developments, the number of tourists and tourism movements is also increasing every day, and it is observed that the national economies have progressed to compete with each other in order to increase their share in the development of the national economy.

A. Economic Effects of Tourism

Considering the world in general, international tourism seems to be the widest account of the world foreign trade. It is the most important source of exports, the industry providing foreign exchange the most and the stimulator of development for most of the countries [11]. Neverthless, it is a known fact that the economic effect of tourism in developed and developing countries differs. That is, considering the international tourism movements, the regions that mostly attract and send tourists are the developed countries. It can be said that tourism is an industry peculiar to the developed countries [12]. Taking into account the ten countries that mostly attract tourists and gain tourism income throughout the world, most of them are the developed western countries. With a comparison among the countries regarding the effects of tourism on an economy, it is a known fact that the economies of developed countries have higher values than those of less developed or developing ones. The foreign exchange income that developed countries gain from tourism is much more than the developing countries get. However, for many developing countries, a certain amount of tourism income gained from tourism is very important in terms of those countries' economic development, growth and progress. From this aspect, in a less developed agricultural country with an economy that is not widespread, the economic benefits and effects of tourism may be more than it is in the economies of developed countries.

For example, while in a developing country each dollar spent for travelling by the tourists is mostly used for creating employment, in a developed country each dollar spent for travelling is used for things such as a hotel construction, renewing investments or telephone calls. The income obtained from tourism in a developing country is used for solving the country's more urgent and important macro economic problems. In other words, while the income attained from tourism in a less developed or developing country is transferred to industries other than tourism, in developed ones it is mostly used within the industry. For reasons such as creating new employment areas, attaining the foreign exchange income that is needed and providing the interregional economic balance, the developing countries have started to give importance to the development of the tourism industry especially after the 1980s. It is possible to understand this importance from the sources transferred to the tourism industry, administrative, financial and legal convenience and opening the regions that have an important tourism potential to the foreign visitors.

An increase in the number of visitors coming to the region increases not only the need for new accommodations, eating, drinking and recreation facilities but some other infrastructure requirements as well. For all these activities, the private sector, the local administration and the government should target new investments toward the studies and policies to be defined. From this aspect, the economic effect of tourism does not only appear within the industry. Since tourism is in close relation with other industries, the investment, employment and income multipliers are dramatically effective. An investmet made in the tourism industry may bring forth an economical activity for other industries and contribute to the creation of new employment areas, while a unit of income has a positive effect on the investments both in tourism and other industries. For example, an investment in tourism may reflect on the construction industry and the tourist inflow to the region freshens the traffic in airline, highway and sea transportation. The positive or negative developments that may occur due to the relationship of tourism with other industries reflect on other related industries in a short time and at changing degrees. Moreover, a tourist accommodates in a hotel, eats at a restaurant, does shopping, utilizes transportation vehicles, communicates with the local people and visits various destinations. Therefore, it should be kept in mind that each individual in direct or indirect relation with the regional tourism industry has a role in the supply of a qualified service [13], and as a result in generating tourist dis/satisfaction or in the tourists' tendency to visit the same destination or not because of the domino effect [14].

As recalled from the previous chapter, the major types of tourism were explained briefly. From this, tourism is a dynamic change from which most large or small organizations, activities, countries, regions and people affect. It is known that different types and movements of tourism brings forth very different kinds of economic effects in different regions. For instance, while vacation/pleasure or coastal tourism affects the economic structure of the seaside regions, business or congress tourism will be related with the economic structure of the metropolitan cities. In this content, the research studies have shown that the economic effects of tourism are examined in three separate ways in terms of the characteristic of the region. The subject matter regions may be listed briefly as [15]:

- Regions less developed in tourism
- Regions developing in tourism
- Regions developed in tourism

While tourism's effect on the economic structure of a less-developed region in tourism and a region that is newly opened is very limited, it has a moderate contribution on the developing region. Yet, the effect of tourism for developed regions may be quite higher. In this scope, considering their current performance, the countries such as Iran, Pakistan or Syria can be considered as regions in the category of less developed regions in tourism, the countries such as China, Russia, Brasil and India can be considered as regions developing in tourism, and finally, countries such as Spain, the USA, France, and the UK as regions developed in tourism.

B. Measuring Economic Effects of Tourism

It is seen that tourism has various micro and macro effects on the economies of the countries. As indicated in the previous chapter, according to the quality of tourism, there are some uncertainties in measuring these effects completely. Since tourism has a share from many industries and it affects some of them directly and some others indirectly, it makes it difficult for this measurement to be made exactly and correctly. Nevertheless, the economic effects of international tourism are attempted to be measured in almost every country by using various methods. In these measurements mainly some certain indicators are taken into consideration. For example, the number of foreign tourists coming to the country, the tourism income attained, tourism spendings, the number/capacity of beds and rooms, the employment situation in the industry, and investment and incentive, and so on.

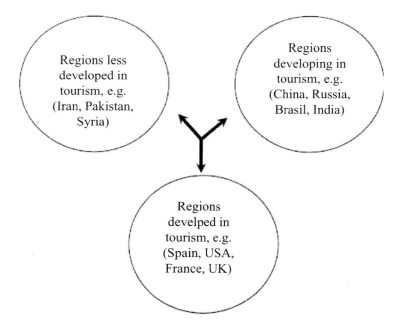

Figure 3.1. Typology of tourism regions.

According to the structure of the tourism industry, it has a combining feature for other industries. A tourist coming to a region not only spends money on accommodations and eating and drinking, but also on shopping, transportation, entertainment, museums, art galleries, sports centers, historical works. As a result, all of this spending constitutes the total tourism income created in the economy. The effect of tourism income in the economy appears in three ways [16]:

- The income attained as a result of the tourists consumption of tourism products at places such as hotels, restaurants, shopping centers operating in the destination and spending directly (primary/direct income).
- The income occuring as a result of trading operations depending on the first stage and attained from spending made for operations among businesses (secondary/indirect income).
- Income attained as a result of the spending of the income gained from tourism by other people employed or profitting from tourism (stimulated/encouraged income).

As pointed out earlier, whereas there is no reliable method to exactly measure or assess the contributuion of these three effects on the countries local economies, it is relatively easier to identify the direct income acquired from tourism. Besides this, several methods are used in order for the secondary and stimulated effects, which are made by the spendings of tourists, to be calculated. Accordingly, the economic effects of tourism occuring as a result of a tourist's visit to a country and spending money are measured in two stages. The first is the method used for the measurement of the primary effects (direct effects); the second is the method used for measuring the secondary and stimulating effects that occurred as a result of tourism activities in the economy as a whole [17] (indirect effects). It is neccesary to emphasize the point that according to the gathering style or form [18] of the data used for the purpose of measuring the economic effects of tourism, the level of truth of the economic effects of tourism may differ. Indeed, since findings attained by judgemental ways result in more correct amd accurate results than the survey method, they are preferred more.

Box 3.1. Effect of Tourism Income within the Economy

Let's suppose a German tourist visiting Barbados spends €1,500 in total after a two-week holiday. Direct or primary income account would occur for the enterprises that take the first share from this amount. However, since these enterprises would have to buy goods or services from other enterprises in order to serve their customers, in return, they would make a certain amount of spending from the income they gained. Assume that the 50% of this income was used for the income – expenditure transfer among the enterprises. While this amount of €750 is the expenditure for the paying enterprises, for the ones accepting the payment it makes up the income. This amount which is still within the economy is called secondary or indirect income, such as the hotels paying the butchers for the meat they bought. Considering that the 20% of the primary income is spent for personnel expenditure, the personnel would have to spend some amount of this for the compulsory expenditures while perhaps spending another part for tourism purposes or else and saving the rest. Here, any kind of spending made by personnel makes up the stimulated or encouraged income account, e.g., paying the rent, spending on food or clothing or paying the education expenses of the children. The circulation of the primary tourism income within the economy will go on like this.

III. METHODS USED IN MEASURING ECONOMIC EFFECTS OF TOURISM

Doubtless to say, the first and the most important economic effect of tourism is tourist spending. It is known that there are some statistical methods using primary and secondary data in the identification of tourist spending. Among these analysis methods the leading one is the method of finding out how much tourists spend on which goods and services by using a survey on a tourist group representing the main population and is chosen by the sampling method. The further step is taking the averages of the obtained figures for each group of goods and services and multiplying them with the number of tourists in the main population.

Another method to measure the economic effects of tourism is the identification of how much change the money incurred in the country's economy by means of tourism causes in the national income, employment, the amount of sales and tax revenues. Questionnaire surveys, analysis of the secondary data obtained by public statistics, economic models, multiplier method, input-output analysis and the satellite estimation method are some of the methods used in measuring the economic effects of tourism [19]. Below, what these methods are and what they refer to are explained briefly [20].

A. Methods Used in Measuring Primary Effects of Tourism

There are various methods used in measuring the primary effects of tourism. As indicated previously, while the primary effects of tourism are measured based on the tourists' spending, there is still no method that everyone compromises on and regards as the best. Therefore, there are nine different methods in measuring the primary effects, which are examined respectively.

1. *Method of Observing Tourist Spending:* According to this method, tourism spending is observed in two ways. First, tourists are monitored in the places they visit in a country and recording the spending they make, it is possible to obtain the exact data so as to calculate the contribution of tourism on the national economy. This method is based on the research made at the place or on tourists upon their travel. Although theoretically, the method seems simple, its application is quite difficult in terms of time, personnel and cost. Second, by gathering all sales figures of the firms operating in the tourism industry, the total income acquired from tourism may be calculated. The data regarding the sales figures may be attained from the records of Ministry of Trade and tax offices rather than observing the sales directly in the sales agencies. Even though the data attained from the areas related with travel and transportation in the tourism industry are identified correctly, it is a general assumption that the data attained from areas such as entertainment, recreation, and accommodation may be sometimes incorrect and not accurate.

2. *Regional Research Method:* This is the most common method for introducing the tourist spending. The basis of the method is the surveys conducted regarding the spending of tourists entering or leaving the country/region from the entrance doors. The average of the primary data obtained from the surveys is taken and multiplied by the main population and the total spending of tourists as well as the industry where these spendings used are estimated. The study findings show that this method has better results especially in the areas related with international tourism but is insufficient in the context of domestic tourism. Considering the survey method in general, the results obtained from the surveys are generalized by an estimation. Therefore, in order for these results obtained from the primary data and hence the survey method to have reliable results, it is necessary that the sample size should be

calculated well and an adequate number of surveys for a correct estimation should be collected. Otherwise, the estimations made as a result of the collected data will not reflect the facts, and the risk may appear where the calculations made out of these regarding the secondary effects of tourism are incorrect.

3. *Method of Consumer Research:* It is possible to predict tourist spending by conducting research among people residing in the tourist generating countries in order to collect a general figure on consumer spending and then subtract tourist spending from this figure. By collecting the spending with regards to business and congress tourism, which are obtained from the tax accounts, reliable data may be obtained regarding the total tourism spending in the tourist generating countries.

4. *Method of Bank Accounts:* It is seen that this method is applied particularly in the countries whose currency is not convertible and that is closed to the outside or restricted. The essence of the system is made up of the calculation of the foreign exchange, which the foreign tourists change at exchange offices and banks in the host country, by the central bank of that country. In other words, the records appearing as a result of tourists' changing foreign exchange are sent to the central bank and there, the total tourist spending is calculated. As explained above, the bank accounts method can be regarded as a successful method in the countries where the foreign exchange is strictly controlled. However, it will not be a correct approach to say that this method gives valid and reliable results in a country whose money is convertible and where tourists may change the foreign exchange wherever they want. Because the foreign exchange purchasing and selling operations had been under the control of the Central Bank, Turkey used to apply this method prior to the 1980s.

5. *Excess Income Method:* In this method, the spending made regionally by tourists visiting the country in terms of cities are calculated as follows: Subtracting the spendings made by the people living in a city from the total shopping income throughout that city as a result of the secondary data, the amount of spending that tourists made is reached. In this approach, the cities or regions where the application is made in general it is needed to record the secondary data very well. Indeed, since identifying the spending of both tourists and local citizens exactly is difficult, this method is regarded as a way which is not helpful. However, the results attained from the method may assist in analyzing the regional economic effects of tourism.

6. *Seasonal Difference Method*: Another method used in calculating tourist spending is the seasonal difference method. In this method, the month when the tourism companies operating in a tourism destination gain the lowest income within the year. The amount of income in that month is accepted to be obtained from that region's local people. The amount of income attained during the following months is subtracted from the income of the lowest month and this amount is accepted as the tourist spending. Similarly, it seems difficult to have an exact and complete measurement of tourist spending in this method because even in a month when the lowest income is assumed to be attained, there may be tourists in the country and spending may be done. Therefore, accepting that the entire income is obtained from the local people will cause unreliable calculations. Moreover, the local people may also spend in the subject matter enterprises and this is added to the calculations as the income obtained from the tourist.

7. *Satellite Estimation Method*: This is a kind of calculation which is made by considering the event in terms of both supply and demand in order to identify the contribution that tourism has on the economy. In other words, the economic effects of tourism are aimed to be measured by the supply and demand approach in this method, and it is foreseen that the calculations are compatible with the national accounts [21]. The satellite estimation method was created in the mid 1990s. Even though it is a new method, many countries are encouraged to use it by various instutions like World Travel and Tourism Council. Reshaping the definitions and concepts in tourism, this method targets to have a standard worldwide. Thereby, it becomes possible to calculate the economic effects of tourism on the countries' national economies more correctly and accurately.

8. *Spending Rate Method:* This method in which the primary and secondary data are used together consits of four stages. At first, the income of the accommodation firms in tourism destinations are determined by means of the secondary data. Second, through the surveys (the primary data) the level of tourist spending is determined. Third, the accommodation income is rated to the tourist spending. Last, in the fourth stage, the obtained rate is multiplied by the accommodation expenses and the total tourist spending is maintained. It may not be possible that a reliable result is reached in practice since it requires a process quite comprehensive and time consuming. The fact that the tourists as well as the companies which may become reluctant to give healthy information may be considered as another adverse factor.

9. *Cost Factor Method:* This method is first based on determining the number of overnight stays in a tourism destination and then on finding out the average daily food cost. The amount of spending is calculated by multiplying the overnight stay by the cost of food. This method is repeated for other groups of tourists and the total spending of that region is calculated in this way. The required data is attained by the household questionnaire and the surveys made in the firms. Moreover, the secondary data can be used under the scope of this method. The weakest side of the method is that the surveys cannot be made in a reliable manner or there may not be secondary data as it is in some developing or less developed countries.

B. Methods Used in Measuring Secondary Effects of Tourism

There are two different methods in measuring the secondary and induced effects that occur in a whole economy as a result of tourism activities. In this part, the indirect effects that occurred as a result of tourist spending are to be measured by the multiplier and the input–output analysis.

1. *Multiplier Method:* Prior to examining the method, it is better to explain the multiplier concept. Multiplier is defined as *"the autonomous increase (independent of income) in total demand causing an increase in national income at a rate of its own size"* [22].

$$\Delta Y = k \times \Delta A_0 \tag{1}$$

$s = 1 - c$

In the above equation, k is the multiplier coefficient. ΔY represents the change in the national income while ΔA_0 shows the autonomous change in the total demand. The autonomous spending components, which are known as A and independent of the income, cover autonomous consumption spendings (C_0), investment spendings (I_0), goods and services spendings of the public sector (G_0) and net exports $(X_0 - M_0)$. The bigger these factors and the smaller the autonomous taxes (T_0) and the autonomous import (M_0) are, the bigger the multiplier coefficient and, hence, the national income will be. Some sources show the multiplier coefficient with the symbol of "k" or "α" [23]. As indicated below, the multiplier has several types and outlooks in respect of the fact that an economy is open or closed to foreign trade. Here, the meaning of multiplier for the closed or open economy is briefly explained. So, it is possible to show a multiplier for a closed economy, as follows [24]:

$$k = \alpha = \frac{1}{1 - c} = \frac{1}{s} \tag{2}$$

Since $(s = 1-c; s = \frac{\Delta S}{\Delta Y})$ in the equation (2) is equal to the marginal savings tendency, it is not wrong to show the multiplier as $k = \frac{1}{s}$. The "c" multiplier here expresses the marginal consumption curve. The marginal consumption tendency showing how much of the change in income goes for consumption can be defined as the ratio of change in consumption spending to the change in the national income $(c = \frac{\Delta C}{\Delta Y})$. Hence, to what extent an increase in the autonomous spendings (ΔA_0) affect the income level (Y) can be said by looking at the marginal consumption tendency. Putting k symbol of equation 2 into equation 1:

$$\Delta Y_0 = \frac{1}{1 - c} \Delta A_0 \tag{3}$$

is obtained. The following should be understood from equation 3: As marginal consumption tendency gets bigger, the multiplier grows as well and as stated previously, this leads to a more increase in the national income. In this respect, a bigger marginal consumption tendency has a bigger multiplier effect.

In an open economy model, as the foreign trade is considered, it is difficult that the import spending is formulized. Since a leakage in the economy occurs in such a situation, the multiplier will reduce the income constituting effect of tourism. For such an economy, the multiplier coefficient is formulized as below:

$$k = \frac{1}{1 - c + m} \tag{4}$$

The "m" symbol shown in equation 4 represents the marginal import tendency. The marginal import tendency can be defined as the ratio of the change in import spendings to the change in national income ($m = \dfrac{\Delta M}{\Delta Y}$).

Figure 3.2 shows that an increase in the autonomous spending raises the balance income level from Y_1 to Y_2 as a result of the multiplier effect. Thereby, the national income increases by $Y_2 - Y_1 = \Delta Y$. To summarize [25]:

- An increase in the autonomous spending leads to an increase in the balance income level.
- An eventual increase in income is "k" as much as an increase in the autonomous spending/
- The larger the marginal consumption tendency, the higher the value of the multiplier effect occuring due to the relationship between consumption and income.

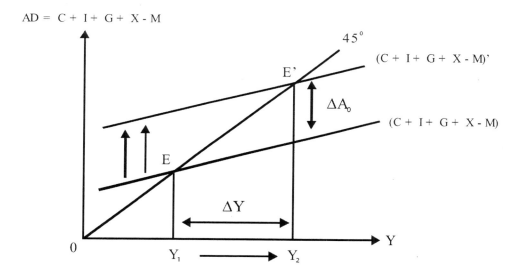

Figure 3.2. Multiplier mechanism.

In assessing the issue in terms of the tourism industry, within the industry, each unit of income obtained from tourists makes up an income for other people and firms by being transferred in the economy. The multiplier that is attained from the ratio of the amount that is formed by the obtained income to the amount at the beginning shows the multiplier effect of tourism income. In other words, that one unit income attained from the tourism industry affects the regional or national economy more than the tourism income at first by being transferred to another person is defined as the multiplier effect in tourism [26].

The multiplier effect is used to measure the effectiveness of tourism income or how much excessive income has been attained in the economy. Tourism makes up a comprehensive income effect due to the multiplier in the economy, because tourists make some consumption spendings to satisfy their needs such as accommodation, eating and drinking, transportation, shopping and recreation. These consumption spending cycle within the economy in some ways, indirectly enabling new income to occur. The firms that are profiting from tourists purchase goods and services from other firms in different industries in order to sustain the service they provide. Hence, some part of the income acquired from tourists turns back into the economy by being spent again. The final tourism income occurring as a result of the operation of this mechanism is relatively higher than the investment spending as well as the consumption spending for tourism purposes made by international and domestic tourists [27]

Box 3.2. Movement of Tourism Income in a Local Economy

Each transfer of income attained in the economy means new tax, savings and spending. For example, a certain proportion of the income (US $2,000) that a hotel has gained is paid to the central or local government as tax, another portion is the payment made to the retailer or wholesaler for purchasing raw material and wages to the personnel working in those companies, and the rest is the amount saved. Retailers and wholesalers will pay tax again, save and will spend the remaining amount again. Meanwhile, spending the money on compulsory needs, the personnel will use the rest for tax and want to save the remaining amount. As seen, each transfer of money means a new tendency of tax, saving and spending. In the economy, some part of income gets out of circulation while the remaining amount will continue operating. In this respect, the amount continuing its operation makes up the part that contributes to the increase in the multiplier coefficient. To illustrate, the multiplier coefficient that Jamaica attains from tourism is lower compared to other countries in the region. Each tourism income of 1 US$ creates a multiplier effect of 1.10 in Jamaica, 1.27 in Barbados and 1.18 in Dominican Republic. Nonetheless, in Europe, this coefficient is seen to be 1.96 in England and 1.27 in Ireland (Source: Eturbonews, February 27th, 2006).

The entire income obtained due to the multiplier effect remains within the economy and a certain amount leaves the economy as a leakage. That is, some leakages from income occur by means of tax, saving and import during the spending made again. The bigger these leakages are, the smaller the income effect occurring by the multiplier effect is. As leakages decrease, the income effect in the economy decreases as well. Here, the longer the income obtained from tourists, namely money stays within the economy, and is transfered (if the ciculation rate of money is high), the higher the multiplier effect.

As explained in Box 3.2, the units in an economy such as travel agencies, hotels, restaurants distribute the income they attain by means of tourists to the other parts of the societies as interest, rent and price of the product purchased. This income taken by the society is spent similarly, causing the other economical units to profit, as well. The total value of the income created in this way is bigger than the first spending that tourists make. Tourists' spending may affect a destination at different levels depending on the structure of the industry or within the industries.

As the multiplier effect changes from region to region, or city to city, if the multiplier value of a country is examined solely, there may be deceiving results. The multiplier values belonging to small residential areas as towns or districts always intend to be smaller than the bigger residential areas. Besides this, as the leakage rates are lower in the economy of developed countries, the multiplier effect in these countries is more than that in developing or less developed countries. If the tendency for importing is very high in a tourism destination and most of the income attained from tourists leaves the country as import spending for various reasons, the multiplier effect will remain lower for that country. For example, 58% of the food and 82% of the meat consumed by tourists were imported from abroad [28] in St. Lucia, Carribean. In such a destination, which imports the majority of raw materials, it is impossible to mention a multipler effect. Therefore, one may highlight that the amount of leakage occurring especially in the countries of "island" nature are high in this respect.

Box 3.3. Other Types of Multiplier Effects

There are various types of multipliers in the literature of economics. In order to narrow down the subject it will be enough to know only their names. There are four kinds of multipliers named as public spending multiplier, tax multiplier, transfer spending multiplier and balanced budget multiplier [29]. Moreover, it is possible to classify the multiplier in two groups a sime and suer multiplier. Besides, the multipliers utilized with regards to tourism are divided into four as employment, operations, output and capital or assets multiplier [30].

Below the way to calculate the multiplier effect is explained briefly with an example. In a closed economy where marginal consumption tendency is 0.75, when an increase in the economy caused by a tourist's spending of 2,000 € is calculated, the result is:

$$k = \frac{1}{1 - 0.75} = 4$$

In other words; the value of multiplier effect is 4.

When the value of multiplier effect is put into the $\Delta Y_0 = \frac{1}{1-c} \Delta A_0$ formula;

Y= 4 x 2,000 = 8,000 € is attained.

Tourism spending of 2,000 € that is made by a tourist visiting a destination leads to an increase of 6,000 € in the economy and causes an increase of 8,000 € to be attained in total. That is to say, 1,000 € of a 2,000 € spending at a destination is spent in the economy again and 500 € of it turns back to the circulation and this process goes on like this. The total value of the income occurred in time as a result of multiplier effect gives the geometrical sum of this progress.

2,000 €+ 1,000 € + 500 € + 250 € ... = 8,000 €

In order for the subject to be understood better, another example is as follows: Considering that the goods and services spending of public sector (G$_0$), which is one of the components of autonomous spending, increases by 3,000 € in a tourism season for a closed economy where the marginal consumption tendency is 0.60 and calculating the amount of income increase caused by this, the result is:

$$k = \frac{1}{1 - 0.60} = 2.5$$

The value of the multiplier effect is $\Delta Y_0 = \frac{1}{1-c} \Delta A_0$

And it is put into the following formula as: Y= 2.5 x 3,000 = 7,500 €

For the same example above, if the marginal consumption tendency is 0.20, then the income increase in the economy will be calculated and the national income attained according to both multiplier values will be compared and interpreted as:

$$k = \frac{1}{1 - 0.20} = 1.25$$

This value is put into the following formula as:

$$\Delta Y_0 = \frac{1}{1-c} \Delta A_0$$

and the result is: Y= 1.25 x 3,000 = 3,750 €

Accordingly, the bigger the marginal consumption tendency and hence the multiplier value is, the bigger the income effect that it causes in national income gets. In this situation, as it is in the previous example, when the multiplier valule is 2.5 according to 1.25, there appears twice as much income effect in the economy. As the marginal consumption tendency (c) grows, multiplier (k) grows as well and as it has been explained before, this causes the national income (Y) to increase. It is important to indicate this issue at this point. First of all, the value of multiplier may change depending on the spending habits and consumption tendencies of the tourists. For example, considering that tourists travelling with a 7 and 14 day package tour and spend approximately 300 – 500 €, whereas a businessman visiting the same country for congress tourism and staying between 2 and 4 days spends approximately 1,000 – 1,500 €, it is a fact that the kind of multiplier will change based on the types of tourism supply and demand. Nevertheless, the multiplier value is said to be based on the various factors such as economic volume and magnitude of the destination, the structure of the industry, whether the demand for goods and services are met by regional/national facilities, the amount of products by tourists visiting the destination and the tendency of the local people for shopping outside the region [31].

2. Input – Output Models: Another method used in calculating the secondary effects of tourism is the input-output analysis. Input-output analysis is a model that is more comprehensive and improved than the multiplier method. While the multiplier coefficient is used in order to measure how much excess income is created in the economy or the effectiveness of tourism income, in input-output analysis the supply and demand interactions among different industry are considered and the reflections of these effects on the economy are put forward. This analysis method developed by Leontief seeks an answer to the question that in what amount should *n* industries in an economy produce so as to meet the total demand for the product [32]. The output of any industry may be needed in other industries or even within that industry. Thereby, the most important feature of the input-output models is that the independency among industries is considered and how much input each industry takes from other industries or conversely, how much output it gives to them is examined [33].

Input-output analysis is made by drawing up input-output tables among industries in an economy at a certain time. The good and services flows (taking input – giving output) within the industries of the economy are demonstrated with these tables. Moreover, the exchange rates among industries are examined in these tables called *input-output tables* by considering imports and exports as well as the three basic industries of economy: agriculture, industry and services. Indeed, the supply and demand interactions of the tourism industry with other industries are shown in these tables. In this method, which is used to measure the secondary effects of tourism, it is seen that tourism is assessed as two separate industries, travel and tourism [34]. For example in Turkey, the State Institute of Statistics used five main industries (e.g. agriculture, mining, manufacturing, construction and services) and 64 subsectors belonging to these in the input-output tables that it produced for the year 1990 [35]. While calculating the input-output tables of 1996, it divided the economy into 205 industries and grouped them in 97 industries [36].

When these tables are examined, although there is no industry named as tourism, it is seen that there are several industries regarding tourism, e.g., hotel business, restaurant and catering business, motorway, airway and sea transportation. The columns in input-output tables show the input requirement needed for a unit production of an industry and the spendings on the value added components, while the lines show which production sectors and final demand elements make what amount of spending for production of each industry [37]. The income and expenses should be equal for each industry. In other words, the sum of lines for each industry must be equal to the sum of columns. The monetary value of the intersectoral operations are used in the subject matter tables. The biggest problem in input-output analysis is to attain the reliable data with convenient time that make up the input-output tables [38]. Since these data are gathered by counting or questionnaires, it may not always be possible to collect the data which would be safe enough.

There are three main components of the input-output model. These are, as indicated in the above paragraph, input-output table (also called operations table), direct needs matrix (direct necessities table) and total direct needs matrix. The direct needs matrix is found by the ratio of the values in the input-output table to the industry production throughout columns. The total direct needs matrix also takes into account the direct and indirect effects in the final demand. These tables have a distinctive importance since they show the money circulation among the industries.

Table 3.1. Input-Output Analysis for an economy with three industries [40]

			Consuming Industries			The Last Demand		Total Output
			A	B	C	Household	Institutions	
Manufacturing Industries		1	X_{1a}	X_{1b}	X_{1c}	C_1	I_1	X_1
	2		X_{2a}	X_{2b}	X_{2c}	C_2	I_2	X_2
	3		X_{3a}	X_{3b}	X_{3c}	C_3	I_3	X_3
Added Value		Workforce	L_1	L_2	L_3	Lc	L_I	L
	Other Added Value		V_1	V_2	V_3	Vc	V_I	V
Total Inputs			X_1	X_2	X_3	C	I	X

In Table 3.1, there is an example of an input-output table designed for an economy of three industries. The figures 1, 2 and 3 represent the manufacturing industries while the letters a, b and c represent the consuming industries. Theoretically, X_{ij} shows the amount of input that the industry j takes from the industry i and hence i represents the lines and j, the columns.

Hence, with "X_{1a}" in Table 3.1, the input amount that the consumer industry "*a*" takes from the manufacturing industry numbered "1" is shown. The added value shown in the table represents the inputs that the tourism industry uses in the economy and the net contribution that the industry has created. In order to produce goods, the tourism industry takes input from other industries and sells this product to the tourist at a certain price at the end of the production process. If the input value gained from other industries is subscribed from the price of the supplied product, the difference gives the value added of tourism [39].

So as to make the input-output model clear in Table 3.2; numerical examples regarding an economy of three industries including manufacturing, construction and tourism are given. Accordingly, it is understood from the mentioned table that the tourism industry constitutes the production value of 15 units (total output) and final demand of six units (business trip, visits of the family and friends) as well as 15/43 of the national income, that is, 34.8% of it. Moreover, the tourism industry supplies two units of input to the manufacturing industry and also, two units of income to the construction industry or gives output. Meanwhile, as an industry, tourism takes five units from the manufacturing industry and four units from construction, creating an added value of one unit for the economy.

In order to see the secondary effects caused by the tourism spending, the direct necessity matrix may be used (Table 3.3). For example, for a 15 unit of output in tourism, the amount of input that should be taken from the manufacturing industry is five units or other industries, the value of input rate. Hence, it is estimated that the input ratio needed for each unit of output is 5/15=0.33 units. When these transactions are made for other industries, the value of the input rates is shown in Table 3.3. Looking at the direct necessity matrix, one can easily see the net impact of 1 US$ increase in tourism spendings.

Indeed, one US$ increase in tourism spending creates 0.33 units of direct spending in the manufacturing industry and 0.26 units in construction. In this respect, an increase in tourist spendings brings out an increase in demand both for manufacturing and construction. With the effect of the increasing spendings in the tourism industry, the production of textile, cotton, plank will increase in order to make more beds and likewise, due to the need for accommodation, there will be more need for cement and iron in construction. Indeed, this new demand that occurred as a result of tourism spending means a stimulated increase in demand. Considering the economy as a whole and the industries as the parts constituting this whole, the economic refreshment or stagnation in any industry may affect other industries and as a result, affect the economy as a whole in a positive or negative way successively.

Table 3.2. An Example of an Input-Output Table

| Industries | Consumption Industries | | | | |
	Manufacturing	Construction	Tourism	Final Demand	Total Output
Manufacturing	10	3	5	2	20
Construction	2	1	4	1	8
Tourism	2	2	5	6	15
Added Value	6	2	1	-	-
Total Input	20	8	15	-	43

Table 3.3. An Example to the Direct Necessity Matrix

| Industries | Consumption Industries | | |
	Manufacturing	Construction	Tourism
Manufacturing	0.50	0.37	0.33
Construction	0.10	0.12	0.26
Tourism	0.10	0.25	0.33

Box 3.4. Input – Output Analysis

Numerical values regarding an economy with the three industries where mining, manufacturing and tourism takes place are given in Table 3.4 and 3.5. accordingly:

What percent of the national income does tourism income constitute solely?

Tourism income constitutes the total output value of 45 units and 45/115= 39.1% of the national income.

As an industry, what amount of input does tourism take from other industries and how many units of added value does it supply to the economy?

The tourism industry supplies five units of added value to the economy by gaining 15 units input from mining and 20 units input from manufacturing industries, which makes a contribution to the economy by five units of added value

According to the direct needs matrix, how many units of increase does one unit increase in tourist spending create on the other two industries?

Looking at the direct needs matrix in Table 3.5, it is seen that one unit increase in tourist spending creates 0.33 units indirect spending in mining and 0.44 units of direct spending in manufacturing. In this respect, this increase in tourist spending brings an increase in demand both for mining and manufacturing.

Table 3.4. An Example of Input-Output Analysis

| Industries | Consumption Industries | | | | |
	Mining	Manufacturing	Tourism	Last Demand	Total Output
Mining	5	5	15	5	30
Manufacturing	10	5	20	5	40
Tourism	2	2	5	36	45
Added Value	13	28	5	-	-
Total Input	30	40	45	-	115

Table 3.5. An Example of Direct Needs Analysis

Industries	Consumption Industries		
	Mining	Manufacturing	Tourism
Mining	0.16	0.12	0.33
Manufacturing	0.33	0.12	0.44
Tourism	0.06	0.05	0.11

In summary, the input-output model is a method that puts forward the demand-supply interactions with other industries, calculates the added value that tourism makes up in an economy, and enables to express its economic relationship with the final consumer with monetary values. However, since tourism is a section of industries and due to the lack of regional data worldwide, the difficulty of this method in practice includes such problems as defining exactly which activities belong to this industry.

SUMMARY

This chapter aims at explaining the importance of the tourism industry as an economic activity in terms of national economies in detail by means of several examples. Taking into consideration from the economic aspect, another issue that is inspected here is some of the main properties of tourism, which is a sub-service industry of economics. There are various methods that are used for calculating the importance of the tourism industry in an economy. These methods, some of which already become traditional and out of practice, are examined by stating their strong and weak points. As well as the methods requiring primary data to calculate the tourism income, a further considerable issue relates to how secondary methods such as multiplier coefficient that is used to put forward the economic benefits that tourism creates in a national economy. Some basic information like the definition of tourism and its properties are under the scope of the following chapter.

DISCUSSION QUESTIONS

1. What can be said regarding the economic importance of tourism? Indicate briefly.
2. What are the industrial properties of tourism? Explain briefly.
3. What are the properties that put forward the close relationship of tourism with economics? Explain briefly.
4. What are the effects that tourism income makes up within a national economy? Explain briefly.
5. Explain by headlines the methods being used to measure the primary and secondary effects of tourism.
6. What is the main difference between the input-output analysis and the multiplier effect analysis? Explain briefly.
7. Give examples from developed, developing and less developed countries in international tourism.

8. Give examples regarding developed, developing and non-developed regions in tourism of your country.
9. Discuss if there is a parallelism between the two by forming a relationship between developed and developing countries in tourism.

REFERENCES

[1] C. Jayawardena, D. Ramajeesingh (2003), "Performance of Tourism Analysis: A Caribbean Perspective," *International Journal of Contemporary Hospitality Management*, 15(3), p. 176; I. Boxill (2004), "Towards an Alternative Tourism for Jamaica", *Journal of Contemporary Hospitality Management*, 16(4/5), p. 269.

[2] WTO (2005), *Tourism Highlights*, Madrid, Spain.

[3] A. Lanza, P. Temple and G. Urga (2003), "The Implications of Tourism Specialisation in the Long Run: An Econometric Analysis for 13 OECD Economies," *Tourism Management*, 24, pp. 315–321.

[4] A. Papatheodorou (1999), "The Demand for International Tourism in the Mediterranean Region," *Journal of Applied Economics*, 31, pp. 619–630.

[5] D. W. Marcouiller, K. K. Kim and S. C. Deller (2004), "Natural Amenities, Tourism and Income Distribution", *Annals of Tourism Research,* 31(4), pp. 1031–1050; K. Göymen (2000), "Tourism and Governance in Turkey," *Annals of Tourism Research,* 27(4), pp. 1025–1048.

[6] Olalı, Timur, pp. 16–17; M. Kozak (2003), *Destination Benchmarking: Concepts, Practices and Operations,* Wallingford: CABI Publishing, pp. 34–36.

[7] V.T.C. Middleton (1994), *Marketing in Travel and Tourism,* Oxford: Butterworth and Heinemann, 2nd edition, p. 4.

[8] Kozak et al., p. 8.

[9] E. D. Lundberg, M.H.Stavenga and M.Krishnamoorthy, *Tourism Economics,* New York: John Wileyand Sons,Inc,1995, p. 4.

[10] P. Callaghan, P. Long and M. Robinson (1994), *Travel and Tourism,* England: Business Education Publishers Ltd., 2nd edition.

[11] C. Lim (1997), "Review of International Tourism Demand Models", *Annals of Tourism Research,* 24(4), pp. 835–849.

[12] Yarcan, pp. 1–31.

[13] J. Jafari (1983), "Anatomy of the Travel Industry", *Cornell Hotel and Restaurant Administration Quarterly*, 24(May), pp. 71–77.

[14] M. Kozak, M. Rimmington (2000), "Tourist Satisfaction with Mallorca, Spain, as an off-season Holiday Destination," *Journal of Travel Research,* 39 (3), pp. 259–268.

[15] D. Foster (1985), *Travel and Tourism Management,* London: The Macmillan Press, p. 42.

[16] B. H. Archer (1976), "Uses and Abuses of Multipliers," *Planing for Tourism Development; Quantitative Approaches,* New York: Praeger, p. 115.

[17] D.C. Frechtling (1994), "Assessing the economic impacts of travel and tourism. Introduction to travel economic impact estimation," in Brent Ritchie, J.R., Goeldner, C.R., 2nd ed. (Eds), *Travel, Tourism, and Hospitality Research. A Handbook for*

Managers and Researchers, John Wiley and Sons, Inc., New York, NY, pp.359-365; D.C. Frechtling (1994), "Assessing the impacts of travel and tourism. Measuring economic benefits," in Brent Ritchie, J.R., Goeldner, C.R. (Eds), *Travel, Tourism, and Hospitality Research*, 2nd ed, John Wiley and Sons, Inc., New York, NY, pp.367-391.

[18] For the detailed information about the ways of data collection and sources in tourism see: D. J. Stynes, "Approaches to Estimating the Economic Impacts of Tourism; Some Examples," http://www.msu.edu/course/prr/840/econimpact/pdf/ecimpvol2.pdf. (Retrieved 07 October 2005).

[19] R. H. Martin (2004), "Impact of Tourism Consumption on GDP. The Role of Imports," http://www.feem.it/Feem/Pub/Publications/WPapers/default.htm, p. 5. (Retrieved 10 June 2005).

[20] Frechtling, "Assessing the impacts of travel and tourism. Measuring economic benefits," pp. 367-391.

[21] For detailed information see: D.C. Fretcling (1999), "The Tourism Satellite Account: Foundations, Progress and Issues", *Tourism Management*, 20, pp. 163–170; A. Blake et al. (2001), "Modelling Tourism and Travel using Tourism Satellite Accounts and Tourism Policy and Forecasting Models," http://www.nottingham.ac.uk/ttri/pdf/ 2001_4.PDF, pp. 1-21. (Retrieved 25 September 2007).

[22] Güran, p. 129.

[23] Yıldırım, Karaman, p. 148.

[24] R.W. McIntosh, C. R. Goeldner, J.R.B. Ritchie (1995), *Tourism Principles, Practices, Philosophies*, New York: John Wiley, 7th edition.

[25] Yıldırım, Karaman, p. 150.

[26] İçöz, Kozak, p. 262.

[27] Bahar, Kozak, p. 13.

[28] For detailed information see: E. D. Lundberg, M. H. Stavenga ve M. Krishnamoorthy (1995), *Tourism Economics*, Canada: John Wiley.

[29] Güran, pp. 133–138.

[30] For detailed information see: A. Mathieson, G. Wall (1982), *Tourism, Economic, Physical and Social Impacts*, Burnt Mill, Harlow, Essex, Longman, p. 65; A. Bull (1995), *The Economics of Travel and Tourism*, Melbourne: Longman, 2nd edition, p. 151.

[31] İçöz, Kozak, pp. 277–280.

[32] For detailed information see: A. C. Chiang (1984), *Fundamental Methods of Mathematical Economics*, McGraw-Hill Publishing Company, 3th edition, pp. 115–124.

[33] Y.Y. Sun (2007), "Adjusting Input–Output Models for Capacity Utilization in Service Industries," *Tourism Management*, 28 (6), pp. 1507-1517.

[34] J. Morrison, R. Powell (1988), "Using Input-Output Methods for Economic Impact Analysis of Tourism," *Proceedings of the Frontiers of Australian Tourism Conference*, BTR, Canberra, pp. 261–268.

[35] DİE (1994), *Türkiye Ekonomisinin Input Output Yapısı 1990*, Ankara: DİE Matbaası.

[36] DİE (2001), *Türkiye Ekonomisinin Input Output Yapısı 1996*, Ankara: DİE Matbaası.

[37] Martin, pp. 13–14.

[38] Lundberg et al., p. 143.

[39] J.E. Fletcher (1989), "Input-Output Analysis and Tourism Impact Studies," *Annals of Tourism Research*, 16 (4), pp. 514-529.

[40] K. Unur (2004), "*Turizmin Ekonomik Etkilerinin Ölçülmesi,*" Dokuz Eylül Üniversitesi S.B.E. Dergisi, 6 (4), p. 133.

Chapter 4

TOURISM SUPPLY AND ITS PROPERTIES

This chapter addresses the concept of tourism supply that constitutes one of the two important components forming the tourism market. In the following parts, the issues of what tourism supply means, its properties peculiar to itself and the elasticity of tourism supply will be discussed. Prior to the explanation of tourism supply, it is essential to inspect the basic concepts such as *market* and *tourism market*. This chapter presents market types, the definition of tourism supply, its properties and the factors affecting tourism supply and its elasticity, and change in the amount of tourism supply.

I. TOURISM MARKET

The organization that is made up by the ones who supply a product so as to sell it and the ones who want to buy that product is called **"*market*"** in economics literature [1]. Shortly, the place where demand and supply meet is defined as the market. However, when market is mentioned, most people think of a physical place at first. Yet, in order to talk about a market, it is not compulsory that sellers and buyers come together and meet. Indeed, market can be a physical place while it can also be a telephone, fax, e-mail or internet atmosphere by which sellers and buyers agreee with each other. Market is an organization enabling buyers and sellers to communicate among themselves. In this respect, there exist different market classifications (see Figure 4.1). Markets can be local, regional, national and international in nature according to their size. Or, according to the status of the product purchased, they can be "*goods and services market*" or "*factor markets*". Finally, as explained in the first chapter, markets can be divided into two groups according to the type of competition as "*perfect competition market*" and "*imperfect competition market*" [2].

Tourism market is the place where demand and supply of the tourism product is met. According to another definition, tourism market is the place where any kind of tourism product is sold and firms and enterprises supplying the tourism product and the tourist or visitors demanding these come together [3]. Under normal circumtances, it is seen that market consists of the three components of other industries in an economy: Supply, demand and price. The tourism industry, besides including these three components, also consists of the other three properties due to its peculiar structure [4]:

1. Tourism market covers a geographical region.
2. Tourism market has continuity owing to its multi-component structure and changes slowly.
3. There exists an internal tourist movement among the geographical regions.

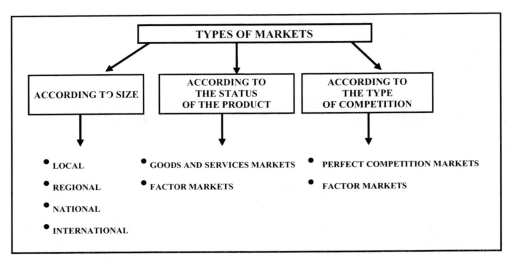

Figure 4.1. Types of Markets.

Following this brief introduction, there can be explanations regarding tourism supply. Here there is a point that should not be avoided. If market and hence tourism market is assimilated to a scissor, tourism supply and demand make up the edges of this scissor. Thereby, the price of the tourism product in the tourism market is formed when the tourism firms that want to sell that good and tourists with a desire of buying it come together. In this context, market or balance price is the price at which the supplied amount and the demanded amount is equal to each other. How the balance price occurs will be inspected in detail following the issues of supply and demand.

A. Tourism Supply

In economics literature, when mentioning supply, it is meant the amount of goods desired to be offered at a certain price in a certain time [5]. The amount that producers present to the market to sell at certain prices at certain times is defined as *supply* [6]. Tourism supply is not so different from the above mentioned definition of supply. That is, *tourism supply* is the entire amount of all tourism assets that a destination (region or country) serves for tourists at a certain price under certain conditions. According to another definition, tourism supply is "all of the material and moral values that a country presents to tourists by means of its firms and institutions" [7]. Finally, another definition states that tourism supply is regarded as "all of the assets, values and facilities that a country, region or an atttraction center have to satisfy the needs of the travelers as well as the flow of goods and services ready for sale at a certain price and are towards travelling and accommodation within a certain time period" [8]. As it can be understood from this definition, when tourism supply is mentioned, all of the activities such as accommodation, eating and drinking, entertainment and relaxation, and transportation, that meets tourists' needs [9].

There are two types of supply concept in the tourism industry: dependent and independent of tourism [10]. Factors such as natural, historical, cultural and traditional values of a country, its forests, sea, sun or historical and artistic values, customs and traditions, remnants, monuments, hospitality, holy places are the components constituting supply elements independent of tourism. All of these properties have no direct connection with tourism, that means, they are independent of tourism. However, that a place has such assets increases, encourages the demand for tourism in that place or region and raises its attractiveness. These mentioned values will appear as tourism supply to the extent that the quality of the facilities and activities, that is the quality of tourism products, established for tourism in that region. Otherwise, in the event that there wasn't sufficient infrastructure, facility and activity, it would be impossible to talk about tourism supply. Dependent tourism supply is supplying the tourism products to tourists by a series of facilities of which infra- and superstructure are completed in order to benefit from the above indicated values. As its name suggests, dependent tourism supply is a kind of supply emerging entirely as a result of the tourism movements to the region. Here, the public and private sector representatives organize every activity from accommodation, transportation, marketing, security to various infrastructure services. In other words, dependent tourism supply can be expressed as supplying the tourism potential existing in a region to the tourist for service purposes by the help of institutions supporting the tourism industry.

Today, several new tourism attraction centers, new markets and a series of new tourism supplies are to be created. What lies under this is the thought of increasing a country's competitive power and market share in the international tourism market, the number of tourists and income by differentiating products in tourism. Thereby, it is economically important that the tourism country analyzes the supply sources that can be used for tourism purposes well and serve those for the benefit of tourists and the country. Taking this as a departure point, inspecting the tourism supply of a country and its properties will be helpful on introducing which regions and attraction centers of the country provide the maximum benefit for that country. Indeed, due to the economic importance of tourism, considering the supply potential of tourism, all countries worldwide try to develop different tourism types and get the desired share from the international tourism income. The detection of the tourism regions in terms of supply will at least assist in identifying the regions available for tourism may be the subject of which type of tourism and contribute to imply effective tourism policies [11].

B. Properties of Tourism Supply

Tourism supply defined above has some properties peculiar to the industry. Examining these one by one will be helpful on stressing tourism supply more clearly and making it comprehensible. In this respect, the properties of tourism supply are listed below [12]:

1. It is obligatory that large scale investments be made in order for the tourism supply to be generated. Beginning to operate, a tourism firm needs a certain time period as

well as lots of spending on constant factors of production, such as the construction of a five-star hotel in the accommodation sector.

2. Tourism supply may differ in terms of the tourism product that is served with the effect of different external factors such as the prices, consumer tendencies and requirements and the quality of goods and services in various destinations. In other words, supply in the tourism industry requires the marketing and presenting goods depending on the structure (socio-economical, demographical and psychological properties) of people in each country.

3. The type of product in tourism supply is service which goes non-stop for 365 days a year and 24 hours a day. Hence, the tourism product has an abstract feature with this aspect. In other words, as a result of the service sale, manufacturing and consumption in the tourism industry the satisfaction of the customer (tourist) is always abstract.

4. That the tourism industry is highly affected by the reasons such as the dynamic structure in the tourism industry, the crises, seasonal fluctuations impedes the hoarding of the goods and services sold. It is for this reason that the goods and services of tourism should be consumed as soon as they are produced. Thereby, due to the structure of tourism supply, it is not possible to hoard the tourism product.

5. That the substitution facilities are high in tourism supply makes the risk distribution difficult. In other words, a tourist with a certain budget can spend his/her money on going on a vacation while s/he may also feel the urgency of buying goods or services that may become a necessity or attract her/him at that moment. In such a case, since the share of tourism industry will be less, a risk may appear.

Box 4.1. Variety of Substitution Facilities in Tourism and Risk Allocation

That there are variety of substitution facilities in tourism makes the risk allocation difficult. For example, an important and costly breakdown in a car of a family who sets off to Paris from London by car, before the journey may hinder the family's intention to buy the tourism product. Such a situtaion is a risk for the accommodation firms such as hotels or guest houses waiting them or others. Or the type of tourist who discards flying and prefers to use highway can be handled within this scope. There is no other way to sell this risk, that is the unsold hotel room or airline seat.

6. Tourism supply is based on production. Tourism supply, which should be handled as a total product, requires a mutual communication in almost every area in the process that begins to be used right after the person decides to travel and ends with his/her return to home. However, due to the fact that technology is used in the tourism industry excessively, it becomes harder to predict its labor intensive nature.

7. Tourism supply is inelastic in short term. More clearly, for many tourism products (such as a hotel room), it is impossible to develop an additional supply. Therefore, if a change in demand occurs, there is no possibility to increase tourism supply in a short term as a result of the project thought by the entrepreneur. Because of the fact that supply cannot meet the demand increasing in a short term, there can be an increase in the price of products.

8. Tourist as the consumer in tourism constitutes a part of the supplied goods and s/he personally has to visit the place where the supply is provided in order to buy and

consume the product and create a positive/negative image about it. This place is at the same time the consumption place. In the manucfacturing and agriculture industries, after the production is made in factories or in the field, it is sent to the places to be sold retail. Most of the time the consumer buys the product from intermediaries, stores, shopping centers at the place s/he lives.

9. Tourism supply has a complex structure that interests very different factors, industries, areas and values. The physical, natural, cultural and human values are regarded as a whole by the tourist visiting the place. In this respect, a tourism destination should be assessed as a whole in the formation of the tourist's holiday experience. In other words, all these factors mentioned above makes up a part of the chain. A negative experience happening somewhere may affect all the impression adversely.

10. The last property of tourism supply is that technological developments change the structure of production in other industries and as a result, increase efficiency and decrease costs. However, due to the structure of tourism, it is possible for only limited areas (booking transactions, visa, passport controls, etc.) to use machines rather than the human capital, that is the people, in this labor intensive industry.

Box 4.2. Relationship of Tourism Products with Each Other [13]

The importance of all geographical and economic components at the host destination in terms of attaining the required quality level in order to serve the tourist visiting the region for holiday and as a result satisfy her/him is evident. According to the structural property of tourism, a tourist will not benefit from only one component; in contrast s/he buys the "final product" that appears as a result of the combination of several geographical, economic and social components, and this product appears as an important factor in the formation of the holiday experience. Numerous motivation factors may be effective in the decision of tourists to choose a specific region. A tourist accommodates in a hotel, eats at a restaurant, does shopping, makes use of the transportatiion vehicles, communicates with the local people and visits some places. So, each individual or firm with a direct or indirect relationship with the tourism industry has a role in presenting high quality service due to the domino effect. Whether a waiter working at a hotel serves in a friendly manner or not is an example. Therefore, a hotel enterprise in a coastal tourism region should not be considered as a separate item from the sea or the beach or even from the staff in charge of service.

Competition in tourism increases every day and appears in different dimensions. If destinations do not serve tourism supply, properties of which are listed above, in line with the expectations and demands of tourists, it is not possible for them to compete in the international tourism industry. First, what the supply potential of a tourism region is, in other words tourism inventory, should be detected. When this is achieved, it is necessary to adopt supply to demand and serve the tourism product in the best way that it meets the tourist's need.

C. Components of Tourism Supply

There are different factors leading individuals to tourism. Some of these are internal and some others are external factors. Examples of internal factors are the ones such as business, curiosity, religion, education, culture, sports, entertainment and relaxation, health, and so on. Similarly, the examples to be given to external factors are introduction, advertising and propoganda as well as new destinations turning up [14]. Hereby, besides these factors leading tourists to travel activity, the country or region's attractive natural assets, values, facilities and socio-cultural assets that increase the motivation of tourists are among the factors affecting tourism supply. While it is known that there are various studies on what is covered by tourism supply, it is possible to summarize them under four or five general headings.

Mill and Morrison (1992), summarize the tourism destination and hence the supply factors creating attractiveness under five headings as the place's attractiveness, facilities and probabilities, infrastructure, transportation and hospitality [15]. Meanwhile, Laws (1995) examines the factors affecting tourism supply as primary and secondary properties [16]. A region's climate, ecological structure, culture and traditional pattern make up that place's primary properties. Similarly, properties such as accommodation and transportation are the components making up the secondary properties. Yet, in another study, the supply conditions of tourism and natural, socio-cultural assets, infrastructure and its transportation situation of a destination are stated [17]. Kozak and Rimmington (1998) divide a region's supply factors that create attractiveness to five groups as the attractiveness of the place, activities and services, infrastructure, hospitality and cost [18]. In another study by Icoz and Kozak (2002), all the factors constituting supply factors are shown in Table 4.1 [19].

Table 4.1. Components of Tourism Supply

Tourism Attractions	
Cultural Environment	Physical Environment
Archeological	*Climate*
Historical Values	*Landscape*
Monuments	*Wild Life Observation*
Structures	Entertainment
Museums	Sports
Traditional Environment	*As a participant*
Political Environment	*As a spectator*
Educational Environment	Parks
Religious Environment	Cinema and Theatre
Modern Environment	Night life /clubs
Cultural Customs	Casinos
Fests	Bars
Art Shows	
Handicraft	
Music	
Natural and traditional life style	
Language	
Religion	
Science	

When the table is examined, there are a lot of different supply sources such as cultural, traditional, political, educational, religious environment, cultural customs, entertainment, sports, cinema-theatre, nightlife among the factors making up the tourism supply.

1. Natural Resources

Natural resources exist in the world in many different patterns, structures and conditions. These are peculiar to a region and arise from nature itself. Natural resources are perhaps the most important of the tourism prosperity because a place without a natural habitat there can be no tourism activity. People want to go to the regions which they cannot see, are curious about and desire to experience. From this aspect, the formations occurring naturally without human effort and contribution are called natural resources or assets. Each region has its own climate, geographical shape, flora, geological structure and beauties.

As an example to these, a landscape of a region, water, flora, mountains, canyon, volcanoes, coral islands, crater lakes, travertines, chimney rocks, lakes, waterfalls, rivers, coasts can be given [20]. All these factors are divided into three as natural beauties, climate, medical and thermal springs [21]. These factors enable the acquisiton of an economic value and creating a demand in tourism, attractiveness and motivation.

Those factors indicated above can generally be summarized under six groups as natural resources, infrastructure, transportation condition, socio-cultural assets, superstructure, and hospitality. Each is briefly examined below.

2. Infrastructure

Today, it is a known fact that the tourism industry is in a rapid improvement both in the national and international area. As a result, a continuous development in tourism supply is attempted to be provided as well as meeting the tourism needs due to the unlimited capacity of demand in tourism. At this point, it is essential to implement a development program that will provide an infrastructure of 15 – 20 years later from now by examining the current demand with the sufficient and future oriented plans and projections of a tourism region. Shortly, it is needed to build and foster the tourism region's facilities and supplies towards infrastructure with sustainable tourism policies in a way that it meets the current and future demand.

When an infrastructure of a region is mentioned, water and drainage systems, communication web, health services, energy sources, waste systems, street, district, subway, tunnel systems as well as security systems appear in mind [22]. These mentioned infrastructure factors are the most important in improving tourism and implementing it successfully. If a tourism destination cannot supply a sufficient infrastructure service to the extent that it can supply the tourism product, a decrease in tourism demand to that region is expected. Occupancy rate and profitability of the facilities built in a region whose infrastructure has not been completed yet and transportation is difficult will be low and hence demand in tourism will decrease.

To summarize, just as it is important for a region to have a certain potential in terms of tourism, it is also an important issue to present an infrastructure service at a quality and standard that will assess this potential. The prequisite of making use of the assets making up tourism supply lies in basing this supply on a suitable infrastructure.

3. Transportation Status

Another important factor constituting tourism supply is the place's accessibility or transportation status. If there had not been the technological developments in public transportation and transportation vehicles throughout history, it would have been impossible for tourism to occur in the present sense [23]. Although a tourism destination has a high level of attractiveness in terms of supply factors, if it were hard to reach that region, there would be major problems in marketing and selling the product to the tourist. Transportation and tourism are the inseperable parts of a whole. Transportation is the most important factor in the development of tourism in a destination. The transportation industry in a tourism destination should be operating well for tourism to gain value. The region's transportation status is one of the most significant reasons that make tourism a necessity and supply added value [24].

Any kind of facility and vehicle used while travelling can be thought within the transportation system, so it is based on the transportation system that the tourism activities start and people change places. The transportation status is an important infrastructure in terms of tourism supply [25]. To reach the target in the market, that is the tourist, tourism centers with high attractiveness should have airline, highway and seaway connection. It is possible to say that transportation is a significant source of supply in terms of the tourism industry since it is expected that long distance overseas journeys in the following years increase and new destinations in various regions of the world become more important. Moreover, the choice of destinations as well as the information technology as an important tool are among the other possibilities [26].

Today transportation is important and customers choose the fastest and cheapest transportation way thanks to the developing technology. In this respect, if the important tourism attraction centers are sensible on this issue, their competitive power in the international tourism market and market share will increase. The success of tourism activities in a region in the rapidly developing global world of today depends on the structure and organizations of the transportation system in that region [27].

4. Socio-Cultural Assets

The values of mankind and society formed socio-culturally are also within tourism supply. These constitute from a region's historical assets as well as its social appearance, and they can be both traditional and contemporary. The socio-cultural assets can be examined in two groups: historical works and museums [28]. Historical and religious monuments, ancient works, historical remains, ruins and excavation places are generally among the tourism assets that attract tourists. These are considered as important by the tourists since they are the works introducing the ancient civilizations and showing a nation's culture, economic-political life and art.

One of the issues that tourists are interested in at the places they visit are museums. Museums are the important factors reflecting the historical formation of a country or a region and hence the ancient civilizations that had existed on Earth. These have an important role in the determination of whether identities belong to a region or a community. From this aspect, it can be said that museums play an important role in the cultural and economic life of any tourism country [29]. Museums making up the significant part of the tourism supply can also be effective in reviving economy in urban and rural areas.

5. Superstructure

The superstructure in tourism consists of important assets such as accommodation, eating and drinking, transportation and entertainment services that make up the tourism industry. Introducing these services to the visitors in high quality standard is important for the competition among destinations [30]. The necessary and sufficient condition for developing tourism in a country or region with tourism potential and the local people's gaining income from tourism as well as benefiting from the external economical impacts of tourism is the existence of infra- and superstructure in that country or region. It is of high importance that there is a structure to maintain all of the expectations and needs of the tourist, such as accommodation, eating and drinking, transportation, health, banking or information services. Naturally, to make the necessary investments for infra- and superstructure, first, the region should have the sufficient economic power. Assuming that this situation is present in the tourism destination, unless the above mentioned factors that lead people to travel are supported by sufficient services of infra- and superstructure, tourism will be meaningless for that region.

Box 4.3. Importance of Supply Sources [31]

Among the tourist groups with different qualifications who are travelling for various purposes, particularly the qualified ones, prefer the natural, historical and cultural places with outdoor activities for holiday. Today, a certain amount of the world population travels in their own countries or abroad to achieve self-fulfillment and experience the life in a natural and cultural environment. Surveys show that the decisions of English and German tourists to travel to the Mediterranean destinations within the scope of coastal tourism are first affected by climate, then the cultural and historical values, and last, by the desire to get closer to nature. In contrast, German tourists differ from English tourists in terms of their interests in cultural values and natural beauties. The fact that tourists participating in nature and culture tourism spend more than the other tourist types show the importance of the assets in terms of tourism economics. For this reason, one should not expect a stable tourism activity in a region without a cultural and natural environment. In other words, in order to maintain sustainability in a regional and natural tourism industry, reach an internal dynamism and get more share from international tourism, it is necessary that natural, cultural and historical values should be perpetuated.

In case there is no facility or building with a level of standard and quality that tourists prefer, it is inevitable that the attractiveness of the region disappears. Some supporting facilities and establishments are the factors complementing the attractiveness of tourism in the region [32], e.g., shopping centers, stores, stations, airports, marinas and harbors, entertainment places as well as the accommodation facilities which are the main component of the superstructure.

6. Hospitality

The hospitality factor differs among the other supply factors since it has a meaning in line with the tourist's moral values [33]. In this respect, the development of the hospitality factor requires a different content than the other factors in terms of tourism supply. However, the natural and physical attractiveness in a country, region or accommodation facility is the negative/undesired attitude or behavior of the local people to tourists may cause the tourist's holiday experience to be negative. So, the tourist may develop a negative image regarding that country. As a result of this, a country may not get the desired share from the tourism industry and may remain idle by not using its resources effectively despite having an important tourism potential. Such a situation brings a great loss in wealth for the citizens and economy of the country as natural, historical and geographical beauties that appear as an economic instrument by means of tourism, cannot be used as well as desired and consequently there appears a waste in sources. In conclusion, today, globalization has been increasing rapidly and service quality and customer satisfaction have become more important. In such a time, it is a fact that treating tourists visiting a country in a proper way will enhance the country's economy.

II. Supply Elasticity of Tourism

The supply elasticity of tourism may be explained as "the level of reaction shown as changing the supplied amount upon the price changes". Supply elasticity of tourism indicates the reaction producers show upon the changes in prices. Supply elasticity is shown with a quotient (e_s). This quotient is equal to the ratio of the percent change in supplied amount to the percent change in price when there is a change in price. According to the supply rules in economics, as the price of a good increases, the supply of that good also increases. For this reason, as price and supply change in the same direction, supply elasticity is always positive despite having some exceptions.

That is, sometimes as the price of a good increases, the supplied amount decreases, e.g., the supply of the agricultural products of working people in rural areas [34]. The event that supply elasticity is 1 is accepted as the criterion. Accordingly, if the quotient of supply elasticity is smaller than 1, it is defined as the inelastic supply; while if it is more than 1, it is defined as elastic supply. If the quotient of supply elasticity is equal to 1, it is unit elasticity. The quotient of supply elasticity is calculated as below:

$$e_s = \frac{\% \ changeinthesuppliedamount}{\% \ changeinprice} = \frac{\Delta q / q}{\Delta p / p} = \frac{\Delta q}{\Delta p} x \frac{P_1}{q_1} = \frac{(q_2 - q_1)}{(P_2 - P_1)} x \frac{P_1}{q_1}$$

Here, e_s is the quotient of supply elasticity, q is the amount and p is the price. Besides, in the above equation $\Delta q = q_2 - q_1$ ve $\Delta p = p_2 - p_1$. For instance, if there is a 25% decrease in the supplied amount due to a 10% decrease in price, the supply elasticity will be:

$$e_s = \frac{\%25}{\%10} = 2.5$$

To give another example, while the number of rooms in a tourism region is 50 and the price of each room is 50 US$ in 1995, assuming that the number of rooms increases to 100 and the price increases to 75 US$ in 2005, the quotient of supply elasticity for this region will be:

$$e_s = \frac{100 - 50 \big/ 50}{75 - 50 \big/ 50} = 2$$

Box 4.4. Reflection of Relationship between Tourists and Local People

> Study results on tourism economics show that the hospitality factor among the supply sources of tourism is an important competitive power for many countries. In a study by Bahar (2004) in order to identify the competitive power of Turkey in the tourism industry, it is found out that the "hospitality" factor, which is assessed within the scope of supply conditions, appeared as the most important factor affecting the competitive power in all analysis [35]. As a result of the indicated study, it appears that in Turkey the attitudes and behaviors towards tourists are much better than they are in Spain, Greece, Italy, Cyprus and France, which are the main competitors of Turkey in the Mediteranean region. Similar results are also attained from the studies by Kozak (2003, 2002, 2001, 1999) in which Turkey was compared to some Mediterranean countries [36]. On the other hand, it is seen that tourists especially face troubles with hassle or harassment and consequently, this directly affects the level of their satisfaction and spending [37].

This can be assessed as follows: if supply elasticity is 2, this means when there is a 1% increase in prices, then there will be 2% increase in the amount of supplied tourism products. That the quotient is positive is due to the supply law, that is, as indicated in the above paragraph, as the price of a product increases, so does the the amount of product supplied. Since the price and amount are in direct proportion with each other, the quotient of supply elasticity should be positive.

In the short term analysis towards the tourism industry, the supply elasticity of the firms individually is to be quite inelastic in the calculations made with the above formula. In other words, supply elasticity is smaller than unit elasticity (smaller than 1). If the market prices are high and there is sufficient demand, supply will be full capacity. This situation is also true if market prices are low and demand is not enough because there would be a pressure on the firm to perform in full capacity under such circumstances. Consequently, this will lead to a low total supply in the industry in the short run. Indeed, it is not possible to increase or decrease supply in the short run against the price changes and accordingly the changes in demand. Regarding the inability to maintain a coherence between supply and demand, firms mostly prefer to have price differentiation and sales development. In the long run, one may decide to stop activities other than sales, as tourism supply is relatively more elastic [38].

A. Tourism Supply Elasticity in the Very Short Run

There is no possibility in increasing the production in the very short term, which is called "market". This period can be defined as the term when no change is made on any of the factors of production [39]. In this short term, the suppliers of the product, that is the sellers cannot increase production to the extent that it meets demand. Within this period production is the stock. For example, during two- or three-day holidays for Christmas day, feast days or national holidays there is a dynamism in domestic or foreign tourism markets. In such periods it is not possible to meet the increasing demand and match the supply to this demad. Hence, supply elasticity of tourism will be equal to zero ($e_s = 0$). The supply elasticity of tourism in such a period is shown below in Figure 4.2. In this case supply is not related with the price. It will cause the prices to increase if demand increases in the very short term.

In Figure 4.2, it is seen that tourism supply indicated with S is zero, in other words, inelastic. As explained above, if it is assumed that the demand increases from D_1 to D_2 due to the reasons such as Christmas holiday, price rises from 50US\$ to 75US\$ as supply is constant. When supply elasticity is calculated for this example:

$$e_s = \frac{100-100/100}{75-50/50} = \frac{0/25}{25/50} = 0$$

As it can be seen in the following chapter, the situation that supply elasticity is zero represents a special case and it is called "zero elastic supply". Detailed information about zero elastic supply will be given below.

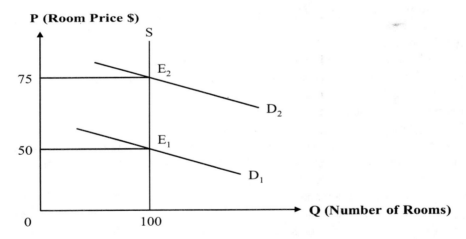

Figure 4.2. Supply elasticity of tourism in the very short run.

B. Tourism Supply Elasticity in the Short Run

Short term represents a time period when a firm is able to change not all factors of the production but only some part in order to change its production volume [40]. In the short run, the firms and enterprises performing in the tourism industry can increase the amount of products they produced to the extent limited to their capacity. In this respect, the short run tourism supply is a little more elastic than tourism supply in the very short run. Short run supply elesticity can change according to the types of goods and services or production conditions, and as a result, supply elasticity may increase. The tourism firm or enterprise may change the amount of some, but not all, of its various inputs that it used in order to attain the tourism product so that it can meet some of the demand increase since it can increase the supply in this period. As a result of the fact that tourism firms change the amount of only a certain part of the factors of production it can meet even a little part of increase in demand. For example, if a hotel increases its bad capacity or number of rooms with some small renewals it shows that there is a short run for the tourism industry.

In the short run the quotient of supply elasticity is smaller than one ($e_s < 1$). So, when supply elasticity of tourism is less than 1, there is a short run. The length of this period when only some part of the factors of production can be changed differs according to the industry in which the firm performs. The firms in an industry using a simple technology (e.g., producing rug with handloom) may change supply in a very short run. However, as the structure of the product gets harder, the time interval of the short run may lengthen (e.g., automobile production) [41]. Although how much time short term covers is not known exactly, it can change according to the industry or the level of difficulty of the produced goods and services. The supply elasticity of tourism in the short run is shown in Figure 4.3.

In Figure 4.3, the tourism supply indicated with S is seen to be more elastic comparing to the one in a very short term. When demand increases from D_1 to D_2 the balance point shifts from E_1 to E_2. As supply is elastic, the amount supplied increases from 100 to 110. However, since increase in demand is more than supply, the price increases from 55US$ to 80US$. When supply elasticity is calculated for the example in the figure:

$$e_s = \frac{110 - 1000 / 100}{80 - 55 / 55} = 0.22$$

As 0.22 is smaller than 1, supply elasticity is not elastic. So, as indicated in the above paragraph, the quotient of supply elasticity should always be smaller than 1.

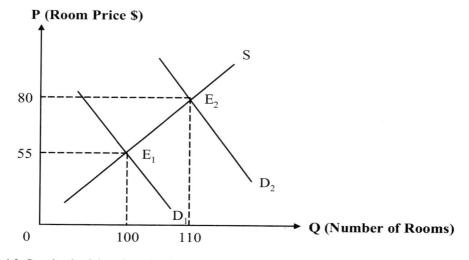

Figure 4.3. Supply elasticity of tourism in the very short run.

C. Tourism Supply Elasticity in the Long Run

In the long run the amount and combination of all factors of production in the production of goods and services can be changed [42]. A tourism enterprise may build a new hotel or increase its current bed and room capacity as much as desired, increase the number of staff and its quality in the long run. Shortly, the firms and enterprises in the industry can change everything regarding infra- and superstructure in this period. Indeed, supply is able to meet any kind of change in the demad. In such a case, the elasticity of tourism supply is larger than one.

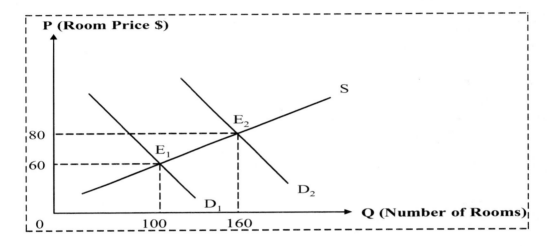

Figure 4.4. Tourism supply elasticity in the long run.

In Figure 4.4, the supply curve in the long run is more inclined comparing to the one in the very short term and short run. As a result of the increase in demand, the balance point shifts from E_1 to E_2. Also, it is understood that the increase in supply is much more than demand in Figure 4.4. Similarly, the supply elasticity for this example is:

$$e_s = \frac{\dfrac{160-100}{100}}{\dfrac{80-60}{60}} = 1.8$$

The value of the quotient of supply elasticty is 1.8. Since this value is bigger than 1, it is clear that supply elasticity is elastic. Therefore, in the long run supply elasticity is generally bigger than 1. Below, all these three terms are shown on the same graph (see Figure 4.5). As the supply curve gets closer to the horizantal axis, it appears to be more elastic. The inclining supply curve is an indicator of its being more elastic. In other words, the supply curve gets more elastic in time. Therefore, it is clear that supply elasticity of a tourism product increases in time and the longer this period is, the more supply reacts to the price changes.

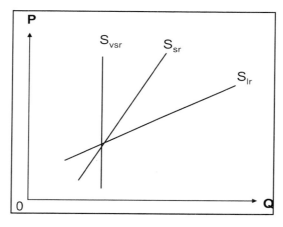

Figure 4.5. Supply elasticity of tourism and time.

D. Conditions When Supply Elasticity of Tourism is the Same

There are three particular conditions when the supply elasticity of tourism is the same on each point. At these points supply elasticity always has the same value. The first of these particular supply curves is the zero elastic supply (Figure 4.6a). The curve is a parallel line to the vertical axis in zero elastic supply and elasticity quotient is equal to zero at each point on the curve. In such a case, it is not possible to change the supplied amount whatever the price level, e.g., private monuments and ancient buildings (Ephesus Theatre, Topkapi Palace, Buckingham Palace, etc.).

The second particular supply curve is unit elastic supply. As seen in Figure 4.6b, the supply curve is a straight crossing the origin and the supply elasticity quotientr on each point on the curve is equal to 1. The amount and price of product in tourism changes at the same rate in unit elastic supply. The utilization frequency, its intensity and level of meeting the needs of products in tourism differ among each other. So, it is difficult to illustrate the unit elastic supply. However, the number of airline seats sold, depending on how well the tourism season is, can be given as an example from the transportation industry. As the amount increases according to the demand, the price increases as well. In contrast, if the number of seats sold decreases, so does the price.

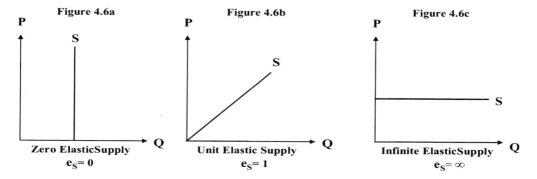

Figure 4.6. Conditions when supply elasticity of tourism is the same.

The third and the last supply curve is the infinite elastic supply. In this type, the supply curve is a parallel line to the horizontal axis (Figure 4.6c). At each point on the curve, quotient of elasticity is equal to infinity ($e_s = \infty$). In this case, the ones producing goods and services in the tourism industry are ready to sell as many tourism products as desired at a certain price. These three conditions are exceptions and special, as indicated above. The supply elasticity is the same at each point on the curve, e.g., a meal eaten at an ordinary restaurant or a pension room rented in small towns.

III. SUPPLY CURVE IN TOURISM

Supply curve is a positively inclined curve, as seen in Figure 4.7. It ascends from bottom left to upper right due to the supply law that asserts the amount supplied and price to be directly proportional. In other words, as the price of the supplied tourism product increases, the supplied amount of that good also increases. In contrast, as prices decrease, the supplied amount decreases as well. When price goes up from P_1 to P_2, tourism supply increases from Q_1 to Q_2. Depending on this price increase, there is an escalation of Q_1Q_2 in supply. In the figure the points S_1 and S_2 show the supply amount of the tourism product, which is ready to be sold at different price levels. The supply curve of tourism is made up of the combination of an infinite number of points like S_1 and S_2.

The below figure (4.7) is the individual supply curve of a single firm performing in the industry. The sum of the supplies of many firms and enterprises performing in tourism is the industry or the market supply. For example, the total supply curve in a tourism area is the one among ministry-registered or non-ministry registered bed capicity served for a tourist's benefit [43]. As another example the number of rooms served by all tourism firms in a destination (country, region, city or district) constitutes the market supply in the industry.

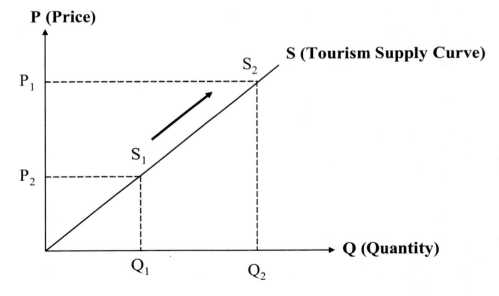

Figure 4.7. Tourism supply curve.

A. Change in Supplied Amount of Tourism

When there is an increase or decrease in the supplied amount of any tourism product, there is a movement on the supply curve. The changes appearing in supply can be dependent or independent of price. When there is a change in the supply amount of tourism, an amount-price combination may be changed with some other one. When there is an increase or decrease in the price of the tourism product, there is also a relative increase or decrease in the supply amount. The supply curve in the Figure 4.7 can be given as an example. The movement from S_1 to S_2 shows the change in the supplied amount of tourism. This change leads to an increase in the supplied amount from Q_1 to Q_2 and consequently supply rises by Q_2Q_1 since the price of the tourism product rises from P_1 to P_2. Apart from the price of the tourism product, the other factors affecting supply (e.g., prices of other tourism products, prices of factors of production or technological level) at the movement on supply curve are constant. Here, there is a change depending on price. This movement on the tourism supply curve reflects *a static change,* which means introducing the price–amount relationship at a specific period (a certain instant, a day, a week, a month, a year, etc.) regarding tourism supply. In such a case, the time factor is not included in the analysis. It is not considered as a separate variable since the change in the amount of tourism supply is within a specific time interval. In short, time is not included as a variable in static analysis [44].

B. Change in Tourism Supply

A change in tourism supply means that the supply curve moves completely to the right or left. Supply change is, shortly the shift of supply curve. The reason of this change is not price, but there is a change not related with price. There is a shift in supply due to the other factor or factors affecting supply apart from the tourism product. For example, a change in the prices of other tourism products or factors of production, an increase at the technological level or a rise or a fall in costs cause shifts in the supply curve. These are called "*factors shifting supply*". Therefore, tourism supply change is a dynamic process as the change in the factors other than price causes to deviate from the period that is assumed as definite, and consequently time factor is included in the analysis. As known, it is not possible to increase supply in the short run. If time gets longer, this makes it possible to increase the supply. The ability to increase or decrease supply depending on time is an indicator of *a dynamic change.* The economic events are examined in time in such dynamic change, and changes that occur within time especially while shifting to a balance from another are considered [45]. Since the supply change in tourism depends on changes in a definite time period, the change in tourism supply is in general a dynamic event economically.

Based on the change in one or all of the factors affecting tourism supply other than price, in Figure 4.8, when supply curve moves from SS to S_1S_1, it shows an increase in supply, and if it moves to S_2S_2, it shows that supply has decreased. An increase in supply causes the curve to shift to the right while a decrease causes a shift to the left. As seen in the figure, a shift to the right or left does not affect the price P_1. In other words, the change in tourism supply is under the effect of factors other than price.

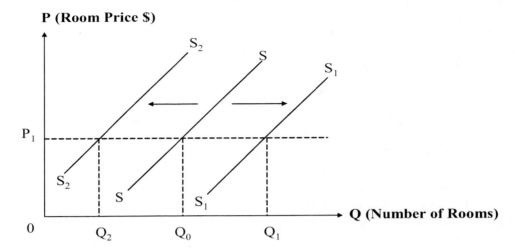

Figure 4.8. Shifts in Tourism Curve.

SUMMARY

As known, in the tourism industry there is a market type where supply and demand are put together, as in every industry. Therefore, the first subject of this chapter is to provide brief information on the tourism market. In the following parts, the concept of tourism supply is examined broadly, which includes some important issues for the tourism industry, e.g., main properties of tourism supply or change of supply elasticity based on different time intervals. Even though the properties of tourism supply are similar to the properties of services industries like health or banking, they differ from the properties of many other industries. Lastly, the supply elasticity of tourism is a subject that is common in practice. The following chapter continues with the discussion of tourism demand and its properties.

DISCUSSION QUESTIONS

1. What is tourism market? Explain by giving an example.
2. Discuss what kind of changes developments in technology can have on the tourism market.
3. Explain briefly the properties of tourism supply.
4. Compare the tourism industry with the health industry and also with the food industry in terms of the properties of supply sources and discuss the similar and different aspects of all.
5. What are the factors constituting tourism supply? Explain briefly.
6. Explain the concept of supply elasticity with an example.
7. Explain the position of supply elasticity of tourism in the very short run, short run and long run.

8. What are the situations when supply elasticity of tourism is the same? Explain briefly.

9. Compare the concept of supply elasticty from the aspect of accommodation, food and beverage and transportation firms.

10. What kind of a change does a time dependent increase or decrease in tourism supply indicate?

REFERENCES

[1]   D. Begg, S. Fischer and R. Dornbusch (2000), *Economics*, England: Mc Graw-Hill, 6th edition, p. 29; Ünsal, p. 54.

[2]   Dinler, pp. 71–72.

[3]   Erdoğan, p. 319.

[4]   Olalı, Timur, p. 194.

[5]   Begg et al., p. 30.

[6]   H. R. Varian (1990), *Intermediate Microeconomics*, London: W. W. Norton and Company, Inc.

[7]   H. Olalı, A. Timur (1988), *Turizm Ekonomisi*, İzmir: Ofis Ticaret Matbaacılık, p. 171.

[8]   Z. Eralp (1970), *Turizm Ekonomisi ve Politikası*, Ankara: Büroteks Yayıncılık, p. 112.

[9]   L. Lickorish, L.C. Jenkins (1997), *An Introduction to Tourism*, Oxford: Butterworth and Heinemann, p. 7.

[10] Dinçer, pp. 21–24.

[11] O. Bahar (2005), "Impacts of Mass Tourism on the Natural Environment," *Quarterly Journal of University of Economics Bratislava*, Volume 4, pp. 433–448.

[12] Erdoğan, pp. 359-361; Olalı, Timur, p. 179; Kozak et al., pp. 42–44.

[13] M. Kozak (2004), "The Practice of Destination-Based Total Quality Management," *Anatolia: An International Journal of Tourism and Hospitality Research*, 15(2), pp. 125-136.

[14] M. Kozak (2002), "Comparative Analysis of Tourist Motivations by Nationality and Destinations," *Tourism Management*, 23 (2), pp. 221-232.

[15] R.C. Mill, A. Morrison (1992), *The Tourism System: an Introductory Text*, Prentice-Hall International Editions, 2nd edition.

[16] E. Laws (1995), *Tourist Destination Management: Issues, Analysis and Policies*, New York: Routledge.

[17] R. W. McIntosh vd. (1995), *Tourism Principles, Practices, Philosophies*, New York: John Wiley and Sons, 7th edition, p. 269.

[18] M. Kozak, M. Rimmington (1998), "Benchmarking: Destination Attractiveness and Small Hospitality Business Performance," *International Journal of Contemporary Hospitality*, 10(5), p. 184.

[19] İçöz, Kozak, p. 37.

[20] M. Çubuk (1981), "Şehircilik Bilim Dalında Bir İnceleme," *Basılmamış Doktora Tezi*, Mimar Sinan Üniversitesi Mimarlık Fakültesi, İstanbul, p. 74.

[21] For additional information see O. Bahar, M. Kozak (2005), *Uluslararası Turizm ve Rekabet Edebilirlik*, Ankara: Detay Yayıncılık, pp. 19–20.

[22] Gee, Fayos-Sola, p. 14; Kozak, Rimmington, p. 184.

[23] B. Prideaux (2000), "The Role of the Transport System in Destination Development," *Tourism Management*, 21(1), pp. 53-63.

[24] J.P. Rodrigue, "International Tourism and Transport," http://people.hofstra.edu/geotrans/eng/ch5en/appl5en/ch5a4en.html, (Retrieved 25 September 2007).

[25] Eralp, p. 128.

[26] http://www.tubitak.gov.tr. "Vizyon 2023 Ulaştırma ve Turizm Paneli," (Retrieved 02 August 2004)

[27] Prideaux, p. 53; A. K. Yağmuroğlu (1998), "Turizmi Teşvik Eden Kaynaklar," *Turizm Bakanlığı 1. Turizm Şurası Bildiri ve Görüş Metinleri,* Ankara, p. 152.

[28] O. Bahar, M. Kozak (2005), *Uluslararı Turizm ve Rekabet Edebilirlik,* Ankara: Detay Yayıncılık, pp. 20–21.

[29] J. Harrison (1997), "Museums and Touristic Expectations," Annals of Tourism Research, 24 (1), pp. 23- 40.

[30] G. I. Crouch and J. R. B. Ritchie (1999), "Tourism, Competitiveness and Societal Prosperity," *Journal of Business Research*, 44, pp.137–152.

[31] Kozak, "Comparative Analysis of Tourist Motivations by Nationality and Destinations," pp. 221–232.

[32] İçöz, Kozak, pp. 39–40.

[33] Erdoğan, p. 373.

[34] Dinler, p. 186.

[35] O.Bahar (2004), "Türkiye'de Turizm Sektörünün Rekabet Gücü Analizi Üzerine Bir Alan Araştırması: Muğla Örneği", *Basılmamış Doktora Tezi*, Muğla Üniversitesi S. B. E., Muğla, pp. 164–165.

[36] M. Kozak (2004), "Measuring Comparative Performance of Vacation Destinations," *In G.I.Crouch vd. (Ed.), Consumer Psychology of Tourism, Hospitality and Leisure,* Wallingford: CAB International, pp. 289–299; M. Kozak (2002), "Destination Benchmarking," *Annals of Tourism Research,* 29(2), pp. 504–512; M. Kozak (2001), "Repeaters' Behavior at two Distinct Destinations," *Annals of Tourism Research,* 28(3), pp. 793–800; M. Kozak, M. Rimmington (1999), "Measuring Tourist Destination Competitiveness: Conceptual Considerations and Empirical Findings," *International Journal of Hospitality Management*, 18, pp. 278–281.

[37] M. Kozak (2007), *"Tourist Harassment: A Marketing Perspective,"* Annals of Tourism Research, 34 (2), pp. 384-399.

[38] İçöz, Kozak, p. 69.

[39] W. Nicholson (1995), *Microeconomic Theory Basic Principles and Extensions*, The Dryden Press Harcourt Brace College Publishers, 8th Edition, pp. 495–496.

[40] Üstünel, p. 162.

[41] Dinler, p. 154.

[42] Dinler, p. 162.

[43] Erdoğan, p. 354.

[44] K. Mortan et al. (2000), *İktisat Teorisi*, Eskişehir: Anadolu Üniversitesi İktisat Fakültesi Yayın No: 36, p. 11.

[45] Mortan et al., p. 11.

Chapter 5

TOURISM DEMAND AND ITS PROPERTIES

As known, the second important factor making up the tourism industry is tourism demand. In this chapter, the issues to be examined relate to the meaning of tourism demand, its peculiar properties and demand elasticity. Besides, there are some other issues like price and income elasticity of tourism demand, demand curve, changes in demand, forming the balance price in tourism, effects of supply and demand change on balance price and meeting tourism demand by tourism supply. Tourism demand is among the significant issues in the literature of tourism economics, so it is important to understand and analyze the structure and nature of demand in order to provide a better quality service as well as customer satisfaction. In today's rapidly developing global environment, obtaining a competitive power is surely based on examining demand in the best way because knowing what the customer or the tourist wants and creating a product variety accordingly will increase the country's success in international tourism.

I. TOURISM DEMAND

Demand is defined as the amount of goods in the market that consumers plan to buy at a particular price [1]. Tourism demand, which makes up the other important part of tourism industry, has quite a similar definition to the demand concept indicated above in general terms. In this respect, tourism demand is defined in tourism literature as follows:

Tourism demand is the flow from a region sending tourists to a region attracting them [2]. According to another definition, "the number of people who have a desire of having a tourism-based movement and can afford such desires" is called tourism demand [3]. In another definition, tourism demand is inferred as "the product as a whole, which a tourist wants to get at a particular price level or exchange rate and actually accepts to get it" [4]. Finally, tourism demand is defined as "the number of people that plan to buy tourism products supported by sufficient purchasing power and spare time in order to meet tourism needs of people" [5].

There are three important properties in almost all of the definitions given above. According to this, for the composition of a tourism demand economically, first of all one has to have a sufficient income and spare time and a desire to travel. When even one of them is missing, it is not possible to talk about tourism demand. So, in order to mention a tourism demand individually, the necessary and sufficient condition is that a person should satisfy his/her basic needs at first, such as food, shelter, education, health and s/he has to have enough income for a tourism activity. The desire to buy without purchasing power is indeed not regarded as demand in economic terms. Another essential point for tourism is, while having a sufficient income, one needs to have spare time out of the working hours, which can be spent for a vacation. Thereby, for a tourism demand to exist, income level, spare time and desire to purchase are three crucial factors.

The world has become smaller due to globalization and the removal of borders. This has led to an increase in people's desire to travel. Nevertheless, thanks to the invention of jumbo jets in the 1970s and rapid technological developments, tourism activities have become intercontinental. Even though there are countries in economic stagnation throughout the world, tourism will keep on growing so long as the mankind exists on Earth. In this respect, the flow of international tourism movements and demand is as below [6]:

From developed countries	→	To developing countries
From America	→	To Europe
From Northern and Western Europe	→	To Mediterranean
From Central Europe	→	To Asia – Pacific region
From industrial regions	→	To coastal regions

Therefore, making reliable predictions for future and knowing the multiplier effect of tourist spending on countries economies will result in benefiting from tourism as an effective tourism factor in economic development and growth. Moreover, the economical units constituting the tourism industry makes tourism supply suitable for demand, which is an important issue for long term planning and estimation for the future of the industry as well as tourism investments [7]. Thereby, three main factors compose demand for tourism (see Figure 5.1) [8]:

1. *Real (Actual) Demand*: This is composed of the people who are actively involved in tourism, demanding tourism, visiting places and buying tourism products.
2. *Potential (under pressure) Demand*: This is composed of the people who have the motivation to travel and utilize tourism services, however, who cannot afford to do these due to financial reasons.
3. *Strained Demand*: This is composed of the people who can travel if motivated but are unable to realize this demand due to the lack of knowledge regarding facilities and activities.

Domestic tourism demand is composed of two different parts as domestic and foreign. Accordingly, the demand for tourism, which people of the country have within the country, is domestic demand. In contrast, the demand for tourism, which people of the country have outside the country, is foreign demand. The foreign tourism demand worldwide is what is constituted by the people who want to travel for tourism purposes [9].

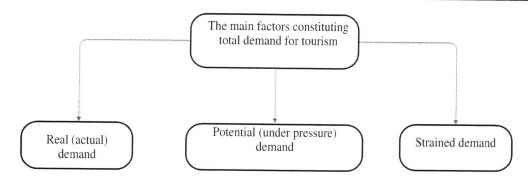

Figure 5.1. Main factors constituting total demand for tourism.

A. Properties of Tourism Demand

Tourism demand has some properties and differences from other goods and services in an economy. These properties are listed as follows [10]:

1. Being an independent demand, tourism demand has several reasons motivating people to travel, which can be economic, social or psychological. These reasons are a result of the desire of a change and some personal effects.
2. Tourism demand is multi-dimensional. People intend to travel with the effect of several motives and participate in various kinds of tourism in different ways. For example, demand can appear as participation to a commercial activity for some people and for some others as a purpose of physical or psychological relaxation [11].
3. Tourism demand is a function of personal disposable income. In other words, there is a positive relationship between tourism demand and personal disposable income. Indeed, that is the reason why tourism demand mostly grows out of developed countries with a high level of prosperity.
4. The tourism product, which is the subject matter of tourism demand, is in competition with other luxurious and cultural goods and services. For example, a family that booked a seaside hotel in winter for a summer vacation may charge their car by a bank credit when there is a decrease in bank loans.
5. There are several substitution facilities in tourism demand and the economic, political and social factors that affect consumption preferences make demand elastic (flexible). A positive or negative change observed in a national economy causes a positive or negative change in the demand of people who want to travel abroad for tourism purposes.
6. Tourism demand is seasonal. Therefore, it is difficult to allocate it to different seasons of the year. As emphasized earlier, it may not be possible to increase tourism supply immediately after the demand increases. The reasons such as employees preferring summer terms for annual leave or school holidays at that period are examples for this property of tourism.
7. Tourism demand may differ according to the development level of the countries. That is, considering the appearance of international tourism movements, one sees

that demand intensifies in the regions of developed countries. Thereby, very few developed countries affect international tourism by having the majority of demand.

Tourism demand should be well analyzed so that the tourism industry contributes to the economical prosperity of countries, the resources are used effectively and efficiently, and as a result, income and share from tourism are attained. The demand analysis in tourism will facilitate better decision making on the prospective investments by looking for an answer for the question of what and how to spend the nation's resources. Therefore, tourism demand is an important indicator on achieving the effective economic growth and development [12].

Box 5.1. European Union Policies and Regional Tourism

> A survey on Euro and tourism demonstrates that the countries in EU with a weak currency are adversely affected by the international tourism movements by the formation of a Euro zone. It is indicated that the income effect and the deterioration in the price competition may increase the import of these countries and this will decrease the demand arising from exporting countries with strong currencies due to the negative flow of prices, since countries were in a cheaper position before the accession to the Union [13]. The results of another Euro survey carried out in Austria suggest that the convenience in travelling and the spending with a standard currency in the member countries of EU seem to be the most important benefits [14]. As known, the commissions and the foreign currency loss while travelling abroad is eliminated and one can spend in Euro at that foreign country. "You needn't always convert your currency into foreign exchange or the foreign exchange back into your currency" states one participant. Similarly, the removal of visa, customs or passport controls at the airports, which used to exist in the previous years, leads to an increase in people's freedom to travel.

B. Factors Affecting Tourism Demand

Estimating tourism demand is an important factor that individuals, representatives of public or private industries and governments require for the future-oriented tourism development plans. To correctly direct pricing, presentation and strategic marketing programs and human resources, natural resources and capital resources, reliable and accurate demand estimation for present time and future is very important for the industry. Estimating the number of tourists visiting a destination, the goods and services they will require and the matching principle, planning several services like infrastructure, accommodation, transportation, attractiveness, advertisement and introducing and coordinating them with the necessary supporting and supplementing industries are significant for the improvement of tourism at that region as well as the competitive power and success in the long run. Besides, by estimating tourism demand, the social and environmental problems are reduced to a minimum level [15].

Tourism demand should be analyzed carefully in order to find answers for the questions, e.g., from where and how often a country mostly receives tourists, the variability of the direct income that this demand creates in comparison with the demand allocation among countries, and the numerical allocation of possible market. Furthermore, it is necessary to examine the foreign tourists visiting a country from different aspects, identify their consumption patterns and preferences and also discuss the results by analyzing demand structurally to rationalize the organizational efforts. Service firms in the tourism industry cannot maintain a balance between demand and supply as demand is unstable and seasonal. In several other industries in the economy, a balance may be maintained by either storing the goods or delaying. However, as explained in the previous chapters in detail, there is no such case for tourism as a service industry.

So, as these explanations suggest, there are numerous and various factors affecting tourism demand. Generally, the demand for all goods and services in an economy depends on the prices of rival/complementing goods and services, tastes and preferences of consumers and technology. Although factors affecting tourism demand are among the ones indicated above, there are also some other variables that have a direct impact on demand. These variables composing of social, political, psychological, legal and other factors apart from economics affect demand at different degrees and hence cause tourism demand to change [16]. The related studies in the literature state that the income per capita, relative prices and relative exchange rates/ratios are among some of the variables explaining tourism demand [17]. According to another study, the population and income level of the tourist generating country, the cost of living in the visited country, presentation and marketing expenses of that country and relative prices of tourism products are some of the factors effective on demand [18].

Box 5.2. EU Policies and Tourism [19]

The most important effect of Euro on tourism in Turkey is related to the price of tourism products. That is, Turkey is regarded as a cheap country by foreign European tourists as a non-member of EU. As indicated above in the economic effects of Euro in the tourism industry, this case creates a price competitive advantage for Turkey in economic terms and results in more tourists coming to Turkey. The most positive side of not converting its currency to Euro is that since the prices in Turkey are lower than the ones in other European Union member tourism countries, many more tourists prefer to visit Turkey. So it is expected that in the coming years the number of tourists visiting Turkey and the income attained as a result are likely to increase.

The previous studies show that almost all of the variables used to explain tourism demand are with an economic content. However, in the recent years there are factors apart from economics that cannot be included in such studies since they cannot be measured statistically in the economic analysis although they are considered as important while explaining tourism demand [20]. In this respect, according to Erdogan, the factors affecting tourism demand can be divided into two as economic and non-economic ones. The non-economic factors are generally the properties that tourists cannot afford but which provide them a psychological satisfaction [21]. According to Yarcan, these factors are divided into three as: a) economic, b) social and c) attractiveness [22]. Thereby, it is apparent that there are several factors generating and determining tourism demand. For this reason, there have been many studies in tourism economics regarding the meaning of tourism demand and the list of factors affecting it [23]. All these factors are summarized in Table 5.1. [24].

Table 5.1. Factors Affecting Tourism Demand

```
A -   Economic Factors
    1.  National Income and Income Elasticity of Tourism Demand
    2.  Income Distribution and Actual Income per Capital
    3.  Relative Exchange Rates
    4.  Distance
    5.  Price of Tourism Produts and Price Elasticty of Tourism Demand
    6.  Accommodation Potential and Supply Capacity
    7.  Advertising and Presentation
    8.  Population and Health
    9.  Transportation
B -   Social Factors
    1.  Fashion, Tastes and Habits
    2.  Spare Time
    3.  Age, Gender and Family Structure
    4.  Occupation
    5.  Level of Urbanization
    6.  Culture and Education Level
    7.  Social Values and Religion
C -   Political and Legal Factors
D -   Psychological Factors
E -   Other Factors
```

II. DEMAND ELASTICITY OF TOURISM

The changes on demand of the factors affecting demand are measured by quotient of the elasticity. These quotients are expressed in different ways. Accordingly, the frequency of the reaction as changing the demand in return of price changes and the frequency of reaction as changing the demand in return of the income changes are both measured by income elasticity of demand. Both cases are examined below respectively.

A. Price Elasticity of Demand

Price elasticity of demand is defined as "the ratio of proportional change in the quantity demanded divided by the proportional change in the price of the same product". As it is in supply elasticity, the elasticity of demand is accepted as a criterion in the tourism industry. Accordingly, if the quotient of demand elasticty is smaller than one, the demand is inelastic; if it is bigger than one, demand is elastic and, finally, if it is equal to one, demand is unit elastic [25]. Considered economically, price elasticities are important for the economic units producing tourism products because the price of products falls down, sales proceeds will increase in the case of elastic demand. If demand is inelastic, the sales proceeds will decrease in return for the price decrease.

Findings of research about the price elasticity of tourism demand demonstrate that the elasticity is quite high and generally it is more than unit elasticity. It can be concluded that the changes in the price of the tourism product may cause a much bigger change in the quantity demanded depending on the gratitue of demand elasticity. Price elasticities may be affected by various factors whose econometrical analysis and modelling are difficult. However, the main indicators are variables such as the accessibility of another tourism product, its relative importance in a tourist's budget, time for identifying price changes, whether the product is necessary or not and its being a luxurious product.

Demand elasticity in tourism shows the intensity of consumers. In other words, the tourist's response to price changes. The quotient of demand elasticity is represented by "e_F". When there is a change in price, this quotient equals the ratio of the percentage change in the quantity demanded to the percentage change in price. According to the law of demand in economics the higher the price of a product, the lower the quantity demanded. Therefore, as price and demand changes are proportionally indirect, the elasticity of demand is always negative, but there are some exceptions. Sometimes although the price of a product increases, the quantity demanded increases as well. For example, people in the tourism industry, due to the effect of free consumption or Veblen effect, may buy the tourism product despite the increasing price only for the status they will gain afterwards [26]. This is called a "snobbish demand". The demand elasticity of such products is positive due to the fact that the price and demand are directly proportional [27]. Examples on this matter include buying an expensive and luxurious car, joining a luxurious cruise or flight or staying at a hotel where not everyone can afford. The quotient of demand elasticity is calculated by using the following formula:

$$e_p = \frac{Percentage\,change\,in\,the\,quantity\,demanded}{Percentage\,change\,in\,price} = \frac{\Delta q / q}{\Delta p / p} = \frac{\Delta q}{q_1} x \frac{P_1}{q_1} = \frac{(q_2 - q_1)}{(P_2 - P_1)} x \frac{P_1}{q_1}$$

Here e_p shows the quotient of demand elasticity, q shows the quantity and p shows the price. Also, the floowing is expressed in the above equation: $\Delta q = q_2 - q_1$ ve $\Delta p = p_2 - p_1$. For example, if a 5% change in price causes a 15% change in quantity demanded, the elasticity of demand is;

$$e_p = \frac{\%15}{\%5} = -3.$$

This can be expounded as follows: *that the price elasticity of demand is -3, which shows there is a 3% increase in the quantity of the tourism product if prices fall by 1%.* The minus sign is due to the demand law, as stated above, which states that when the price of a product increases, the quantity demanded will decrease. In other words, as price and quantity are indirectly proportional, the sign of the price elasticity of demand must be minus. The only exception for this is the consumption based on prestige. The demand for and elasticity of the previously mentioned products for prestige are with a positive sign.

To illustrate, suppose the number of rooms demanded by tourists in a tourism region is 75 and each room costs 40US$ in 2000 but in 2005 the demand is 125 and the price falls to 30US$. The value of the demand elasticity for this region is:

$$e_p = \frac{125 - 75/75}{40 - 30/30} = -2$$

What is important here is that the quotient of price elasticty of demand is always negative, in other words with a minus sign. Rising of the demand as a resut of the fall in price or falling of the demand as a result of the rise in price enables the value of the elasticity quotient to be negative. It is for this reason that the minus sign is mostly not written in econometrical analysis or research as it is already known.

B. Income Elasticity of Demand

The income elasticity of demand is also called income elasticity and it is the percentage change in the quantity of the tourism product demanded divided by the percentage change in income. As the income increases, the demand for tourism also increases. Indeed, the higher a person's income is, the less luxurious it seems to him/her to travel. Because of this reason, there is a direct proportional relationship between income and tourism demand. An increase in people's income level leads to an increase in the number of people travelling.

Just as it is in the price elasticity of demand, when demand equals 1, demand is unit elastic. If it is greater than 1, the income elasticity of demand is soft and if it is smaller than 1, the income elasticity of demand is rigid.

$$e_i = \frac{percentagechangeinquantitydemanded}{percentagechangeinincome} = \frac{\Delta q / q}{\Delta i / i} = \frac{\Delta q}{\Delta i} x \frac{i_1}{q_1} = \frac{(q_2 \quad q_{1)}}{(i_2 \quad i_1)} x \frac{i_1}{q_1}$$

In this equation, e_g represents the quotient of income elasticity of demand, q is the quantity and g is the income. Also in the above equation, $\Delta q = q_2 - q_1$ ve $\Delta g = i_2 - i_1$. For example, if a 15% increase in income causes 30% increase in quantity demanded, the income elasticity of demand is:

$$e_i = \frac{\%30}{\%15} = 2$$

This can be expounded as follows: *that the income elasticity of demand is 2 shows there is a 2% increase in the quantity of the tourism product if income rises by 1%.* The positive sign shows that as people's income rises, the quantity demanded increases. In other words, income and quantity are directly proportional so the income elasticity of demand is positive.

The only exception for this occurs in the demand for inferior goods. The goods that people consume more as their income rises are called *normal goods*. In contrast. the goods people consume less as their income rises are called *inferior goods* [28]. In economics, this is called *Giffen Paradox*.

C. Cases in Tourism when Demand Elasticities are the Same

Just as it is in supply elasticity in tourism, the demand elasticity in tourism can be the same at every point in three particular cases. At these points the value of the elasticity of demand is always the same. The first of these particular supply curves is the zero elastic demand in Fig. 5.2a. In zero elastic demand, the curve is parallel to the vertical axis and at every point on the curve the elasticity is zero. In such a case, it is impossible to change the quantity demanded whatever the price may be. This means the quantity demanded remains constant regardless of an increase or decrease in prices.

The second particular situation is the unit elastic demand. As seen Fig. 5.2b the unit elastic demand is an isosceles hyperbolic curve and at each point on the curve the demand elasticity is one. In unit elastic demand, the quantity of the product and its prices change at the same rate.

The third and final particular situation is the infinite elastic demand. Demand curve is a line parallel to the horizontal axis (Figure 5.2c). At each point on the curve the elasticity is infinite ($e_p = \infty$). In such a case, a good or service produced in the tourism industry, that is the tourism product will have buyers at any price. As emphasized before, all of these three situations are exceptions and particular. The demand elasticity is constant at each point on the curve.

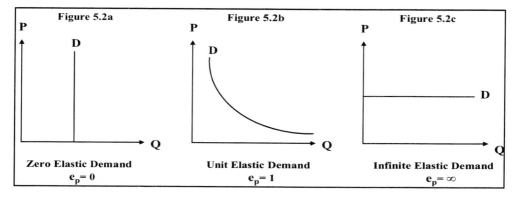

Figure 5.2. Tourism demand curves with constant elasticity.

III. DEMAND CURVE IN TOURISM

As it is seen in Figure 5.3, demand curve has a negative slope. The curve slopes downward since, according to law of demand, quantity demanded and price have a negative relationship. In other words, as the price of the tourism product increases, the quantity demanded from that good decreases. In contrast, as the price falls, the quantity demanded increases.

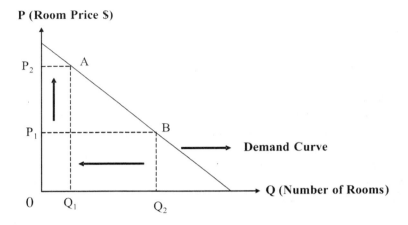

Figure 5.3. Tourism demand curve.

When price rises from P_1 to P_2, the quantity demanded falls from Q_1 to Q_2. Depending on this price increase, there is a decrease of Q_2Q_1 in the quantity demanded. The A and B points indicated on the figure show different combinations of price and quantity. According to this, as we move from point A to point B on the demand curve, the quantity demanded increases depending on the decrease in price (from upper left to bottom right). Hence, tourism demand curve occurs from the combination of the infinite number of points like A and B that are assumed to be on the curve.

The below figure shows the individual demand curve of a single consumer (tourist) who wants to buy a room at a particular price. The market demand is the cumulative demand of all tourists in a country's tourism market for a particular tourism product.

A. Change in Quantity of Tourism Demand

An increase or a decrease in any tourism product means a movement on the demand curve. The changes in demand can be divided into two as price dependent and independent of price. When there is a change in the tourism product's demanded quantity, one moves from a quantity – price combination to another. When there is an increase or a decrease in the price of a tourism product, in return for this there will be an increase or a decrease in the quantity demanded. To illustrate, considering the demand curve in Figure 5.3, a movement from point A to point B shows the change in the quantity of the demanded tourism product. This change causes the quantity of the demanded product to rise Q_1 from Q_2 as a result of the price decrease from P_2 to P_1 and there is an increase of Q_2Q_1. The other factors (level of tourist income, price of other tourism products, tastes and preferences of tourists etc.) other than the price of the product, which affect demand, are constant in the movement on the demand curve. In other words, there is a change dependent on price. Since time is not included into the analysis in the movement on the demand curve, a static change exists here.

B. Change in Tourism Demand

Shortly, a change in demand means that there is a shift in the demand curve. In other words, demand change is the movement of the demand curve to the left or right. The reason for this change is not the price. That is, there is a change independet of the price. Demand curve shifts depending on the other factor or factors affecting demand other than the price of the tourism product. For instance, change in the income level of a tourist, who is also a consumer, or his/her taste and preferences cause a shift in the demand curve. Indeed, these factors indicated above, are called "factors shifting demand". Therefore, a demand change in tourism represents a dynamic change because the change in the factors other than price results in moving beyond a period assumed as definite, which means the involvement of time factor in the analysis. This is an indicator of a dynamic change.

Figure 5.4 shows that depending on the change in one or all of the factors affecting tourism demand other than price, demand curve's moving from DD to D_1D_1 demonstrates an increase in demand while its moving from DD to D_2D_2 shows a decrease in demand. If demand increases, the curve shifts to the right and if demand decreases, it shifts to the left. As shown in the figure, the price P_1 is affected by the demand's shift to the left or right. In other words, the change in tourism demand is affected by the factors other than price.

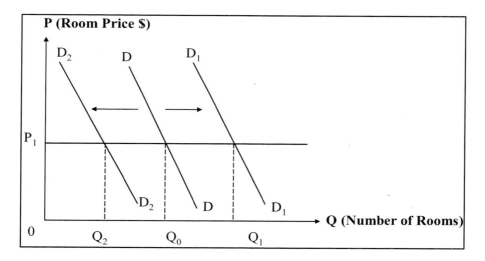

Figure 5.4. Shifts in Tourism Demand Curve.

IV. EQUILIBRIUM PRICE (MARKET PRICE) IN THE TOURISM MARKET

As indicated in the previous chapter, the tourism market is the place where demand and supply meets. Therefore, tourism demand and supply are both like a tool making up the tourism market. In this respect, the price of tourism products in the market is made up of the firms (who want to sell) and tourists (who want to buy) coming together. Hence, market or equilibrium price is the price at which the quantity demanded equals the quantity supplied.

As shown in Figure 5.5, the market or equilibrium price is the price level at the point where supply and demand curves intersect. At the equilibrium point D, the quantity of supplied tourism product and the quantity of demanded product are equal to each other. The market equilibrium price at this point is P_1 and equilibrium quantity is Q_1. At the E point at which tourism supply and demand curves intersect, all tourism products supplied to the tourism market have a buyer at the price P_1. There is neither excess demand nor excess supply at the level of equilibrium price-quantity. In Figure 5.5, at point M, the equivilant of price P_2, the tourism supply is greater than tourism demand so, there is a supply excess. Due to this excess supply, there will be competition among the producers supplying the tourism product and the price will decrease and reach equilibrium price level. Likewise, at point N, the equivilent of price P_3, tourism demand is greater than tourism supply so, there is a demand excess in the market. When there is excess demand, the competition among the consumers (tourists) who want to buy the tourism product will raise the price and the price will reach the level of equilibrium price.

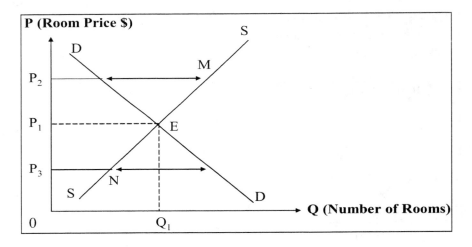

Figure 5.5. Equilibruim (Market) Price in Tourism Market.

In a tourism market where there is excess tourism supply or demand, there will be no equilibrium price because as indicated in the above paragraph, when equilibrium price exists, the quantity of supplied tourism product and the quantity demanded are equal to each other and there is no excess supply/demand in the market.

A. Effect of Demand/Supply Changes on Equilibrium Price

Equilibrium price changes only when there is a shift in tourism supply and demand curves. In this case, a new equilibrium price occurs, at which a new quantity–price combination exists. In other words, when supply is constant demand may change and when demand is constant supply may change or both of them may change all together. As a result of these, there appears a new equilibrium price and quantity. These three cases will be examined below respectively:

1. Demand Change When Tourism Supply is Constant

When tourism supply is constant, the equilibrium price will increase if demand increases somehow and the equilibrium price will decrease if demand decreases. In Figure 5.6, initally, the equilibrium price is at the equilibrium point E_0 at which the supply and demand curves intersect. In such a case the equilibrium price is P_0 and the equilibrium quantity is Q_0. The equilibrium price rises from P_0 to P_1 and the equilibrium quantity rises from Q_0 to Q_1 when tourism demand moves from DD to D_1D_1. Similar to this, the equilibrium point shifts from E_0 to E_1. When tourism demand moves from DD to D_2D_2, that is, when demand decreases, the equilibrium price moves from P_0 to P_2 and the equilibrium quantity decreases from Q_0 to Q_2. Thereby, the equilibrium point shifts from E_0 to E_2.

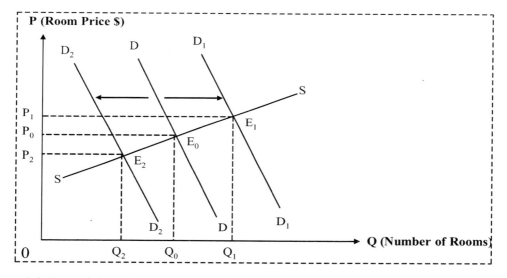

Figure 5.6. Demand change when tourism supply is constant.

2. Supply Change When Tourism Demand is Constant

If supply increases somehow while demand is constant, the equilibrium price falls down. If supply decreases, the equilibrium price rises. In Figure 5.7, initially, the equilibrium price is at the equilibrium point E_0 at which the supply and demand curves intersect. In such a case the equilibrium price is P_0 and the equilibrium quantity is Q_0. The equilibrium price falls from P_0 to P_1 and the equilibrium quantity rises from Q_0 to Q_1 when tourism supply moves from SS to S_1S_1. Similar to this, the equilibrium point shifts from E_0 to E_1. When tourism supply moves from SS to S_2S_2, that is, when supply decreases, the equilibrium price moves from P_0 to P_2 and the equilibrium quantity decreases from Q_0 to Q_2. Thereby, the equilibrium point shifts from E_0 to E_2. From here, when tourism supply increases, the equilibrium price falls whereas, when tourism supply decreases, the price increases and the quantity decreases.

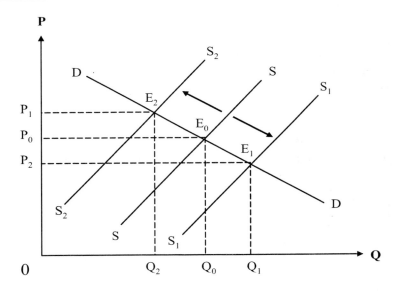

Figure 5.7. Supply change when tourism demand is constant.

3. Common Change in Tourism Demand and Supply

In case of tourism demand and supply change together, the equilibrium price will change, as well. However, the increase or decrease of price will be based on the level of the change of supply and demand size. In other words, if the demand and supply changes at the same size/rate, then the price does not change. If the increase in demand is greater than the increase in supply, price increases accordingly. On the other hand, if the increase in supply is greater than the increase in demand, price decreases accordingly. In Figure 5.8, the level of equilibrium in case of the common change in demand and supply is shown. According to this, initially the equilibrium point is E_0, equilibrium price is P_1 and equilibrium quantity is Q_1. By demand's moving from DD to D_1D_1 and supply's moving from SS to S_1S_1, that is, both of them increase, it is seen that the equilibrium price does not change. The equilibrium quantity rises from Q_1 to Q_2 and there is an increase of Q_2Q_1. Hence, the equilibrium point shifts from E_0 to E_1. Likewise, if both of them decrease at the same rate, price will not change and there will be a decrease only in the quantity.

As indicated in the above paragraph, if the rate of increase in demand is greater than the rate of increase in supply, then there will be an increase in the equilibrium price. The initial equilibrium point is E_0 and the second one is E_1. According to this, depending on the proportional change in demand, it is seen that price increases from P_1 to P_2 and quantity increases from Q_1 to Q_2 (see Figure 5.9). If supply increases more than demand, price will decrease accordingly. The proportional increase in supply will cause the equilibrium price to fall from P_1 to P_2 and quantity to increase from Q_1 to Q_2. In Figure 5.10, it is seen that eqilibrium point shifts from E_0 to E_1.

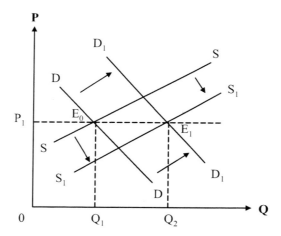

Figure 5.8. The common change in tourism demand and supply.

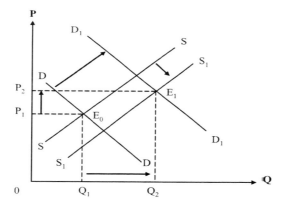

Figure 5.9. Rate of increase in tourism demand is greater than supply.

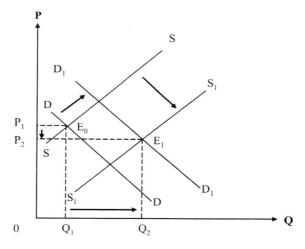

Figure 5.10. Rate of increase in tourism demand is greater than supply.

V. TOURISM SUPPLY MEETING TOURISM DEMAND

As indicated before, long term planning and estimation regarding tourism investments, the future of the industry, and that economic units forming the industry enable tourism supply to meet demand are important issues. Tourism supply should be planned by taking into consideration the necessities of the tourism destination in line with the estimated tourism demand. In order to meet the foreseen demand, there are two different statsitical methods to determine the supply.

1. Occupancy Rate Method

In this method, in order to determine tourism supply, there are two different approaches. These are: (1) determining the required number of rooms and (2) determining the required number of beds. The following equation is used to determine the required number of rooms:

$$T \ x \ P \ L$$

$$D = \frac{T \ x \ P \ x \ L}{365 \ x \ N \ x \ O}$$

Here:

D = Demand for room at a particular date,
T = Expected number of tourists,
P = Percentage of tourists staying at the hotels,
L = Average accommodation period of a tourist,
N = Average number of tourists staying at a room,
O = Average occupancy rate.

The second approach used in this method is determining the supply of beds according to the most occupied month. The following equation is used for this:

$$D = \frac{T \ x \ P \ x \ L \ x \ S}{M \ x \ Q}$$

Here:

D = Demand for bed for the most occupied month,
T = Expected number of tourists,
P = Percentage of tourists staying at the hotels,
L = Average accommodation period of a tourist,
S = Percentage of tourists staying at hotels in peak-seasons,
M = Number of days in the peak month,
Q = Occupancy rate in the peak month.

To illustrate, let's suppose Ds in both situations above. In other words, the demand for room increases to 150. Considering the supply capacity in this region is 125, the additional number of rooms and beds will be 25 for both situations.

2. Quotient of Tourism Potential Method

The mentioned quotient in this method shows the relationship between the total accommodation potential that a region may serve to the tourists at a "t" time and the permanent population of the region. Tourism potantial is calculated as follows:

$$F_t = \frac{D}{R} \, x\,100$$

Here:

F = Quotient of Tourism Potential
D = Number of beds
R = Permanent population in the region.

The rate F_t is defined as below. If:

- F = 0, there is no tourism potential.
- $0 < F \leq 1$, there is very little and ineffective tourism potential
- $0 < F < 35$, there is little tourism potential
- F = 35, the condition required to be accepted as a tourism destination
- $35 < F < 100$, it is an average tourism destination
- F = 100, it is a attractive tourism destination,
- F > 100, it is a very attractive tourism destination,
- $F = \infty$, there is no local people and the destination is open for tourism only in certain seasons.

For example, let's assume that the quotient of the tourism potential is found as 50, as a result of the calculations made for a tourism destination. It should be understood that the tourism potential here is above an average level since the value is among the average values between 35 and 100. This quotient is important in terms of the investments of accommodation supply planned in accordance with the structure of the destination population. As it is in the above example, the industry representatives either determine to invest in that destination or make no investment at all according to the estimated value of the quotient.

SUMMARY

Today, the tourism industry is often seen to be sensitive and fragile to the various natural, economic and social changes and so, the main idea of this chapter is to review the properties of tourism demand. Tourism demand made up of people with adequate time, income and desire, has a sensitive structure for the income changes as well as the price changes occurred in the regions to be visited. Therefore, it is important to examine the economic conclusions that the tourist receiving destinations arrive by means of the tourism industry from the aspect of both positive and negative sides. This may at least affect tourism income or the number of tourists to be generated. The following chapter discusses the positive or negative effects of such conclusions over the national economy.

DISCUSSION QUESTIONS

1. What is tourism demand? Explain with an example.
2. What are the properties of tourism demand? Explain briefly.
3. What do the concepts of real, strained and potential demand mean? Explain briefly.
4. How do the price and income elasticty of demand occur? Explain briefly.
5. What is Giffen Paradox? ? Explain briefly.
6. What are the demand shifting factors in tourism demand? Explain with a proper example.
7. How does the equilibrium price occur in tourism demand? Explain briefly.
8. How will the equilibrium price change in case of a common change in demand and supply? Explain briefly.
9. What are the statstical methods in determining tourism supply to meet tourism demand? Give an example.

REFERENCES

[1]   Begg et al., p. 29.
[2]   C. Lim (1997), "Review of International Tourism Demand Models," *Annals of Tourism Research*, 24(4), pp. 835–849.
[3]   Bahar, Kozak 2005.
[4]   Olalı, Timur, p. 195.
[5]   Usta, p. 114.
[6]   İ. Ç. Oktay (1997), *Güncel Turizm Yönelimi İçinde Kongre Turizmi ve İstanbul*, Ankara: Turizm Bakanlığı Yatırımlar Genel Müdürlüğü Araştırma ve Değerlendirme Dairesi Başkanlığı Yayını, p. 4.
[7]   C.J.S.C. Burger, M. Dohnal, M. Kathrada and R. Law (2001), "A Practitioners Guide to Time-Series Methods for Tourism Demand Forecasting," *Tourism Management*, 22, p. 404; V. Cho (2003), "A Comparison of Three Different Approaches to Tourist Arrival Forecasting," *Tourism Management*, 24, p. 323.
[8]   Yarcan, *Turizm Endüstrisinin Yapısı*, p. 29.
[9]   Dallı, p. 6.
[10] Olalı, Timur, p. 196.

[11] http://rptsweb.tamu.edu/courses/331/RPTS%20331/homepage/class/Class%20Notes/Ch apter%204/TourismBehavior_Spring%202006.PPT#258,3,Why is This Important? (Retrieved 22 October 2007).

[12] M. Uysal, M.S. El Roubi (1999), "Artificial Neural Networks Versus Multiple Regression in Tourism Demand Analysis," *Journal of Travel Research*, 38 (November), p. 111.

[13] E. Smeral, A. Weber (2000), "Forecasting International Tourism Trends to 2010," *Annals of Tourism Research*, 27(4), pp. 982–1006.

[14] K. Meier, E. Kirchler (1998), "Social Representations of the Euro in Austria," *Journal of Economic Psychology*, 19, pp. 755–774.

[15] Gee, Fayos-Sola, p. 200.

[16] H. Kazdağlı (1995), *Türkiye'de Turizm Talebi ve Talebi Etkileyen Faktörlerin Analizi*, Ankara: Turizm Bakanlığı Yayını, p. 1.

[17] Uysal, Roubi, pp. 111–118.

[18] R. Law (2000), "Back-Propagation Learning in Improving the Accuracy of Neural Network-Based Tourism Demand Forecasting",*Tourism Management*, 21, pp. 331–340.

[19] O. Bahar, M. Kozak (2006), "Potential Impacts of Euro on Destination Choice." *TOURISM: An Interdisciplinary Journal*, 54 (3), pp. 245-253.

[20] H. Song, K.K.F. Wong and K.K.S. Chon (2003), "Modelling and Forecasting the Demand for Hong Kong Tourism," *International Journal of Hospitality Management*, 22, p. 438.

[21] Erdoğan, p. 342.

[22] Yarcan, *Turizm Endüstrisinin Yapısı*, p. 29.

[23] Olalı, Timur, *Turizm Ekonomisi*, pp. 197-213.

[24] For detailed information about all factors influencing tourism demand see; O. Bahar, M. Kozak (2005), *Uluslararası Turizm ve Rekabet Edebilirlik*, Ankara: Detay Yayıncılık, pp. 27-42.

[25] http://neutralsource.org/content/blog/detail/519/ (Retrieved 22 October 2007).

[26] Lundberg et al., p. 34.

[27] Dinler, p. 80.

[28] Dinler, pp. 72-74.

Chapter 6

ECONOMIC EFFECTS OF TOURISM

It is not always possible to state that tourism has completely positive effects on a country's national economy. More clearly, the improvement of tourism in a country/region causes some positive or negative societal and economic changes, too. For the purpose of limiting the subject, this chapter will examine only the macro- and microeffects of tourism on the national economy by emphasizing the positive and negative aspects. Hence, within the scope of macroeconomic impacts, the effect of tourism on balance of payments, income generation, employment, and balanced development among the regions and other industries as well as on developing infra- and superstructure can all be assessed. Similarly, prices and salaries, opportunity costs and internal and external economies are the examples of microeffects of tourism.

I. POSITIVE ECONOMIC EFFECTS OF TOURISM

In today's global world of hard economic conditions as well as harsh competition, the importance of tourism on countries' economies is clear. Regardless of the level of a country's development, tourism has a certain share on the economic development of countries. Particularly in developing countries, it makes tourism even more important that the savings are inadequate, there is lack of exporting possibilities and sources, the necessary foreign exchange income is gained cheaply and without any effort. Indeed, to illustrate the economic importance of tourism in such countries, the deficit of savings and foreign exchange in less developed and developing countries gets bigger each day. As a result, the need for foreign exchange, the amount of foreign debts, the investments needed for development, and new job opportunities are among the examples. Today, throughout the world in both developed and less developed countries, no matter scientific or not, it is emphasized in all meetings, conferences and seminars that tourism is necessary and it has a great importance in terms of the country's economy and the discussions on its economic effects are stil ongoing.

The service industry worldwide is developing day by day and it makes up approximately the 70 – 80% of a country's gross domestic product. According to the 2005 data, tourism, which is a part of service industry and considered as one of the biggest industries of the world, constitutes 30% of the world's trade of services by itself [1]. Here, tourism as a labor intensive industry, has become a distinctive industry in many countries with its property of creating an inflow of foreign exchange and new income, and increasing employment opportunities especially in rural areas in order to achieve the economic development of developing countries. Moreover, considering the cost of foreign exchange attained from tourism income that it is consumed at the production location and it is a privileged industry in terms of subsidies, the cost of exporting a tourism product is less than the cost of exporting products in other industries. For this reason travel/tourism is second in international trade following computer, information, communications, and other commercial services [2]. Hence, tourism is an important tool of economical policy in the resolution of economic problems that countries face, finishing the foreign trade deficits and overcoming the troubles. Following this brief introduction, the priority is given to the positive economic effects of tourism discussed below.

A. Effect of Tourism on Balance of Payments

As emphasized in the previous chapter in detail, the foreign balance of payments is an account showing the status of the external economic and financial relationship a country has with the rest of the countries in the world [3]. In international tourism, it has a direct effect on the balance of payments account as an invisible exporting item. Therefore, it exists in the international services part of the mentioned balance sheet. The balance of payments is important in terms of its showing the amount of tourist spending (the foreign exchange inflow to the country) and the expenditure that the citizens travelling abroad make (the outflow of foreign exchange), and as a result, the absolute contribution that tourism has on the economy. If the foreign exchange income that a country gained from tourism is greater than the outflow of foreign exchange, this affects the balance of payments account positively. In contrast, if the foreign exchange income that a country gains from tourism is smaller than the outflow of the foreign exchange expenditure, this affects the balance of payments account negatively. In this respect, it is necessary that the country's foreign exchange income gained by tourism should be more than the foreign exchange loss in order for tourism to have a positive effect on the balance of payments.

FOREIGN EXCHANGE INCOME >	FOREIGN EXCHANGE EXPENSE	⇒	POSITIVE EFFECT
FOREIGN EXCHANGE INCOME <	FOREIGN EXCHANGE EXPENSE	⇒	NEGATIVE EFFECT

These two effects mentioned above are in fact the essence of the balance of foreign trade. That is, the tourism income and expenses arise from the relations of a country with other countries in a certain time are shown in the balance of foreign exchange (see Table 6.1) [4]. There is no generally accepted model that is determined regarding the arrangement of the tourism balance sheets. In this respect, while it is known that countries make various arrangements and models, regarding the tourism balance sheets it is seen that two basic models are used as wide and narrow perspectives [5].

- *Tourism balance sheet as a narrow perspective* is composed of the tourism income and expense accounts arisen as a result of the direct expenditure tourists make.

- *Tourism balance sheet as a wide perspective* is composed of the income and expense accounts arising from the transactions that tourists as well as the other people and firms in an economy make for tourism and travel purposes.

As seen in Table 6.1, the balance of foreign tourism is made up of two different accounts as active foreign trade at which tourism income is shown and passive foreign trade at which the tourism expense is shown. Therefore, the excess in the balance sheet will surely reflect in the balance of payments positively.

Table 6.1. Balance of Foreign Tourism [6]

Active Foreign Tourism (Tourism Income)	Passive Foreign Tourism (Tourism Expenses)
Payments tourists made such as accommodation, food and drinks, transportation, etc.	Citizens' expenses made abroad such as accommodation, food and drinks, transportation, etc.
Exporting the consumption goods for tourism purposes and souvenirs	Importing the consumption goods for tourism purposes and souvenirs
Foreign investment for tourism purposes and profit transfers to the country	Foreign capital investment for tourism purposes and profit transfers outside the country
Rental income of the facilities for tourism purposes by foreigners	Payments made for the rented facilities in foreign countries for tourism purposes
Payments made for the training of foreign staff	Payments made for the training of the staff abroad
Advertisement and promotion expense in the country	Advertisement and promotion expense abroad
Commissions taken from the foreign tourism firms	Commissions given to the foreign tourism firms
Payments made to the banks and insurance companies in foreign countries	Payments made by the local firms to the banks and insurance companies abroad
Other foreign exchange income	Other foreign exchange outflow
Net Expenditure	Net Income

The "Ideal Balance of National Tourism," which is prepared on this subject by the international tourism instutions like World Tourism Organization and some economists, is shown below in Table 6.2. The purpose here is to show what the ideal balance of tourism that enables an effective and complete economic efficiency of all the tourism activities and transactions. However, in order to identify the ideal balance of tourism perfectly and put forward the real effect of tourism activities, it is crucial to record the statistical data of countries very well.

Tourism has another positive impact on the balance of payments as a stabilizing factor in international trade. For example, some developed countries like Japan, whose balance of payments has surplus, encourage their citizens to travel and spend abroad in order to prevent this surplus from creating an inflation effect. In contrast, many less developed or developing countries with a deficit in their balance of payments, try to increase the number of foreign tourists visiting the country.

Table 6.2. Ideal Balance of National Tourism

PART	TOTAL	PART	TOTAL
Tourism Costs (Expenses of citizens abroad)	----	Tourism Income (Expenses of foreign tourists)	----
Importing Goods (Main food and intermediary goods)	----	Exports (Durable or semi-durable goods, handicrafts)	----
Transportation (Travelling share provided by citizens)	----	Transportation (International travel provided by except for citizens)	----
Tourism investments abroad	----	Foreign tourism investments in the country	----
Expenses regarding foreign investments and paying back the capital	----	Income gained from investments made abroad	----
Income returns from the payments made to foreign tourism workers	----	Paying back the income paid to the national tourism workers abroad	----
Public Relations, advertising	----	Public Relations, advertising	----
Balance of Debt	----	Balance of Credit	----
= Deficit	----	= Surplus	----
Total		Total	

Hence, their national income to prevent this deficit from causing some negative conclusions in income distribution, employment, salaries and prices, the convenient and effective sustainability of the external debts and other macro and micro sizes. In so doing, countries try to attain a balance of payments by the income gained by tourism or the expenses [7].

Box 6.1. Tourism and Balance of Payments

In Table 6.3. the balance of payments of Turkey is shown by years in summary. The table points out that the country's balance of foreign trade is constantly negative. In other words, there is a deficit and this deficit increases year by year. On the contrary, it is seen from the table that the tourism balance has a tendency to increase. Considering the 2005 data, we see that an income of 15.048 billion US$ has been gained from tourism. So, it is clear that tourism's share in closing the foreign trade deficit of Turkey rose from 9.3% in 2004 to 51.9% in 2005 (see Table 6.4). Taking into consideration that this rate is 139.8% in 2001 and 110.1% in 1988, it is evident that tourism has a great impact on the balance of payments. In other words, the tourism industry has an important role in compensating the foreign trade deficit of Turkey.

Table 6.3. Balance of Payments in Turkey (Million US$) [8]

YEARS	1990	1995	2000	2001	2002	2003	2004	2005
I-CURRENT ACCOUNTS	-2625	-2339	-9819	3390	-1524	-8035	-15604	-18740
I-a-Balance of Goods , Services and Investment Income	-6990	-6737	-14593	-413	-3960	-9062	-16731	-20039
I-b-Balance of Goods and Services (A+B)	-4482	-3532	-10591	4587	596	-3505	-11094	-14807
I-A-BALANCE OF FOREIGN TRADE	-9448	-13152	-21959	-4543	-7283	-14010	-23878	-28968
I-Aa-Total Goods Export	12959	21636	30721	34373	40124	51206	67047	69326
I-Ab- Total Goods Import	-22407	-34788	-52680	-38916	-47407	-65216	-90925	-98294
I-B-BALANCE OF SERVICES	4966	9620	11368	9130	7879	10505	12784	14161
I-Ba-Income	8083	14939	20364	16030	14783	19025	22928	24657
I-Bb-Expense	-3117	-5319	-8996	-6900	-6904	-8520	-10144	-10496
I-B2-TOURISM	*2705*	*4046*	*5923*	*6352*	*6599*	*11090*	*13364*	*15048*
I-B2a-Income	3225	4957	7636	8090	8479	13203	15888	17602
I-B2b-Expense	-520	-911	-1713	-1738	-1880	-2113	-2524	-2554

On the other hand, as well as being a stabilizing factor in terms of equal income distribution and economic development among the less developed geographical regions of a country, tourism also helps to overcome the economic instabilities among countries. Because many less developed and developing countries bring in an important amount of foreign exchange income to their countries by marketing the values peculiar to their countries/regions, goods, services and events to the foreign tourists besides the natural, geo-economical, socio-cultural and historical assets by means of tourism.

The income that the above mentioned countries have gained and will gain from tourism as an economic activity has more advantages than the economic aid, debts and supports because this type of income is far away from the political and economic limitations and they cannot be used as a restriction tool at all. Besides this, the less developed and developing country can determine the price of the tourism product itself. Therefore, the tourism product could be sold to foreign tourists at a higher price than in the domestic tourism market compared to the other exporting goods in the economy [9]. Moreover, with this aspect tourism makes a diversification and product differentiation in terms of exporting and the fluctuations that might arise accordingly can be minimized or abolished. As stated earlier, since foreign exchange income attained from tourism is less costly than the foreign exchange income attained from other industries, the income gained from this industry is used for improving the other industries, mostly production, sustaining development, alleviating the load of external debts, creating new job and employment opportunities and as a result increasing the national income per capita.

B. Income Creating Effect of Tourism

It is doubtless that the most important economic and positive effect of tourism is its income generating effect. The expenditure that tourists make at the countries/regions they visit for tourism purposes to meet their necessities such as accommodation, entertainment, travel, eating and drinking, transportation, shopping, and gifts cause the economical income to increase at that location. To express it more clearly, these types of expenditures with tourism purposes will make up the local people's wages or prices. According to a research by the WTO, in at least 38% of the countries participating in the research, tourism is the main source of income. According to the same research, tourism exists among the first five industries among the 83% of the participant countries in exporting categories.

However, income effect is not limited with this. As indicated before, as the tourism industry is related with many industries of different sizes, the tourism income gained will cause the production, consumption, import or export of many goods and services in these industries. Hence, the tourism income attained will be spent by the owners of factors of production (the ones providing construction, agriculture, industry and other services) in other industries and transfered to other people in the economy in various ways, so it will be an income for other owners of the factors.

Table 6.4. Share of Tourism in Closing Foreign Trade Deficit (Turkey) [10]

Years	Balance of Foreign Trade (Million US$)	Balance of Tourism (Million US$)	% of Tourism in Compensating Foreign Trade Deficit
1984	-2.910	271	9.3
1985	-2.976	770	25.9
1986	-3.018	637	21.1
1987	-3.206	1.028	32.1
1988	-1.813	1.997	110.1
1989	-4.190	1.992	47.5
1990	-9.448	2.705	28.6
1991	-7.290	2.062	28.3
1992	-8.076	2.863	35.5
1993	-14.081	3.025	21.5
1994	-4.167	3.455	82.9
1995	-13.152	4.046	30.8
1996	-10.264	4.385	42.7
1997	-15.048	5.286	35.1
1998	-14.052	5.423	38.6
1999	-10.185	3.732	36.6
2000	-21.959	5.923	27.0
2001	-4.543	6.352	139.8
2002	-7.283	6.599	90.6
2003	-14.010	11.090	79.2
2004	-23.878	13.364	56.0
2005	-28.968	15.048	51.9

As known, this relationship of income and expenditure, which is called the "*multiplier mechanism*" is discussed in Chapter 3 in detail, so the issue is kept in short here.

Besides this, it is important to consider the following issue. In economics literature the more money changes hands, or in other words, the greater its circulation rate, the more the income effect it creates [11]. In this respect, in economics the expenditure made by tourists, first of all, creates an income effect as much as their own size. Later, a part of this income is transfered in various ways as investment, savings, spending and tax and causes new income to occur. Tourism firms gaining income from tourists buy goods and services from other firms of different industries in order to sustain the services they have and hence, a part of the income gained by tourists return back to the economy by means of spending. The final tourism income occurred as a result of this mechanism and is quite higher than the consumption and investment expenditure that the foreign and domestic tourists make.

These economic events occurring subsequently continue until the amount of income is equal to zero (0) [12]. As a result, the expenditure made with tourism purposes in a tourism country causes an income increase of a much bigger size by means of multiplier effect [13]. To illustrate, the experience gained from the Europe tourism year shows that the multiplier effect of the expenditure the EU made for tourism is high because the multiplier effect of only a 20% expenditure is 500%, which is an indicator of an important impact [14]. According to the findings of research in which Turkey exists as well, the value of the multiplier effect of the country is indicated as 1.96 [15]. Here, we can conclude that the tourism spending made by the tourists result in an income increase almost double the size of itself.

Box 6.2. Income Impact of Tourism

Table 6.5 shows the income impact of tourism in Turkey. While in 1980 the income obtained from tourism is 327 million US$, it rises by 5288.5% (almost 53 times) in 2005 and reaches to 17.602 billion US$. Similarly, the expenditure per tourist in 1980 is 253.6 and it rises by 167.7% reaching 69 US$. Moreover, according to the tourism income data of WTO in 2005, Turkey is the eighth country in the world with 15.9 billion US$ income and a 2.6% share [16]. In the same table, it is seen that the subsidies in Turkey have such a big increase in the amount of tourism income that there are significant increases between 1980 and 1990 as well as between 2000 and 2004.

Table 6.5. Income Effect of Tourism [17]

Years	Number of Tourists (Million)	Rate of Increase (%)	Tourism Income (Billion US$)	Rate of Increase (%)	Expenditure per Tourist (US$)	Rate of Increase (%)
1980	1.288.060	-	326.654	-	253.6	-
1985	2.614.924	103.0	1.482.000	353.6	566.7	123.4
1990	5.389.308	106.0	3.225.000	117.6	621.3	9.7
1995	7.726.886	43.3	4.957.000	53.7	684.0	10.0
2000	10.428.153	35.0	7.636.000	54.0	764.3	11.7
2004	16.826.000	61.3	15.888.000	108.0	661.0	-13.5
2005	20.522.621	21.9	17.602.000	10.7	679.0	2.7

C. Employment Generating Effect of Tourism

Tourism is an important industry in generating income as it has a labor intensive structure. Tourism is a labor intensive industry and hence, it makes up one of the main sources of employment in the rural or less developed region that is suitable for tourism. As indicated before, tourism is the combination or intersection of many industries. Therefore, while the tourism expenditure for tourism purposes creates a direct employment effect at first stage, an indirect employment effect is provided by spending the income again at the second stage. As a final stage, an additional employment effect is attained by the local people's spending the income that they gained from the tourism firms again [18].

Parallel with the development of tourism and the increasing tourism demand in a region, new facilities and investments should be made. Hence, the first of income attained from tourism is spread to the other industries in the country's economy and cause a spending and income flow. New investments and employment fields occuring as a result of the development by tourism result in an increase in employment by increasing demand [19]. As known, due to the tourism's seasonal structure, the increase in employment in the summer time when tourism is intense is much more than the increase in winter. It also creates a benefit for employing under qualified employees in this term. However, with implementing the investmenst and operating the facilities, the number of qualified employees in this industry will increase accordingly.

Here, the main properties of employment in the tourism industry are summarized as follows [20]:

- The number of female employees in the tourism industry is higher compared to some other industries. The reason for this is that there are more career opportunities for women. However, the level is not as desired for the tourism industry.
- Seasonal employees are common in the places where mass tourism is intense (e.g., Turkey, Greece, Eygpt, Spain, Thailand, etc).
- The work force in the tourism industry has a mixed structure in terms of socio-economical and socio-demographical aspects. The people of different ages, occupations (student or the ones with a previous job), income and culture may be hired in the same firm in tourism.
- Due to the intensity of the seasonal workforce employed as full time or part time, it may not be possible to find out statistically how many people are employed by the tourism industry at the national and international level.
- It is difficult to determine how much of the workforce employed in tourism serves directly to that industry or to the local people.
- The turnover rate in tourism is high. The hard working conditions, seasonal structure, low wages and other attractive offers increase this rate.
- It is sometimes possible to employ foreign workforce in tourism at the international level in changing amounts based on the nature of the countries, e.g., tourist guides working in travel agencies, reception clerks at hotels.
- The staff working in the sub-sectors like transportation and accommodation serve 24 hours a day. Therefore, service has a continuous structure and consumers may demand for 24 hours a day in the tourism industry.

- Many sub-occupations created by the tourism industry require high level of skills, e.g:, gardening, cleaning work, purchasing and stocking.
- The power of unionization and negotiation rights are low in the tourism industry. This is because tourism has not still had a systematical structure with a strong bargaining share.
- There are differences between the property of the work force employed in developed and developing regions and the level of the salaries paid. Developing or less developed countries can employ the personnel with lower salaries than do the other countries.

Box 6.3. Tourism and Employment

The tourism industry constitutes 8% of the total employment of 200 million people in the world. In line with this improvement, for the last 30 years, more than 8.5 million people have been employed in new fields. It is estimated that until 2010, it will have made up the 12% of the world GDP and 9% of the total work force [21]. In EU, approximately 8 million people are employed directly, which is 6 % of the EU total employment. Furthermore, it is predicted that tourism will employ 3.3 million people in the next 10 years [22]. This number may exceed 20 million including the indirect employment as well [23]. Considering the employment report of Turkey's tourism industry, according to the Association of Turkish Travelling Agencies (TURSAB), direct employment in the industry exceeded 1.2 million people by 2003. This number, in addition to the indirect employment (total employment) is expected to have exceeded 3 million [24]. Considering the family members in terms of direct and indirect employment, it is seen that approximately 10 million people earn money in the tourism industry in Turkey [25].

D. Effect of Tourism on Interregional Economic Development

International tourism has a direct contribution on the balance of payments account as an invisible export item and therefore, there is an outflow to the country's economy. As it is in other exporting industries, public revenues, income per capita and employment increase and new job opportunities are created by means of spending this income again within the economy. In this respect, international (foreign) tourism brings in foreign exchange to the country, domestic tourism provides this income to be allocated again in the country and has a positive effect on the economic wealth and development among the regions [26]. The tourist spends in the available places of the country for tourism as a result of tourism activities and the income gained there is spent in other parts of the country as investment and new fields for employment.

Besides this, tourism introduces more effective alternative opportunities for the economic development in the less developed regions of the country. In fact, in such less developed regions tourism has important effects. That is, people in these regions like farmers or fishermen work for the sake of tourism or earn income by means of this industry and can increase their total income at a greater extent. The widespread growth of tourism in such regions will lead to the development of traditional activities such as handicraft or gifts.

Hence, this situation will provide a monetary source and incentive to the region and local people will have the chance to introduce peculiar goods and services to the hotels/firms in the region. This will contribute to the development of the region's economy. So, the widespread growth of tourism in less developed regions of a country may have a more positive impact on the economic prosperity of the local people than it does on developed regions.

On the other hand, tourism has an aim of minimizing the developmental differences among regions by economic and social development. As the countries/regions that do not have the sufficient resources and developmental opportunities in terms of agriculture and industrial activities but have substantial tourism potential implement a planned and effective tourism policy, a balanced development is achieved in tourism. As a conclusion, the economies of the mentioned countries get a certain amount of share from the international tourism revenues. In this respect, although tourism policies and in general the foreign exchange income are not merely enough to accomplish economic development, they are important since they accelerate development and have complementary effects [27].

Another effect of tourism on interregional economic development is that the people of that region get the chance to reach high standards in public transportation, infrastructural systems or shopping centers, which perhaps could not have been attained in another way, due to the tourist mobility to the country. Hence, tourism provides opportunities for reducing the unemployed human sources that do not have certain qualifications, diversifying the economic structure and developing other industries in the country/region [28].

Tourism's continuous development in a region increases the demand for the scarce resources in that region. For example, if the demand for land increases, the price of land increases as well. In such a situation local farmers and land owners living in the less developed region begin to sell the land they have. Although they get an important and sufficient amount of revenue in the short run, they work for low salaries in the long run. The ones who benefit most from this situation are the speculators trying to attain great revenue and profit from those lands.

E. Effects of Tourism on Other Industries of the Economy

Just as the other industries in an economy, the tourism industry also directly or indirectly affects the other industries in the place/region or country where the tourism activities take place. This effect can be important for both the place where the tourism activity happens and the region or national economy as well. The effects tourism have on the agriculture, manufacturing and services industries are discussed below respectively.

1. Effect of Tourism on the Agricultural Industry

The development of tourism in a region causes the awakening and brisking of that region's economy because in a tourism region there are several hotels, apart hotels and holiday villages. In order to satisfy the foreign tourist's eating and drinking needs, the local people, especially the ones working in the agricultural field, have important missions. In order to meet the increasing demand in tourism season the population working in the agriculture industry should produce more. This is an important factor for serving more qualified goods and services to the tourist and also, this is an invisible source of export for the agricultural industry. The price of the agricultural goods provided for tourism in this way, is by far more than it is in the normal season. In the agricultural industry that tourism affects indirectly, the price of the agricultural goods that is low at normal times will increase and an additional value added will be attained by this way.

The increasing demand for the agricultural goods in the tourism season will cause the national income and income per capita to increase as well as the local people's supply of more qualified goods and services. The mentioned increase in production will also enable new employment opportunities as well. People will tend to a structural change from the agriculture industry to tourism in order to benefit from the added value increase in the tourism industry. In other words, for the sake of attaining more income from tourism, local people operate, rent or sell their agricultural land, and farm themselves. Hence, the level of wealth of the local people increases by means of tourism.

However, there is an issue to be pointed out in particular. As tourism develops especially in the rural areas, there are construction and purchasing of the second residences, which has negative effects on the local economies in tourism regions. The decrease in the agricultural land of the local people who cannot compete with large tourism investments and do not have any tourism income cause them to fall into a marginal situation economically [29]. In this respect, while the tourism industry has a positive effect on the agricultural industry, it has also some negative effects. Tourism is not only an income creating activity but it is also an industry that has a role in eliciting the economical wealth and happiness of the future generations [30].

2. Effect of Tourism on the Manufacturing Industry

The effect of tourism on the manufacturing industry differs according to the sub-sectors of production of the manufacturing industry. As known, the production of manufacturing industry is divided into three sub-sectors as consumption goods, intermediary goods and investment goods. Therefore, the effect of tourism on these sub-sectors also differs. Accordingly [31]:

- Effect of tourism over the industries producing consumption goods (food, beverage, etc.) is more.
- Effect of tourism over the industrial section producing such intermediary goods such as leather, leather products and ceramic is high.
- Effect of tourism over the industries producing investment goods is not much.

The major effect of tourism on the manufacturing industry appears on the industrial goods used in tourism investments when they increase [32]. These cover the heavy industrial goods at most and the infrastructural facilities like cement, steel, computer technologies, highway, marinas and airports, other industrial goods like construction, forest and ceramic. Thanks to tourism, in the industries supplying these equipment and the industrial branches as well as the necessary infra- and super-structure, there is a liveliness and activation. There are production increases in such industrial branches as a result of the increasing demand especially in the tourism season.

3. Effect of Tourism on the Services Industry

From the aspect of balance of payments, international tourism is assessed within the international services account. From the aspect of tourism, the most important account in the balance of payments are the international services. Therefore, tourism also affects the services industry in which it is included at a great extent and contributes to its development in various ways. In this respect, the effect of tourism on the services industry is examined as follows:

1. It assists the third production industry regarding current consumption to develop (e.g., bread that is sold in markets, meat, milk, fruit and vegetables, etc.)
2. It assists the third production industry regarding equipment arts (e.g., electrician in charge of various technical services, painter, blacksmith, construction workers, etc.)
3. It helps the third production industry regarding comfort (e.g., designers, sports equipment, perfumery, news agency, florist, patisserie, café, etc.)
4. It helps the third production industry regarding the assistance and security services (e.g., health facilities, banks, insurance, police)
5. It helps the third production industry regarding the services considered as luxiruous by the society (e.g., jewelry shop, night club, sauna, monitors, antique shops)
6. It supports the maintenance of the servcies for holiday and tourism (e.g., transportation, petrol stations, historical places, entertainment places and recreational activities)
7. As tourism develops in a region, the public services in that region also develops (e.g., road, water, electricity, communication, drainage systems and social residences).

Hence, it is possible to state that the tourism activities in a region cause some qualitative and quantitative developments on the above mentioned activities in the service industry by means of expenditure–income relationship.

II. NEGATIVE ECONOMIC EFFECTS OF TOURISM

Generally, the positive economic effects of tourism are discussed. On the other hand, it is not always possible that the tourism movements in a region have positive economic effects. The development of tourism in time shows that the benefit from tourism may decrease and cause negative developments. In this respect, the negative economic effetcs that tourism cause can be listed in Figure 6.1 [33].

A. Importing Effect

Although it is stated that the international tourism movements encourage the economic independence of countries, it is accepted that tourists travelling from the developed countries to the less developed countries are dependent on the region's economy. Because of the effects with consumption purposes or the ones that tourism may introduce in the consumption habits of the society, some goods (food, construction equipment, kitchen equipment, etc.) need to be exported.

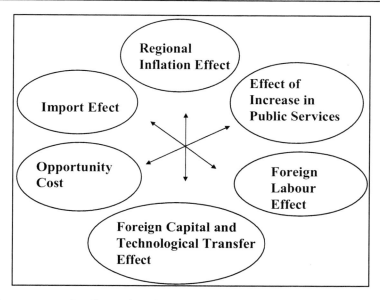

Figure 6.1. Adverse economic effects of tourism.

As tourism will bring technological investments, the importing tendency of the country may increase. This is so especially for the less developed countries without the necessary capital and technology. Developing tourism in certain regions in the country, the demand for goods and services that domestic and foreign tourists as well as the people coming to that region for work may cause the prices of the goods and services in the destination to rise (regional inflation). Mostly, the local people inhabiting there for so long are influenced by this situation.

The tourism industry may require a higher priority at some periods than other industries. Fostering only the tourism investments may cause an instability among industries. If the qualified work force cannot be attained from the region, the work force can be transfered from either outside the region or abroad. This will cause the multiplier effect to decrease since the income gained is off the region. The same property is valid for the foreign capital investments. International tourism investors transfer the investments they make in other countries to the center.

The positive or negative effects of foreign capital on the tourism industry of the region are still under discussion and it may not be possible to reach an exact conclusion on this issue especially after the developments regarding EU [34]. Finally, tourism may increase the load of current public services in the region. Additional spending should be done in order to make new infra- and super-structure investments and renew the existing ones. For example, for 92 fire alarms reported by the tourism firms, 63,300£ was spent (among the total of 2,767,000£). Therefore, it is useful to investigate some of the important negative effects listed above.

B. Effects of Tourism on Inflation

In economics, inflation occurs if there is rise in the general level of prices when the aggregate demand exceeds the aggregate supply. In other words, inflation means the continuously rising prices and as a result, the decrease in the purchasing power of money [35]. Inflation has three main properties:

- Prices of all goods and services in the economy will increase
- Such an increase in prices will be permanent
- Such an increase in prices will be at important levels.

When an impact arises in a country's economy so rapidly, it is considered inflation. The reasons why inflation is important are explained as follows: Due to inflation, the income per capita in a country is allocated unfairly. Inflation has an effect of distorting the income distribution. Moreover, it causes the production to decrease, the external value of money to fall faster than its internal value and hence, exporting to increase and exporting to decrease. The inflation brings forth the concept of "running away from money" and it enables an increase in constant capital investments and economic activities to stop. Last, confidence in the government is damaged and the entrepreneurs do not want to invest any more [36].

The inflation effect that tourism makes on a country's economy is seen in different ways. To indicate these briefly: Considering the fact that the purchasing power of foreign tourists is much more than that of the local people, with the tourism season in less developed and developing countries such as Turkey, Tunisia, Eygpt, and Morocco, tourists accept to buy the high priced goods and services and pay the higher price. The retailers operating in the region want to increase their profit by rising the price of tourism products [37]. An increase in prices will lead to an increase in costs and hence the inflation rate will tend to increase even in regional terms.

As a result of these general price increases, in tourism season, especially real estate and land prices will increase in line with this. In this case, because of the fact that the value of the real estate increases, people will pay more rent and tax and hence, the inflation pressure will increase [38]. In this respect, whether an inflationist pressure occurs in a tourism region/country with the development of tourism or not depends on the following three conditions [39]:

1. First, in a country if the foreign exchange attained by means of tourism, in other words, net foreign exchange income is less than the foreign exchange expenses, it may be said that tourism creates an inflationist pressure because the foreign exchange income gained in such a situation will have been absorbed by the "tourism product" that the industry produced. To express it more clearly, should the balance of foreign exchange demand and supply in the country be destroyed, the price stability in the economy cannot be maintained and hence this will cause inflation. It is inevitable that inflation occurs when the amount of foreign exchange demanded in the country by means of tourism exceeds the foreign exchange supply.

2. Tourism is considered as an important industry by the employees working in other industries in the national economy, and if they choose tourism as a pilot activity, it is inevitable that tourism has an inflationist pressure. As it is in developing countries, in the tourism industry there is a chance of attaining more income in comparison with the other industries. So there is a flow of jobs and entrepreneurs from other industries towards tourism. There is a possibility that this situation creates an inflationist

pressure on the tourism industry and the national economy in general. This possibility is in line with the economic share of tourism in the country's economy.

3. Whether tourism will create an inflationist pressure or not can be defined according to the actual foreign exchange value. That is, if the amount of national money spent with the effect of tourism is greater than the sum of net foreign exchange income, tourism will create an inflationist pressure since, in such a case, the actual foreign exchange will be less than 1. However, for situations when foreign exchange equals 1 or is bigger than it, there is no such inflationist pressure.

Maintaining the price stability is a monetary policy important for every country's economy today. Therefore, it is important for the representatives of the private and public industries to take the necessary precaution on time for the sudden and unexpected price movements and the fluctuations in the tourism industry in terms of the citizens of the country affected by inflation in the minimal and most harmless way. This is because inflation surely causes some economic distortions and recession such as cost and price of inputs, decrease in tourism demand, price instability, lack of investments in efficient areas and a negative impact on the country's trade [40].

C. Opportunity Cost of Tourism

As known, the existence of economy-related problems require choosing among the alternative utilization facilities of the scarce resources. In this respect, as indicated in the first chapter of the book in detail, the *opportunity cost* of a good is foregoing the production of another good in order to increase the production by one unit. In other words, the opportunity cost means choosing among various scarce resources, sacrificing or not using the unchosen alternatives.

To review in terms of the tourism industry, comparing the income countries gain by investing in the tourism industry with the predicted income gained by investing in other industries is briefly called the opportunity cost of tourism. In this respect, the concept of opportunity cost can be perceived as the maximum amount of outcome that can be attained by directing the scarce economic sources of the country in the best way. For example, assume that an amount of resource has been saved from the budget for investment. Considering that this monetary resource can be used in building a five star hotel, constructing the water and drainage systems or building a school, the current resource sould be directed to the best alternative that can meet the needs in that region and increase the income per capita as well as the national income in the long run. If, according to the cost and benefit analysis, the sources are spent on the hotel construction although the most efficient area is the school construction, it means the sources are not used effciently and effectively and hence, there is a loss of economic wealth in terms of the country's citizens.

So, while a country/region has a need of constructing a school, directing the resources to the hotel construction causes both the inefficient usage of the resources and a loss of economic prosperity in terms of the citizens. Therefore, it is a very important decision that the scarce resources in an economy should be allocated in the best way and sent to suitable places for increasing the wealth of the citizens and minimizing the sacrifice. Indeed, this situation of high importance for the less developed and developing countries will cause the scarce resources to be invested in relatively less efficient tourism fields and to be wasted, although there are probably more efficient areas. So policy makers, especially in the countries with tourism potential, should consider these points in order to direct the country's resources to the efficient and effective areas. No matter which industry they are made in, the investments having good results, in other words, the investment efficiency expressed as input/output ratio is a significant issue for the country's economy.

D. Seasonal Effects of Tourism

The property of seasonal fluctuations in tourism is mostly seen in the accommodation industry. That is, that the tourism movements are intense in certain periods of the year and made mostly by the foreign visitors make it difficult for the accommodation supply in tourism regions in the summer terms when demand is very high. Comparing to the summer terms when demand is higher, the capacity rate in tourism regions decrease in winter terms and this causes supply to be idle. This situation causes the accommodation facilities to stop operating at certain periods in a year or continue operating with even a little income. The most important drawback it creates is that the facilities in the accommodation industry have to live with the income they gain during the season for the whole year. Hence, some of the workforce in the off season will be fired and so, the problem of seasonal unemployment will occur. Meanwhile, since the return on the investment seasonality effect is long in the tourism industry, it will hinder the investors to invest in this industry with the effect of seasonality as well. As a result of the seasonality property of tourism, the capacity increase in the industry will be prevented, unemployment will increase, and there will be a decrease in both national income as well as in income per capita.

In order to overcome these difficulties arising due to seasonality, the duration of the tourism season for 12 months and producing strategies and policies regarding the development of some other tourism forms, except for coastal tourism, will cause the efficient and effective utilization of the capacity to idle in the winte time and so the resources. If these mentioned points are taken into consideration by the public and private industry representatives there will be an increase in tourism demand for that region and the number of tourists coming to country with tourism income. Besides this, the tourism regions that can minimize the product differentiation will protect and develop its market share and enter new markets, and as a result attain a more competitive power than their competitors by minimizing this seasonal effect of tourism.

Considering that the investments in tourism have a return rate to the low capital [41], it can be said that seasonal fluctuations harm the industry because from the moment the idle capacity increases, the capital's duration for the return decreases. When this is the case, many investors, no matter domestic or foreign, give up investing in the tourism industry. Scientists still research in order to solve the seasonality problem with some studies [42]. Solving this important problem means that the economies of countries with a tourism potential gain more income.

E. Effects of Tourism over Foreign Capital

Before talking about the negative effects of the tourism industry regarding foreign capital, it is necessary to point out what this concept means. "Foreign private capital" or "private foreign capital investment" can be defined as *"private capital investments from developed countries to the less developed countries and cooperation or partnership constructed technologically* [43] *"*. According to another definition, foreign capital is defined as "the financial or technological resources that a country may add up to its economic power in short time by supplying from other countries on the condition that it pays it back in different ways in the future" [44]. When *foreign capital* is mentioned, it is meant that due to the fact that the less developed and developing country has inadequate internal savings and hence cannot have an accumulated capital, they fill this gap with external funding sources. The economic benefits of foreign private capital in general terms are summarized as below. The foreign capital:

- Contributes to the capital accumulation
- Contributes to the technological development
- Enables the real wages to increase
- Makes up external economies
- As a result of the technology, production becomes cheaper and quality increases and so some new goods can enter the market
- Income generated through the government tax increases
- Brings a dynamism to the internal market by enabling the recovery of the competitive power and duration and also creating an awareness of the market economy.

In a general assessment in terms of the tourism industry, it is seen that the foreign capital investments have some negatives besides the benefits, because foreign investment finally goes to another country with an intention of profit [45], which is the nature of economics. In this respect, people go wherever there is an expectation of highest profit and aim to maximize their own profit due to the property of "homo economics". In light of above-mentioned points, the possible economic effects of foreign capital on the tourism industry can be summarized as below [46]:

- The vertical and horizontal cooperations that foreign firms make among each other cause the formation of a monopolistic power and hence the local firms lose their market share and competitive power as they cannot compete.
- That the foreign equity owners bring the employees that they will hire from their own countries will have negative effects on the employment structure of the host country. Moreover, there will also be an outflow of foreign exchange as the wages of the personnel they hire abroad.
- By the arrival of foreign capital to the country, importing the necessary machines and equipment for investments will also cause a loss of foreign exchange.
- That the foreign companies are far beyond the local companies in terms of know how, capital structure and technological facilities make it harder for the local firms to

compete with them in time. In this respect, against this destructive power of the foreign capital that aims to develop the local tourism industry and provides some economic benefits, the local firms may have to stop their operations in time.

- The firms of foreign capital change the consumption patterns in the tourism industry and this results in an increase in the dependency on imports, causing a decrease in the multiplier effect of tourism.
- Foreign companies cause to increase the investment costs and as a result local firms cannot buy real estate at high prices due to the inadequate capital structure and give up entering the tourism industry.
- Another negative effect that foreign private capital generates on the nation's economy as well as the tourism industry is that the foreign firms import some intermediary and investment goods, which in fact can be provided within the country.
- Similarly, most of the income that foreign firms gain is transferred to their own country prior to being used as an investment within the country and this creates a deficit in the country's balance of payments, affecting the country's economy in a negative way.

These negative points indicated above reduce the positive effects of foreign capital on the tourism industry and the country's economy. Therefore, the improvement of the capital market of the country and the existence of a well operated financial system mean overcoming most or all of these problems indicated above. If this is not possible, it is needed to control the foreign capital entering the country continuously and sufficiently by the necessary units. However, it is another important point that the incentives and legal arrangements should be made by paying attention not to scare the foreign private capital. As a result, although it has some negative sides, the amount of foreign capital entering the country has an importance on developing the industry, establishing new facilities, know–how, administrative knowledge, marketing strategies and transfer of new technologies, etc.

Box 6.4. Foreign Capital and Tourism

In financing development after World War II, in most of the less developed and developing countries, foreign capital has been used as an important financial source and it is still being used. According to the 2000 data, in general, Turkey has only 1/1000 of the world inflow of foreign capital with an amount of approximately 1 billion US$ foreign capital investment. On the other hand, South Korea has 9.3, Poland has 9.4, Argentina has 11.7, Italy has 13.2, Denmark has 34.2, China has 38.4 and Germany has 189.2 billion US$ foreign capital investment, which shows that the foreign capital that Turkey attains is so little [47]. In this respect, the countrywise allocation of the foreign capital allowances in the tourism industry, according to 2000 data, the Netherlands is the first with 696 million US$. Following it, Germany has 583 million US$ and the third one is Italy with an amount of 271.7 million US$ [48]. The amount of foreign capital of Turkey in the tourism industry is almost 50.2 million US$.

SUMMARY

This chapter has first examined the positive effects of the success attained in the tourism industry on a national economy. Among the first positive effects there are positive contributions on the balance of payments, supporting the improvement of GDP, fostering the creation of new job opportunities, having a stimulating effect on other industries due to the multiplier effect. Besides, it is not wise to disregard the negative effects by only concentrating on the positive ones for the development of future policies. Therefore, depending on the countries level of development, it should not be forgotten that sometimes there are some negative conclusions of the development in tourism on the national economy, e.g., increasing cost of living, increasing amount of leakage after importing, not calculating the opportunity cost well enough. All of these variables show a national economy's competitiveness level at the international level. So, the following chapter will discuss a general analysis of the competition models and their applicabilities.

DISCUSSION QUESTIONS

1. What are the positive economic impacts of tourism? Explain briefly.
2. What should be the direction of the relationship between the foreign exchange income and expenses in order to mention the positive effect of tourism on the balance of payments? Explain briefly.
3. Find different examples showing the share of tourism in GDP and compare them with each other.
4. Define the concept "balance of tourism" both in narrow and wider terms.
5. List the main properties of employment in the tourism industry.
6. Give brief information regarding the effects of tourism on other industries.
7. What are the negative effects of tourism on a country's national economy? Explain briefly.
8. Give examples from your country about the positive and negative effects of tourism.
9. State your own opinion about the dependancy of a particular island on tourism.

REFERENCES

[1] WTO (2005), *Tourism Highlights*, World Tourism Organization, Madrid.
[2] Worldbank (2006), *World Development Indicators*. http://devdata.worldbank.org/wdi2006/contents/Section4.htm. (Retrieved 22 September 2007).
[3] P. R. Krugman, M. Obstfeld (2000), *International Economics Theory and Policy*, America: An Imprint of Addison Wesley Longman, Inc., pp. 313-316.
[4] O. Bahar, M. Kozak (2006), *Turizm Ekonomisi*, Ankara: Detay.
[5] K. Unur (2000), "Turizmin Türkiye'nin Ödemeler Dengesine Etkisinin Analizi," *Dokuz Eylül Üniversitesi S.B.E. Dergisi,* 2 (3), pp. 1–18.
[6] O. İçöz, M.Kozak (2002).
[7] İçöz, Kozak, pp. 198–199.

[8]   This is a summary table. Source: www.tcmb.gov.tr. (Retrieved 09 February 2006).

[9]   R. R. Croes (2006), "A Paradigm Shift to a New Strategy for Small Island Economies: Embracing Demand Side Economics for Value Enhancement and Long Term Economic Stability," *Tourism Management*, 27(3) pp. 453-465.

[10]   O. Bahar (2006), "Turizm Sektörünün Türkiye'nin Ekonomik Büyümesi Üzerindeki Etkisi: Var Analizi Yaklaşımı," *Celal Bayar Üniversitesi İ.İ.B.F. Yönetim ve Ekonomi Dergisi*, Cilt 13, Sayı 2, pp. 137–150; www.tcmb.gov.tr. (Retrieved 09 February 2006).

[11]   R. T. Froyen (1990), *Macroeconomics Theories and Policies*, New York: Macmillan, pp. 68–71.

[12]   Çakır, p. 66.

[13]   For detailed information see multiplier effect given in Chapter 3.

[14]   Bahar, Kozak, p. 186.

[15]   İçöz, Kozak, p. 270.

[16]   http://www.world-tourism.org/facts/eng/pdf/highlights/2005_eng_high.pdf. (Retrieved 14 February 2006).

[17]   http://www.turizm.gov.tr

[18]   J. Lea (1988), *Tourism and Development in The Third World,* London: Routledge, p. 46; Kozak et al., pp. 83–85.

[19]   M.M. Coltman (1989), *Introduction to Travel and Tourism: an International Approach*, New York: John Wiley, p. 230.

[20]   İçöz, Kozak, pp. 237–238.

[21]   P. Rita (2000), "Tourism in the European Union," *International Journal of Contemporary Hospitality Management*, 12(7), pp. 434–436.

[22]   Rita, p. 434.

[23]   http://europa.eu.int/comm/enterprise/services/tourism/tourismeu.htm. (Retrieved 30 April 2004).

[24]   http://www.tursab.org.tr/content/turkish/istatistikler/akrobat/GENEL/04mtIstih.pdf. (Retrieved 15 February 2006).

[25]   A. Çımat, O. Bahar (2003), "Turizm Sektörünün Türkiye Ekonomisi içindeki Yeri ve Önemi Üzerinde Bir Değerlendirme," *Akdeniz Üniversitesi İ.İ.B.F. Dergisi*, 3 (6), p. 14.

[26]   B. Archer, C. Cooper (1994), "The Positive and Negative Impacts of Tourism," *Global Tourism The Next Decade*, W. F. Theobald (Ed.), Oxford: Butterworth and Heinemann, p. 75.

[27]   M. E. Erkal (1990), *101 Soru 101 Cevap Bölge Açısından Azgelişmişlik*, İstanbul: Der Yayınları, pp. 234–236.

[28]   İ. S. Barutcugil (1989), *Turizm İşletmeciliği*, İstanbul: Beta Yayıncılık, p. 30.

[29]   O. Bahar (2005), "Impacts of Mass Tourism on the Natural Environment," *Quarterly Journal of University of Economics Bratislava*, 4, pp. 433–448.

[30]   Bahar, p. 446.

[31]   Olalı, Timur, p. 113.

[32]   See detailed information about the related effects see: O. İçöz, M.Kozak (2002), p. 255.

[33]   Kozak et al., p. 89.

[34]   Kozak et al., pp. 89–90.

[35]   J. C. Fuhrer (1994), *Goals, Guidelines, and Constraints Facing Monetary Policymakers*, Federal Resrve Bank of Boston: Conference Series No: 38.

[36]   Gürkan, p. 189.

[37] Erdoğan, p. 293.

[38] Lea, p. 50.

[39] Erdoğan, pp. 294–296.

[40] Kozak, et al., p. 91.

[41] Erdoğan, p. 300.

[42] See: J. Cunado et al. (2005), "The Nature of Seasonality in Spanish Tourism Time Series," *Tourism Economics*, 11(4), pp. 483-499; E. Koç, G. Altınay (2006), "An Analysis of Seasonality in Monthly Per Person Tourist Spending in Turkish Inbound Tourism from a Market Segmentation Perspective," *Tourism Management*, 27(1), pp. 1-11.

[43] Han, Kaya, p. 94.

[44] Krugman, Obstfeld, pp. 169-177.

[45] H. Çeken (2003), *Küreselleşme, Yabancı Sermaye ve Türkiye Turizmi*, İstanbul: Değişim Yayınları, p. 176.

[46] Çeken, pp. 176–178.

[47] Eren, p. 165.

[48] Çeken, p. 93.

Chapter 7

TOURISM AND INTERNATIONAL COMPETITIVENESS

According to the figures of WTO, throughout the world the number of people participating in the international tourism movements in 2005 was 763 million and international tourism income was 623 billion US$ [1]. These figures are expected to reach 1.6 billion people and 2 trillion US$ in 2020 [2]. Hence, as income and transportation facilities increase in line with the increasing level of wealth, it is estimated that the competition among the tourism countries and regions (destinations) will increase accordingly. The success of tourism regions in the international tourism market is determined by their competitive power. Sustaining and increasing the tourism income depends on a region's competitive power in serving the tourism product [3]. As the concept of tourism and competition is among the important subjects of today's economies, their relationship with tourism economics is examined in this chapter.

I. IMPORTANCE AND STRUCTURE OF COMPETITION

World economy has experienced rapid and big changes with the liberalization of the capitalistic movements, especially after the 1980s. The removal of obstacles and restrictions in liberalizing foreign trade, in other words the concept of globalization, oblige the domestic industries to cooperate with the world markets no matter which industries they are. Thereby, it can be said that the world seemed to have become a single market with the liberalization efforts of world trade as well as the improvements in areas of communication, technology and transportation. From the aspect of tourism economics, air jet travels started in 1958 had a big impact on the rapid development of tourism especially after World War II.

The globalization activities and fast technological developments worldwide bring in concepts such as competition, competitive power and international competition power. These transformations occurred in the world economy after 1980 that caused an increase in the international competition power to intensify in terms of countries, regions, firms, organizations and economica unions. In other words, the globalization process has increased the competition among countries and firms. This situation is valid for the tourism industry as well. There has been an enlargement in the tourism industry with the effect of globalization and getting the desired share from the excess market share has led to an increase in the competition among the tourism destinations [4].

In light of these, several factors such as the success in rapid adaptation to the changing market structure and reacting, customer satisfaction, service quality, prompt delivery, changing the demand profile of consumers in the age of information, adaptation to the environment, flexibility and innovation make it difficult to make an exact definition of the competitive power and identify the factors affecting competitive power either positively or negatively. Indeed, it is not easy to make an exact and apparent definition of this concept that is widely used today. The reason for this is that competition has a very dynamic and multi-dimensional feature and it cannot be explained with a single variable or indicator [5].

A. Concept of Competitive Power

The effort to increase the market share due to the rapid growth and development in the travel and tourism industry shows how important competition is in terms of tourism destinations [6]. Indeed, competition and obtaining competititve power have an important role in the long-term success of the destinations and marketing of tourism products [7]. As indicated in the previous chapter, it is not possible to find a definition which everybody agrees on and has an exact and perfect content because integrating the properties of many countries and firms and distinguishimg some of them are almost impossible. Therefore, competition can be defined in five different classes as firm, industry, cluster, national and international. Defining the competitive power in terms of the firms is easier than its definition in terms of regions or countries.

So, in terms of firms, the competitive power can be expressed as "the ability of maintaining the sustainable preference for the supplied goods and services against their alternatives" [8]. Meanwhile, the competitive power in terms of industries is "the ability of an industry to reach an efficiency level that is equal or higher than its competitors and sustain this level or produce and/or sell products at an equal or lower cost than do its competitors" [9].

Examining the regional competitive power by forming clusters at an industry level brings forth the concept of competitive power in terms of clusters. Accordingly, an increase in the efficiency of clusters forms and develops the competitive power by degrees [10]. The national competitive power is, on the other hand, "the ability of a country to gain a considerable amount of profit from the resources it has, differentiating products in foreign trade, creating new products and being able to distribute it through distribution channels" [11]. The international competitive power can be defined as follows: "Having an international competitive power in domestic and foreign markets means for a local firm that it is more advantageous than the competitor domestic and foreign firms on factors such as the product price and/or product quality, prompt delivery and after sales services now and in the future" [12]. Therefore, the concept of international competitive power can be considered apart from the firm/industry and national competitive power because the below properties are the ones separating international competition from the other definitions about competition [13]:

- Countries have different factor costs
- There are different conditions in foreign markets
- Countries have different structure

- Abilities on observing the sources, purposes and rivals are different.

As a general definition of the competitive power, it can be said that "it is the relative industrial efficiency and relative cost advantage of the industry and industrial firms occurred based on quality, cost, rapid delivery, marketing power and financial power, as well as the working atmosphere" [14]. Besides this, all definitions state a main property of the competitive power that a relative power is needed to compete with the rivals. In other words, the essence of the competition theory is shortly to exist in a market and survive. Those firms organizations, regions and countries that cannot survive in the free market conditions will surely lose the competition game with their rivals.

B. Examining Competitive Power from a Theoretical Perspective

Throughout history, the countries in the world have always been and are still competing with each other in order to get the underground and surface resources, under which lies the aim to reach economic prosperity. For example, the intervention of the USA to Iraq to get the utilization right for petrol and enhance the energy resources in the region.

As explained in the first chapter in detail, the reason for the existence of the branch of economics is that the resources that can satisfy the necessities are scarce. Since the basic problem of economics is scarcity and preferences, the history of economic events and mankind can be regarded as equal. Hence, it can be asserted that the competition for sharing the economic assets among people and countries dated back to the past. However, the scientific literature regarding the issue states that at first competition used to be explained with the absolute advantage approach of Adam Smith covering two countries and two products. Later on, David Ricardo examined the competitive power by his concept of comparative advantage, which is the more improved version of the former one. Thereby, first it is necessary to review what these two main theories express because in order to make these clear, their content should be examined.

Adam Smith defines the competitive power as the absolute advantage in production and absolute advantage based on the geographical allocation, in other words factors of place of establishment and division of labor in activities. According to Smith, as a result of the economic activities that will be made nearby the raw material and energy sources, the countries specialized and this results in an increase in their production, which also causes the increase in international trade as well as the competitive power [15]. Moreover, in the theory of absolute advantage it is stated that a country should specialize in the production of goods that are produced at a lower cost than another country and it should export them and import the ones that it is costly to produce [16].

David Ricardo put forth the comparative advantage theory 40 years later than Adam Smith by developing his theory. Hence, the competition and trade power among countries are explained based on this theory. Shortly, the essence of Ricardo's theory is that countries will be successful by transfering their resources particularly on the production areas on which they are most efficient [17]. According to Ricardo, absolute advantage is not important in production, the important thing is the level of this advantage. In other words, a country should specialize on the products on which it has a higher level of advantage in production [18]. That a country has a proportional advantage in producing any product means that it produces that good more efficiently but at a lower cost than the competitiors do. The comparative advantage theory is also valid for the tourism industry. That is, some countries are more advantageous over the other in terms of their natural, historical and cultural resources as well as their attractiveness [19]. That a country has such resources means that the tourism demand for that country and accordingly the competitive power in the international tourism market increase.

Table 7.1 shows other theoretical approaches underpinning the competitive power. As indicated above, the theoretical approaches explaining the competitive power have also changed with the rapid developments worldwide. According to the traditional approach during the period beginning with Richard Cantillon and ongoing until the 1960s when there was no technological developments at all, the production advantage was the most important factor explaining competitive power. The way to be advantageous in production and hence obtain competitive power grows out of the close relationship between the geographical location and transportation costs [20]. In this respect, the regions close to the sea and rivers have an advantage and competitive power due to their transportation convenience and by producing easily and cheaply, they experience a rapid development process. In the following years, in order to protect their market share and attain an advantage over their competitors producing similar goods, countries/firms have been directed towards new policies. Hence, in the 1970s the most important factor explaining competitive power is seen to be the production approach based on the cost advantage.

Table 7.1. Theoretical Approaches to Explaining Competitive Power [21]

Theoretical Approaches	Properties	Theorists
Traditional Approaches (until 1980)	• Production Advantage • Geographical Location and Transportation Costs • Cost Advantage	R. CANTILLON A. SMITH D. RICARDO
Porter's Approach (1980–1995)	• Internal Variables – Factor Conditions – Demand Conditions – Affliliated and Support Industris – Firm strategy, construction and competition • External Variables – Opportunities – The Role of State	M. E. PORTER
Efficiency Approach (1995 and afterwards)	• Clusters • Cluster Economies • Small and Medium Enterprises	World Bank World Economical Forum

A new and different approach, which expounded the competitive power, was put forth by Michael Porter between the years 1980 and 1995. The essence of Porter's theory called "The Competitive Power of the Nations" is shortly towards transforming the comparative advantages to the competitive advantages. Because of the gradual decrease in the natural resources that countries have, the structure of comparative advantages change and differentiation of costs, quality and product, new products, technological differences, focusing and market structures have become the factors explaining competitive power. According to Porter, competition will not be the production advantage or price competition anymore but it will be product differentiation, innovation, modern design and appearance [22].

Box 7.1. Competition Status of Turkey in the Mediterranean Region

In a research carried out in 2004 with a purpose of putting forward what might be the factors explaining competitive power in terms of the tourism industry in Turkey, the closest competitors of the country in the Mediterranean region are found out as Spain, Greece, France, Italy and Cyprus. Turkey, compared to these five important competitors in the region, is said to be more competitive than Cyprus; in other words, its level of competition is much higher than that of Cyprus. It is seen that Greece, Spain, France and Italy follow respectively [23]. Moreover, it is confirmed by the analysis that the supply conditions of tourism, cost, investment, incentive and financial arrangements; sustainable tourism and environment, service quality and customer satisfaction, efficiency and effective usage of resources, differentiation of tourism products, image and innovation, tourism competition strategy as well as other factors affecting competitive power positively influence Turkey's competitive power in the international tourism industry. In the same way, factors such as human capital and education, information technologies and technological development, supply conditions of tourism, demand conditions of tourism, cost, investment, incentive and financial arrangements, marketing strategy of tourism and its market share and bureaucracy negatively influence the country's competitive power [24].

After 1995, it is seen that the Efficiency Approach developed by the World Bank and the World Economic Forum in explaining the competitive power. The basis of the industrial region approach based on efficiency is known to be dated back to the famous economist Alfred Marshall. This has been improved and adopted to the present time. According to this approach, the most important factor over the competitive power is the externality shaped according to the economic structure of the region as a result of the concentration in economic activities and the interaction of these externalities. The geographical residence area or region provides economic benefits to the firms and these benefits are called cluster economies [25]. Hence, the cluster economies attained depending on the geographical residency increase the competitive power by causing an increase in efficiency [26].

II. CONCEPT OF COMPETITION IN TOURISM

Price was the first important factor to recall when competition was referred before the 1990s in the international tourism industry. However, the concept of competition has become a more complex and multidimensional structure which requires more of a struggle with the competitiors [27]. It is not possible to explain the competition and competitve power issues by the traditional competition theories in today's economic circumstances. As indicated before, the demand structure of consumers has changed with the effect of globalization and technological developments. That is, tourists in the future will be the consumers improved multidimensionally who want to benefit from the technological facilities as much as possible and who take care of the environmental issues. The increase in the new production technologies covering other industries of economics as well as the tourism industry cause the behavior and desire of consumers and hence the market competition to change as it is in the production methods.

In line with this, it has become difficult to explain the competitive power with the traditional competition theories in the tourism industry because the most important factor affecting competitive power is not the price itself. Many factors may be effective on the competitive power besides the price of the tourism product. Below in Figure 7.1 the factors affecting the competitive power of tourism destinations are shown. Here, the competitive power of destinations is affected by individual, regional, countrywise, qualitative and quantitative factors with complex structures that interact with each other. For the destinations to attain the competitive power, all these factors should perform at the same time and in harmony. A delay in any of these factors constituting the competitive power will be a big obstacle in the country's attaining full competitive power.

A. Definition of Competition in Terms of Tourism Destinations

Destination competitiveness is the first thing that comes into mind when competition is referred in tourism economics. As explained in the second chapter in detail, *destinations,* "are the geographical regions serving integrated services to the consumer group called tourists and are composed of the combination of the tourism products" or "the places that have distinguished natural attractivenesses and properties that may be appealing to the tourists" [28]. The destination may be a country or a continent, city, town, an island, places with natural and outstanding landscapes and so on.

There is no general theory, an explained definition or a model whose validity is tested regarding the issue of competition. The reason that a definition or model for tourism is the difference between the traditional production of goods and services and the production of the tourism product. In contrast to a certain category of production goods, the competition in tourism regions has a structure including many industries and sub-industries, some of which are larger and some others are smaller, serving individual goods and services in terms of the combination of all experiences in the visited destination [29].

Competition can range from the competition among nations and firms to the competition among the destinations due to the effect of globalization. However, there are not any determining factors of the destination competition. In this respect, various definitions regarding tourism (destination) competition in literature are presented here to clarify the issue. Tourism competition (see Table 7.2) is defined as "facilities providing a high standard of living to the inhabitants of a tourism destination, the competitive power of that place" [30]. Another definition states that the tourism competition is the ability of a destination to attract the potential tourists to its region and satisfy their needs and expectations [31]. According to a final definition, the tourism competition is "the ability of a destination to sustain its market share and power, protect and develop it in time" [32].

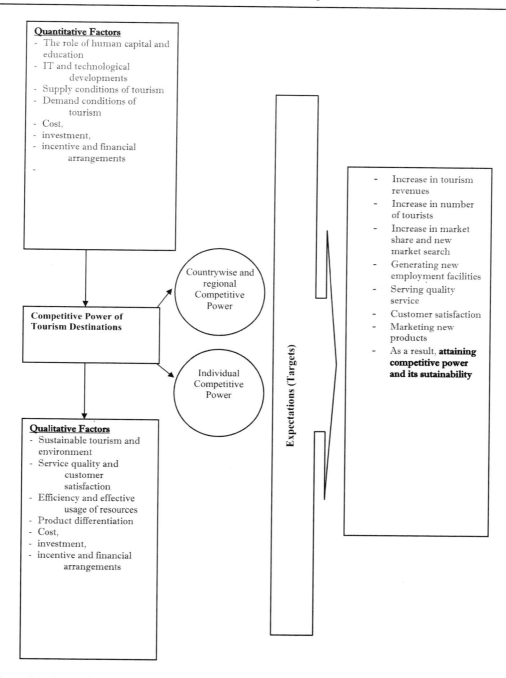

Figure 7.1. Competitive power of tourism destinations.

The common features of the definitions above emphasize that they refer to a period through which competition is ongoing; that is they cover a dynamic period, that the level of prosperity of the local people can be increased and customers (tourists) can be satisfied at the highest level. Unless a destination has three of these properties, it may not be possible that a country or a region gets the competitive advantage in the tourism industry because whether it is a tourism country or not, a country can increase its economic prosperity and improve its economic growth and development depending on its ability to develop its competitive power in almost every industry. Indeed, in the global economy of the third millenium, the importance of concepts such as globalization and competitive power are increasing each day.

Table 7.2. Definitions of Competitiveness in Tourism

No	Authors	Definitions
1	Ritchie and Crouch 2003	Facilities that provide high standard of living to the inhabitants of that region.
2	Enright and Newton 2004	The ability of a destination to attract the possible tourists and satisfy them.
3	d'Hauteserre 2000	The ability of a destination to sustain its market share and power, protect and develop it in time.

B. Models Regarding Destination Competition

It is necessary that the competitive models in the literature be investigated in order to analyze the factors that the competitive power is under the influence of. These models allow the handling of the competition and the determinants of the competitive power in a different way. Therefore, giving a little information about these models will be helpful in clarifying and presenting the issue in a better way.

Porter's Model that redefines the boundaries of strategic management is composed of the combination of four basic internal and two external factors that are called "dynamic diamond" [33]. These are listed as in Figure 7.2 [34]. According to Porter, the dynamic diamond composing of four internal and two external determinants have a structure that fosters each other with mutual interaction. It is not possible for only any one of them to have enough impact on the competition power [35]. In order for the competition power in the tourism industry to develop, there is a need for a system in which all determiners are together and in mutual interaction. Something missing or wrong in any of these factors is an obstacle for the formation of an effective competition power.

The most elaborate study on tourism competition made so far is the one that belongs to *Crouch and Ritchie.* They try to explain the destination competition by developing a conceptual model. According to them, the most competitive destination is the one that can provide its citizens the best prosperity conditions in the long run. In this respect, the facilities that can form a high standard of living for the people in a tourism region are expressed as the competition power of that place. However, the policies implemented to attain the competitive power should be on the sustainable basis. Competition power cannot be thought apart from the sustainability principle. If this principle were underestimated, the generated competitive power would reflect a deceiving situation. Moreover, Ritchie and Crouch investigate the factors affecting destination competition in four groups. These are qualitative determinants, destination management, basic source and attractivenesses, and support factors and sources [36]. The destinations using its resources in the most effective and efficient way are the ones who are the most advantageous in attaining competitive power.

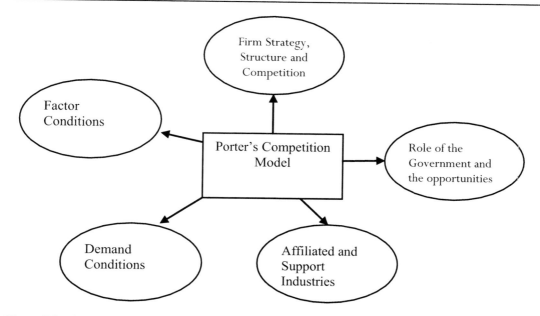

Figure 7.2. Elements of Porter's competition model.

The integrated competitive model was developed by Dwyer and Kim in order to make a comparison between the tourism industries and countries. The purpose of the model is to identify the factors affecting the competition power of destinations. The integrated competitive model also puts forth the factors constituting the weak and strong points of tourism destinations. Dwyer and Kim explains the tourism competition with five factors as natural and improved resources, destination management, demand conditions, regional conditions, and destination competitiveness [37]. According to this model examining the indicators and structure of the destination competitiveness comprehensively, it is possible to make a comparison among different countries and tourism industries.

The tourism competition model of Kim was developed in order to explain the competition industry has four dimensions in this model, which are the primary, secondary, third and fourth sources of competition. He also states that the effect of each industry on the competitive power is different. In a study conducted in Korea, Kim indicates that the fourth source of the competition has much more of an effect on the competitive power of destination than the others [38].

After this short explanation of the most important four models developed on the issue of competition, their scope, advantages and limitations are shown as a summary in Table 7.3. That these four models are demonstrated in the table altogether makes it easier to understand the issue well and to put forth the relative strengths and weaknesses of the issue. All these four models have a similarity giving no empirical investigation on the measurement of destination or tourism competitiveness. On the other hand, Kim's model differs from the others to the extent that it shows an example for practice.

Table 7.3. Models Developed on Tourism Competitiveness [40]

Authors	Scope	Advantages	Limitations
M. E.Porter (1980-1995)	1)Factor conditions 2) Demand conditions 3) Affiliated and support industries 4)Firm strategy, structure and competition 5) Role of opportunities and governments	1) Investigates competitiveness in macro and micro terms and handles the issue from the aspect of firm/industry and country. 2)Theoretically serves a valid structure for all industries and sectors in economy and while explaining the competitiveness of countries, examines the issue as a whole by combining the other countries' economies and competition strategies. 3) It is a dynamic model and explains the factors which the competitiveness among the national tourism economies is influenced by.	1) The model mostly explains the competitiveness of the developed countries and in terms of the developing countries it should be reviewed again. 2) Concepts such as foreign exchange rates that is one of the main factors affecting competitiveness, cost of labor, technology, foreign capital investment and globalization shaping the world economies are not taken into consideration. 3) Human capital, which is a very important factor in improving the destination competitiveness is not examined sufficiently, either.
J.R.B.Ritchie, G.I.Crouch (1999-2000)	1) Basic sources and attractions 2) Support factors and sources 3) Destination management 4) Qualitative determinants	1) The model is the most comprehensive study on destination competitiveness. It examines the destination competitiveness conceptual wise. 2) The issue of competitiveness regarding the service industry is applied to tourism destinations in a country, industry, enterprise and product basis.	1) The model, having a linear structure, cannot examine many factors affecting competitiveness comprehensively. Therefore, integrated competitive model is developed. 2) The biggest limitation of the model, which has a very elaborative conceptual context is that its empirical tests have yet to be undertaken.
L. Dwyer - C. Kim (2001)	1) Natural and developed sources 2) Destination management 3) Demand conditions 4) Regional conditions 5) Destination competitiveness	1) Comparison is made among the tourism industries and countries and the factors affecting competitiveness are put forth. 2) By identifying the weak and strong points of different tourism destinations, the model presents policies and strategies for the authorities to implement in terms of competitiveness.	1) The biggest limitation of the model is that there is no test for the practice in the measurement of the destination competitiveness. 2) Another point that makes it difficult to measure the destination competitiveness is that several factors investigated in the model seem to become qualitative.
C. Kim (2000)	1) Primary sources 2) Secondary sources 3) Third sources 4) Fourth sources	1) 17 factors among the four main determinants identify the destination competitiveness.	1) It is not clear that based on what scientific basis the factors affecting destination competitiveness are divided as primary, secondary, third or fourth sources in the model. 2) It is a linear model. Therefore, it is inadequate in clarifying the interaction among different factors and sources that affect the destination competitiveness.

C. Other Approaches to Competitiveness in Tourism

There are some other approaches in the tourism literature regarding competition that is an important factor in the success of tourism destinations. These studies are said to have a multidimensional structure such as from economic, socio-cultural and environmental perspectives [39]. In this respect, especially in tourism, competition is a complex concept which is hard to be measured by the secondary data and includes several visible and invisible factors.

In the tourism industry, which is included in the international services account of the balance of payments, it is not possible to measure some variables or factors statistically as it is in the other industries of an economy. This is because of the fact that the tourism competition is measured by the experience of customer groups called tourists. In other words, the tourism product is perceived by tourists in a different way. From this aspect, there seems to be some trouble in presenting the problems like competition and competitive power in tourism. It may become hard to see some theoretical issues in practice and testing them with primary data cannot always be an easy task. Despite this, some approaches developed by various scholars are investigated below respectively.

1. Poon's Approach

Adrian Poon divides the tourism industry into two groups as new tourism (flexible, segregated – sliced, integrated crosswise and environmentally sensitive) and old tourism (mass, standardized, stable rigid packets) and states that both of them affect consumers, enterprises, technology, production and main conditions. As well as this, he emphasizes that the new tourism has changed the rules of the game and there is a need for new strategies to generate and sustain competitiveness. The faster a firm or enterprise changes the tourism region/environment that it activates, the easier it is to implement the new strategies, because the tourism and travel industry is changing every day [41]. The essence of Poon's theory is brief, making innovation and creating a new tourism product.

According to Poon, the destination competitiveness is regarded as a continuous innovation and as a continuous change. Poon points out that there are four main principles for the destinations to attain competitive power: (1) the environment's being the policy of first priority, (2) tourism's being the leading industry, (3) fostering the distribution channels in the marketplace, and (4) the existence of a dynamic private industry in tourism [42]. These four properties are the competitiveness strategies that a destination needs to implement at macro levels. Moreover, Poon states the four main principles that enterprises at the micro level and the assigned personnel within the industry need to implement. These are, first, putting oneself in a tourist's place, being a leader in quality, accomplishing drastic innovations and finally, the need for fostering the strategic position of the enterprise [43]. As a result, it is pointed out that the competitive power in the industry can be attained when a tourism enterprise at a micro level or a tourism destination at a macro level think and implement the eight principles mentioned above.

2. WES Appoach

WES (Westvlaams Economisch Studiebureau) approach has arisen from the idea of the Inter American Development Bank to analyze the competitive power of the countries in the

Caribbean Region. The essence of the approach is how to put forth the competitive power of the countries in the subject matter region, to explain the differences of the competitiveness of destinations and focusing on the idea of competitiveness in the long run. According to this approach, competitiveness is defined as the ability to reach the goals defined more effectively than the international and regional averages in the long run. In this respect, the destination competition is attaining a profitability above the avearge with the lowest social cost without harming the environment and current resources [44].

According to the WES Approach, there are five different factors affecting the destination competiveness as macro–economic, supply, demand, transportation and tourism policy. These factors and the variables regarding these factors are shown in Table 7.4. As seen in the table, there are a total of 19 variables within the scope of factors affecting destination competitiveness. In this respect, the approach puts forth what may be some variables and factors affecting competitiveness. When these factors are examined (Table 7.4), it will be seen that these are mostly related with the macro economic indicators of the country. Hence, this approach emphasizes that in order for a destination to attain a competitive edge in the future, the above mentioned macro economic factors should be focused on.

Table 7.4. Factors Affecting Compettive Power According to WES Approach [46]

Factors	Variables
Macro-economical Factors	• Income generating countries • Actual rate of foreign exchange • Cost and existence of capital • Financial policies Import tax Taxes increasing cost price Tourism tax Tax of cruise travel
Supply Factors	• Tourism products Attractiveness Accommodation Prices • Labor Existence Cost Quality and education • Infrastructure Transportation Public services
Demand Factors	• Market dependence • Functionality in distribution channels • Marketing efforts • Product existence in a market that will grow in the future
Transportation Factors	• Existence of regular services • Existence of charter services • Existence of cruises and boat trips
Tourism Policy Factor	• Public structure • Generating policies • Planning capacity • Trading • Support from public budget

3. Price Competition Approach

The core of this approach is related with the most common factor of the competitive power. In other words, it states that the ones determining the most suitable price of a product in the long run will be the ones having the competitive power in the tourism market. So, the tourism competitiveness is explained as a general concept, covering the price differences occurring by the foreign exchange movements coming together, the efficiency level of various components of the tourism industry and qualitative factors affecting the attractiveness of a region [45].

4. "Bordas" Model

The model dwells on two important factors for marketing tourism destinations in the long run in a sustainable way. The first is the "perceived value", which refers to the image of a destination created among potential tourists. Image is a concept which is dependent on other tourism activities. If the destination's image is negative, it is quite difficult to change it. The advertisement and promotion activities are mostly insufficient in changing this negative image. Therefore, destination authorities that want to gain competitive power in the tourism industry should take care of the studies to develop a positive image.

The second important factor in Bordas' model is the "perceived cost". By this, costs that are abstract and hard to measure such as economic costs, the physical exhaust or cost of travelling, physiological costs, e.g., the waiting period at the airport, stress, tiredness due to the time gap and the time zone difference while flying (jet lag), the inactive waiting period during the flight and all of the possible risks and costs of hygiene and care. The more such troubles that tourists may encounter during their travels are reduced by the destination, the higher will be the competitive power to be attained in the end [47].

5. Classification of Competition by WTTC

In a study on competitiveness by the WTTC, there are eight different categories explaining competitiveness [48] (see Table 7.5).

The price competition index is composed of hotel price index, parity of purchasing power index, corrected consumer price index and tax index on goods and services. The human tourism index measures the human development on tourism activities. Accordingly, the index covers several indexes consisting statistical data about tourism such as the index of participating in tourism, tourism influence index, international tourism arrival and departure index, tourism annual import and export index, tourist spending and rate of annual growth of tourism index. The human resources index covers some demographical variables such as the average life, the literacy ratio, rural population – total population, labor, participation in workforce, unemploymemt rate, employment rate in agriculture, manufacturing and services industries, increase in the amount of population and growth rate. Finally, the infrastructure index includes road index, development of health conditons index, railway transportation, access to the sources of drinking water.

Table 7.5. Competitiveness Index of WTTC [49]

No	Indexes	Scope
1	Price Competition Index	Hotel price index, parity of purchasing power index, corrected consumer price index and tax index on goods and services
2	Human Tourism Index	Participating in tourism, tourism influence index, international tourism arrival and departure index, tourism annual import and export index, tourist spending and rate of annual growth of tourism index
3	Human Resources Index	Average life, literacy ratio, rural population – total population, labor, participation in workforce, unemploymemt rate, employment rate in agriculture, manufacturing and services industries, increase in the amount of population and growth rate
4	Infrastructure index	Road index, development of health conditons index, railway transportation, access to the sources of drinking water
5	Environment Index	Population density, environmental sensitivity, emission rate of CO_2
6	Technology index	Rate of utilizing mobile and hot line (in 1000 people), exporting of advanced technology, internet access
7	Extraverison Index	International trade and extraversion in tourism and the custom tax rates
8	Social Development Index	Number of newspapers read daily, rate of computers and TV per person, the rate of committed crimes that are recorded.

The environmental index covers some factors such as the population density, environmental sensitivity, and emission rate of CO_2. The technology index shows the development in the modern technology and includes some variables such as the rate of utilizing mobile and hot lines (in 1000 people), exporting advanced technologies and internet access. With the extraversion index, the extraversion in the international trade and the custom tax rates on these are meant. The social development index suggests the number of newspapers read daily, rate of computers and TV per person, rate of committed crimes that are recorded.

III. IMPORTANCE OF DESTINATIONS IN TERMS OF COMPETITIVENESS

As in every industry and enterprise, in order to attract more tourists to its region and increase its tourism revenues, many tourism destinations in the international tourism and travel industry are in a harsh competitive environment. It depends on their adoptation to the demand structure amd global conditions of today that the destinations develop their competitiveness and protect their market share in these increasing competitive conditions. In sustaining the market share, which is one of the important indicators of competitiveness in international markets, the issues of primary importance are utilization of tourism sources and transportation capacities as well as the protection of the peculiar properties and attractivenesses. It is for this reason that the significance of destinations in the international tourism and travel industry is to become higher [50].

Destinations in terms of the tourists participating in international tourism movements loom large according to the hotel, holiday village, pension serving individual services or activities like the sea, sun or sand and become a trademark. In other words, the future of tourism and travel industry depends on the existence of destinations because the people participating in tourism movements travel not only for the reasons stated above (spending time at a hotel /holiday village, benefiting from the sea, sun or the sand, etc.) but also to learn about the region as a whole, see new places, experience new things and spend most of their time in the destination. This is due to the fact that the demand in the tourism market is directed towards the tourism variation that satisfies new interests, enables learning new things and being more active and that the tourist profile of the information age has changed. Tourists are, today, more experienced and informed, able to benefit from technology well, able to speak different languages, travel by various vehicles, plan their vacations in advance and most importantly, able to visit the same destination more than once.

Finally, the competition among the destinations in the international tourism is intensified. The new destinations, products and tourist profiles appear in the tourism market and the ones satisfying tourists by finding customer focused strategies become advantageous in gaining the competitive advantage. Meanwhile, destinations leave the traditional products approach and focus on the items that enable the highest customer satisfaction. Destinations, so as to act against competition or overcome it, are in a search of product variation, advertising, marketing, image development, service quality and customer satisfaction as well as different technologies.

Box 7.2. Results of the Competition Index of WTTC [51]

According to the 2004 data of the WTTC, the value of Turkey's price competition index is 84.77. This value is said to be high to the extent that it is close to 100. Turkey's competitiors in the Mediterranean region have lower values, for example, Spain has 54.28, Italy 47.06, Greece 54.41, France 51,34 and Cyprus 71.51. Here, according to the price competition index of WTTC, one could understand that Turkey is a tourism country that has a high level of competition power. Similarly, according to the infrastructure index, while Turkey has a value of 45.69, Spain has 46.32, Italy has 55.72, Greece has 55.72, France has 56.53, and Cyprus has a value of 72.06. The environment and technology indexes show that Turkey is far beyond its competitiors with the respective values of 65.52 and 25.17. That is, it is known that the environment index for Spain is 94.84, Italy (86.72), Greece (75.07), France (80.71), Cyprus (66,09) whereas the technology index for Spain is 55.20, Italy (66.64), Greece (57.29), France (62.64), and Cyprus (52.46).

SUMMARY

It is important to examine the concept of competitiveness in order to measure the success of every economic activity. Indeed the investments on an industry do not contribute to a firm's or enterprise's competitiveness internationally, the result would be a failure. Therefore, in this chapter the importance of international competitiveness in the tourism industry is emphasized, and some approaches about how to compete particularly in terms of destinations are discussed. In this scope, several models developed by the scholars like Michael Porter are examined in terms of the tourism industry and some other models developed for tourism are presented with a comparative analysis. Additionally, the chapter is concluded by reviewing the capacity of some countries in their ability to compete in international tourism by means of the data attained as a result of the models developed by the WTTC.

DISCUSSION QUESTIONS

1. Why is competitiveness important today? Explain.
2. Define the term "competitiveness" in terms of firm, industry, cluster, national and international.
3. What are the main properties of conceptual approaches explaining competitiveness? Explain briefly.
4. Why is competitiveness important in the tourism industry? Explain briefly.
5. What factors are needed for tourism destinations to attain competitive edge? Explain by giving examples.
6. Define competitiveness in terms of tourism destinations.
7. Give brief information about the models and approaches explaining competitiveness in the tourism industry.
8. Why are destinations important in gaining the competitive advantage? Explain briefly.
9. What do your think about the factors affecting the competitiveness of your country in international tourism? Explain.
10. Discuss the competitive advantage/s of the two countries you would prefer.
11. Discuss the competitive disadvantage/s of the two countries you would prefer.

REFERENCES

[1]   WTO (2005), *Tourism Highlights*, Madrid, Spain.
[2]   Cho, p. 323.
[3]   D.O. Gomezelj, T. Mihalic, "Destination Competitiveness—Applying Different Models, The Case of Slovenia," *Tourism Management*, (Article in Press); J. R. B. Ritchie, G. I. Crouch (2003), *The Competitive Destination*, England: CABI Publishing, pp. 5-23; V. Patsouratis, Z. Frangouli and G. Anastasopoulos (2005), "Competition in Tourism Among The Mediterranean Countries," *Applied Economics*, 37, pp. 1865–1870.
[4]   J. Falzon (2003), "The Competitive Position of Mediterranean Countries in Tourism: Evidence from the Thomson Brochure," http://www.erc.ucy.ac.cy/english/ conference2003/Falzon%20Paper.pdf, (Retreived 31 March 2006).
[5]   K. Chon, K. J. Mayer (1995), "Destination Competitiveness Models in Tourism and Their Application to Las Vegas," *Journal of Tourism Systems and Quality Management*, 1 (2/3/4), p. 228.

[6]   N. Gooroochurn, G. Sugiyarto (2006), "Competitiveness Indicators in the Travel and Tourism Industry," http://www.erc.ucy.ac.cy/english/conference2003/Guntur.pdf, (Retrieved 21 March 2006).

[7]   D. Gürsoy, K. W. Kandall (2004), "A Competitive Positioning of Mediterranean Destinations", *CD Proceedings of Euro-Chrie Congress*, 3–7 November, Ankara; E. P. Lopez, M. M. Navarro and A.R. Dominguez (2004), "A Competitive Study of two Tourism Destinations Through the Application of Conjoint Analysis Techniques: the Case of the Canary Islands," *Pasos. Revista de Turismo y Patrimonio Cultural*, 2 (2), pp. 163–177.

[8]   O. Çoban (2001), "Teknolojik Gelişme ve Rekabet Gücü", *Verimlilik Dergisi*, 4, p. 27.

[9]   C.C. Aktan (2005), "Rekabet Gücü Kavramı," http://www.canaktan.org/yeni-trendler/yeni-rekabet/kavram.htm, (Retrieved 30 March 2006).

[10] Bakımlı, p. 13.

[11] P.J. Buckley, C.L. Pass and K. Prescott (1988), "Measures of International Competitiveness: A Critical Survey," *Journal of Marketing Management*, 4 (2), pp. 175–200.

[12] A. Kibritçioğlu (1996), "Uluslararası Rekabet Gücüne Kavramsal Bir Yaklaşım," *Verimlilik Dergisi*, 3, p. 111.

[13] M.E. Porter (1998), *Competitive Advantage*, New York: The Free Press, p. 276.

[14] E. Ertürk (1991), *Ekonomik Entegrasyon Teorisi ve Türkiye'nin İçinde Bulunduğu Entegrasyonlar*, Bursa: Ezgi Kitabevi, p. 208.

[15] A. Smith (1999), *The Wealth of Nations*, England: Penguin Books, p. 30.

[16] Seyidoğlu, p. 17.

[17] R. Lynch (1997), *Corporate Strategy*, Londan: Putman Publishing, p. 144.

[18] Seyidoğlu, p. 18.

[19] İçöz, Kozak, p. 11.

[20] A. Smith (1999), *The Wealth of Nations*, England: Penguin Books, p. 30.

[21] R. Lynch (1997), *Corparate Strategy*, London: Putman Publishing, p. 145; N. Ayaş (2003), "Bölgesel Rekabet Gücünü Geliştirmeye Yönelik Alternatif Bir Yaklaşım: Yeni Endüstriyel Bölgeler Yaklaşımı (Denizli Örneği)," *Basılmamış Doktora Tezi*, M.Ü.S.B.E., Muğla, p. 7; O. Bahar (2004), "Türkiye'de Turizm Sektörünün Rekabet Gücü Analizi Üzerine Bir Alan Araştırması: Muğla Örneği," *Basılmamış Doktora Tezi*, Muğla Üniversitesi S. B. E., Muğla 2004; N. Gooroochurn, G. Sugiyarto (2006), "Competitiveness Indicators in the Travel and Tourism Industry," http://www.erc.ucy.ac.cy/english/conference2003/Guntur.pdf, (Retrieved 21 March 2006).

[22] M. E. Porter (1998), *Competitive Strategy*, New York: The Free Press.

[23] O. Bahar, M. Kozak (2005), "Türkiye Turizminin Akdeniz Ülkeleri İle Rekabet Gücü Açısından Karşılaştırılması," *Anatolia: Turizm Araştırmaları Dergisi*, 16 (2), p. 145.

[24] Bahar, Kozak, "Türkiye Turizminin Akdeniz Ülkeleri İle Rekabet Gücü Açısından Karşılaştırılması," p. 147.

[25] M. Novelli, B. Schmitz, T. Spencer (2006), "Networks, Clusters and Innovation in Tourism: a UK Experience," *Tourism Management*, 27 (6), pp. 1141–1152.

[26] N. Ayaş (2003), "Bölgesel Rekabet Gücünü Geliştirmeye Yönelik Alternatif Bir Yaklaşım: Yeni Endüstriyel Bölgeler Yaklaşımı (Denizli Örneği)", *Basılmamış Doktora Tezi*, M.Ü.S.B.E., Muğla, pp. 6–45.

[27] R.D. Ireland, M.A. Hitt (1999), "Achieving and Maintaining Strategic Competitiveness in the 21st Century: the Role of Strategic Leadership," *Academy of Management Executive*, 13 (1), p. 43.

[28] Buhalis, p. 97; T. Duman, A.B. Öztürk (2005), "Yerli Turistlerin Mersin Kızkalesi Destinasyonu ve Tekrar Ziyaret Niyetleri ile İlgili Algılamaları Üzerine Bir Araştırma," *Anatolia: Turizm Araştırmaları Dergisi*, 16 (1), p. 9.

[29] N. Vanhove (2006), "A Comparative Analysis of Competition Models for Tourism Destinations," *Progress in Tourism Marketing,* M. Kozak, L. Andreu (Ed), London: Elsevier, pp. 101–114.

[30] Ritchie, Crouch, *The Competitive Destination,* pp. 5–23.

[31] M. J. Enrigth, J. Newton (2004), "Tourism Destination Competitiveness: A Quantitative Approach," *Tourism Management*, 25 (6), p. 778.

[32] A. M. d'Hauteserre (2000), "Lessons in Managed Destination Competitiveness: The Case of Foxwoods Casino Resort," *Tourism Management,* 21 (1), pp. 23–32.

[33] A.M.Gonzalez, J.M.Garcia-Falcon (2003), "Competitive Potential of Tourism in Destinations," *Annals of Tourism Research*, 30 (3), pp. 720–740.

[34] J. L. Furman, M. E. Porter ve S. Stern (2002), "The Determinants of National Innovative Capacity," *Research Policy*, 31(6), pp. 899–933.

[35] M. E.Porter (1990), *The Competitive Advantage of Nations*, New York: The Free Press.

[36] See: Ritchie, Crouch (2003).

[37] L. Dwyer, C. Kim (2001), "Destination Competitiveness: A Model and Determinants," www. ttra.com./ pub/uploads/007pdf. (Retrieved 08 August 2003).

[38] A. Kim (2000), "A Model Development for Measuring Global Competitiveness of the Tourism Industry in the Asia-Pacific Region," http://www.kiep.go.kr/project/ publish.nsf/243478c1734df97649256d9900210480/546cfe150257031a49256db200487 ea8/$FILE/APEC00-03.pdf, (Retrieved 19 August 2003).

[39] N. Vanhove (2006), "A Comparative Analysis of Competition Models for Tourism Destinations," *Progress in Tourism Marketing*, M. Kozak, L. Andreu (Eds.), London: Elsevier, pp. 101–114.

[40] O. Bahar, M. Kozak (2005), *Uluslararası Turizm ve Rekabet Edebilirlik*, Ankara: Detay Yayıncılık, pp. 112–113.

[41] A. Poon (2002), *Tourism, Technology and Competitive Strategies*, Wallingford: CAB International, p. 24.

[42] Poon, pp. 236–237.

[43] Poon, p. 24.

[44] Vanhove (2006).

[45] L. Dwyer, P. Forsyth and P. Rao (2002), "Destination Price Competitiveness: Exchange Rate Changes Versus Domestic Inflation," *Journal of Travel Research*, 40, p. 328; L. Dwyer, P. Forsyth and P. Rao (2000), "The Price Competitiveness of Travel and Tourism: a Comparison of 19 Destinations," *Tourism Management*, 21 (1), pp. 9–22.

[46] Vanhove (2006), p. 112.

[47] Vanhove (2006).

[48] http://www.wttc.org/frameset3.htm. (Retrieved 10 April 2006).

[49] http://www.wttc.org/frameset3.htm. (Retrieved 10 April 2006).

[50] M. Kozak, *Destination Benchmarking: Concepts, Practices and Operations,* pp. 37–38.

[51] http://www.wttc.org/frameset3.htm. (Retrieved 24 May 2006).

INDEX

D

E

J

K

L

M

Q

R

S

W